GEOFFREY CHEW
Architect of the Bootstrap

GEOFFREY CHEW
Architect of the Bootstrap

Edited by

Lars Brink
Chalmers University of Technology, Sweden

Richard C Brower
Boston University, USA

Carleton DeTar
University of Utah, USA

Chung-I Tan
Brown University, USA

K K Phua
Founding Director Emeritus IAS NTU, Singapore

 **World Scientific**

NEW JERSEY · LONDON · SINGAPORE · BEIJING · SHANGHAI · HONG KONG · TAIPEI · CHENNAI · TOKYO

Published by

World Scientific Publishing Co. Pte. Ltd.

5 Toh Tuck Link, Singapore 596224

USA office: 27 Warren Street, Suite 401-402, Hackensack, NJ 07601

UK office: 57 Shelton Street, Covent Garden, London WC2H 9HE

Library of Congress Cataloging-in-Publication Data
Names: Brink, Lars, 1945–   editor.
Title: Geoffrey Chew : architect of the bootstrap / edited by
    Lars Brink, Chalmers University of Technology, Sweden, Richard C. Brower. Boston University, USA,
    Carleton DeTar, University of Utah, USA, Chung-I Tan, Brown University, USA,
    K.K. Phua, Founding Director Emeritus IAS NTU, Singapore.
Description: New Jersey : World Scientific, [2022] | Includes bibliographical references.
Identifiers: LCCN 2021032431 | ISBN 9789811219825 (hardcover) |
    ISBN 9789811219832 (ebook) | ISBN 9789811219849 (ebook other)
Subjects: LCSH: Chew, Geoffrey F. | Nuclear physicists--United States--Biography. |
    Bootstrap theory (Nuclear physics)
Classification: LCC QC774.C495 .G46 2022 | DDC 530.092 [B]--dc23
LC record available at https://lccn.loc.gov/2021032431

British Library Cataloguing-in-Publication Data
A catalogue record for this book is available from the British Library.

The editors and publisher would like to thank the following publishers of the various journals for their assistance and permissions to include the selected reprints found in this volume: The American Physical Society (*Phys. Rev., Phys. Rev. Lett., Phys. Rev. D*); University of Chicago Press (*Isis*); Pauline, Frank and Pierre Chew (*arXiv preprints*). We thank the following organizations and people for permissions of reusing various photos in this volume: UC Berkely, AIP Emilio Segrè Visual Archives, Ling-Lie Chau, and Basarab Nicolescu.

While every effort has been made to contact the publishers of reprinted papers prior to publication, we have not been successful in some cases. Where we could not contact the publishers, we have acknowledged the source of the material. Proper credit will be accorded to these publications in future editions of this work after permission is granted.

For any available supplementary material, please visit
https://www.worldscientific.com/worldscibooks/10.1142/11815#t=suppl

Contents

Preface   ix

Photos   xi

Part I: Recollections   1

1. The Analytic S Matrix and the Ideal Mentor   3
   William R. Frazer

2. How Geoff Got Started   12
   James S. Ball

3. Interactions with Geoff Chew   15
   Owen Chamberlain

4. Interactions with Geoff Chew   18
   Steven Weinberg

5. On the Uniqueness of Physical Theories   33
   David J. Gross

6. My Postdoctoral Years at Berkeley with Geoff Chew   42
   Steven Frautschi

7. Salesman of Ideas   48
   John Polkinghorne

8. The Bootstrap: Still a Relevant and Prolific Idea   51
   Gabriele Veneziano

9. Geoffrey Chew and Regge Poles   59
   A. Donnachie and P. V. Landshoff

10. Memories: Geoff Chew, S-matrix, Bootstraps and Dual Topology   64
    Carl Rosenzweig

11. Geoff Chew and the S-Matrix   69
    Alan White

12. Geoff Chew of Illinois   70
    Jerrold Franklin

13. From the S Matrix to String Theory   72
    John H. Schwarz

14. Geoffrey F. Chew (1924–2019): A Passion for Physics and the
    Ph.D. Thesis Professor for Me and Seventy-plus Others   84
    Ling-Lie Chau

15. The Reggeon Field Theory, Describing Financial Markets in
    Crises, and Predicting Crises   92
    Jan W. Dash

16. Pomeron — A Bootstrap Story   101
    Chung-I Tan

17. Back to the Future: Reflection on Geoffrey Chew's Legacy   111
    Richard C. Brower

18. Recollections of Life as a Berkeley Chew PhD Student in the Late 1960s 118
    Carleton DeTar

19. Recollections of Geoff   126
    R. Shankar

20. A Passion for Physics   128
    Marvin L. Goldberger

Part II: Architect of the Bootstrap   **133**

21. Recollections of Geoffrey Chew   135
    David Kaiser

22. Bootstrap Physics: A Conversation with Geoffrey Chew   178
    Fritjof Capra

23. The Bootstrap Principle and the Uniqueness of Our World   216
    Basarab Nicolescu

Part III: Reprints of Selected Articles   **231**

24. Effective-Range Approach to the Low-Energy p-Wave
    Pion-Nucleon Interaction [Phys. Rev. **101**, 1570–1579 (1956)]   233
    G. F. Chew and F. E. Low

25. Application of Dispersion Relations to Low-Energy
    Meson-Nucleon Scattering [Phys. Rev. **106**, 1337–1344 (1957)]   243
    G. F. Chew, M. L. Goldberger, F. E. Low and Y. Nambu

26. Regge Trajectories and the Principle of Maximum Strength for
    Strong Interactions [Phys. Rev. Lett. **8**, 41–44 (1962)]   251
    G. F. Chew and S. C. Frautschi

27. Hadron Bootstrap Hypothesis [Phys. Rev. D **4**, 2330–2335 (1971)]   255
    G. F. Chew

28. Multiple-Production Theory via Toller Variables
    [Phys. Rev. Lett. **19**, 614 (1967)]   261
    N. F. Bali, G. F. Chew and A. Pignotti

29. An Integral Equation for Scattering Amplitudes
    [Phys. Rev. Lett. **22**, 208–212 (1969)]   266

    G. F. Chew, M. L. Goldberger and F. E. Low

30. Single-Surface Basis for Topological Particle Theory
    [Phys. Rev. D **32**, 2683–2697 (1985)]   271

    G. F. Chew and V. Poénaru

31. Unification of Gravity with Electromagnetism
    [arXiv:1209.6290 [gen-ph]]   286

    G. F. Chew

32. Extended-Lorentz Quantum-Cosmology Symmetry Group
    $U(1) \times SD(2,c)_L \times SL(2,c)_R$ [arXiv:1308.4366 [gen-ph]]   295

    G. F. Chew

Preface

Geoffrey F. Chew (1924–2019) was a prominent scientist, teacher, and mentor. He is best known as the architect and passionate champion of the bootstrap concept, sometimes called nuclear democracy. This memorial volume collects scientific commentary and reminiscences from many of his former PhD students and associates. It also includes a small sampling of reprints from his extensive scientific writing.

His firm vision of the unity and self-consistency of physical law has inspired a generation of students, which totaled more than seventy. A previous volume celebrating and honoring Geoff, entitled "A Passion for Physics", was also published by the World Scientific in 1985, at his sixtieth birthday. We have reprinted here several noteworthy articles from that volume. The title "A Passion for Physics" was taken from Murph Goldberger's after-dinner talk, which is also included here.

Chew grew up in Washington, DC, graduating from high school at the age of 16. He attended George Washington University, worked with George Gamov, and received his BS degree in Physics in 1944. He was recruited to do calculations related to the development of the hydrogen bomb with Edward Teller's group in the Los Alamos Manhattan Project. In 1946 he began PhD studies with Teller at the University of Chicago, but later joined Enrico Fermi's group there and completed his doctorate with Fermi in 1948. Subsequently, he took a postdoctoral position and then assistant professorship at the University of California at Berkeley. Controversy over a new requirement that all faculty members take an oath of loyalty to the Constitutions of the United States and the State of California led several physics faculty members including Chew to resign their positions in protest. Chew joined the physics faculty of the University of Illinois in Champaign-Urbana. Seven years later in 1957 he returned to Berkeley where he remained on the physics faculty until his retirement in 1991.

Geoffrey Chew is best known for developing the bootstrap concept as embodied in S-matrix theory, wherein no elementary particles are assumed to be fundamental, but all particles and interactions arise in a self-consistent manner in keeping with the requirements of analyticity (causality), unitarity (conservation of probability), and crossing symmetry. Further requirements were added — for example, that particles lie on Regge trajectories. Chew's realization of the bootstrap evolved over the course of his career as he improved it often by assimilating new developments.

The bootstrap concept was central to his all-encompassing vision of elementary particle physics. His passion for physics was an inspiration for his many students and associates.

In recognition of the profound impact of his work on the progress of elementary particle theory, Chew was the recipient of many honors, including memberships in the National Academy of Sciences and the American Academy of Arts and Sciences and prestigious prizes, including the Hughes Prize of the American Physical Society (1962) and the Ernest Orlando Lawrence Award of the Atomic Energy Commission (1969).

The Editors

Photos

Geoff discussing bootstrap ideas in his office at the Lawrence Berkeley Laboratory (photos from *A Passion for Physics: Essays in Honour of Geoffrey Chew, including an Interview with Chew*, eds. C. DeTar, J. Finkelstein and C.-I. Tan (World Scientific, 1985)).

(Top) Geoffrey Chew and Steven Weinberg. Courtesy of the AIP Emilio Segrè Visual Archives. (Bottom left) Geoffrey Chew and Chung-I Tan. (Bottom right) Photo: Courtesy of the UC Berkeley.

(Top left) Photo: Courtesy of the UC Berkeley. (Top right) Photo: Courtesy of Ling-Lie Chau. (Bottom) Basarab Nicolescu and Geoffrey Chew. Courtesy of Basarab Nicolescu.

Part I
Recollections

The Analytic S Matrix and the Ideal Mentor

William R. Frazer

*Department of Physics, University of California, Berkeley,
CA 94720-7300, USA*

1. Geoffrey F. Chew, Ideal Mentor

Looking back on my graduate student days, I realize that having Geoffrey Chew as my advisor was a great privilege. Geoff was always available, always supportive, always generous, always showing a deep sense of ethics.

- **Available**: Geoff carried availability to an extreme! His door was always open and his attitude welcoming. More remarkably, I often had the experience of being called as I walked by his office: "Bill! Come in, I have an idea I want to try out on you!" Try out on me, a mere graduate student? What a remarkable learning experience! [One amusing event: one day as I walked by Geoff's door, I saw him, as usual, at the blackboard staring at an equation and tossing a piece of chalk up and down. Being a well-coordinated, athletic type, he was not looking at the chalk. But at this moment, he looked down at his hand, with puzzlement — no chalk! After looking around, baffled for a few seconds, he looked up. Roof repairs were in progress, and a panel was missing from the ceiling. One of the remaining panels held the missing chalk!]

- **Supportive**: All Geoff's students benefitted from generous support. For example, there was a weekly private seminar at which one of us gave a talk, always with helpful questions from Geoff. He also prodded us to give external talks, ranging from a ten-minute talk at an APS meeting to an evening seminar at Panofsky's house. A very supportive experience indeed, with Sid Drell tossing me softball questions!

- **Generous**: A few months after I received my PhD, Jose Fulco and I published a paper that was widely noted and quoted. We predicted the existence of what is now known as the rho meson on the basis of fitting the nucleon structure data of Hofstadter *et al.* Frankly, we could not have done this work without Geoff's basic idea. Therefore, Jose and I wrote up a draft with authors Chew, Frazer, and Fulco. Geoff helped edit the draft; one of his edits was to cross off his name. Why? I asked incredulously! He answered, "If it is published as Chew, Frazer, and Fulco, people will refer to it as Chew *et al.*; if published as Frazer and Fulco, it will jump-start your careers." It did so, and also gave a boost to the nascent field of analytic S-matrix theory. Ironically, ours was one of the very few quantitative predictions the theory ever produced. Some years later, Sidney Coleman introduced me to give a seminar at Harvard as "Bill Frazer, whose work set back the progress of physics by ten years."

- **Ethical**: It should not be necessary to teach a young scientist that you do not lie, cheat, or steal! Nevertheless, many universities do teach science ethics courses. Those of us fortunate enough to have Geoff for an advisor learned by emulating a physicist who took the ethics of the profession very seriously. We did not rush into print at the cost of getting it wrong. We, of course, did not plagiarize. On the contrary, we spent a lot of time ensuring that we had given credit to those on whom our work was based. Moreover, Geoff had an extraordinary commitment to the science he pursued. When Geoff declined to be a coauthor with Fulco and me, he gave a second reason: If he put his name on the paper, it would become "his problem." He would feel a need to keep track of further developments, and to contribute when appropriate. It was part of his commitment to the profession.

Again. What a privilege to have had Geoff as a thesis advisor, and friend. He greatly influenced, and greatly facilitated my physics career.

2. The Analytic and Unitary S Matrix[a]

This talk, on the occasion of the Chew Jubilee, is a collection of observations and reminiscences on the development of the theory of the analytic S matrix. It does not presume to be a history of that development, because to write such a history would impose a burden of completeness and fairness which I am not prepared to assume! I shall select the developments I describe, simply because 1 was fortunate enough to observe them, and not because I make any claim that they are more important than others I do not describe.

The method I shall use is a diagrammatic method, a sort of "intellectual history" diagram. A typical diagram is shown in Fig. 1.

The diagram rules are as follows: A solid line represents the Chew worldline (see Fig. 2).

---

[a]Reprinted from *A Passion for Physics: Essays in Honour of Geoffrey Chew, including an Interview with Chew*, eds. C. DeTar, J. Finkelstein and C.-I. Tan (World Scientific, 1985).

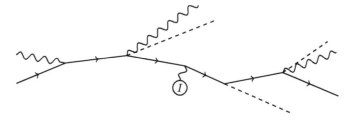

Fig. 1.   A typical intellectual history diagram.

Fig. 2.   Chew world-line.

Those of you who, along with Geoff, have played a fundamental role in the development of S-matrix theory, please do not feel slighted at not seeing your world lines — this is not your jubilee! Another diagram element is shown in Fig. 3:

Fig. 3.   Emission of a student or postdoc.

The emission of a student, which occurs so remarkably frequently along the Chew world-line, illustrates the profound analogy between our diagrams and other more familiar diagrams; namely, the emission of a student is quite analogous to bremsstrahlung. Students are, with high probability, emitted in the forward direction, and tend to remain moving along the direction Geoff taught them, even after Geoff himself has turned in another direction!

Another important diagram element, the absorption of a collaborator, is shown in Fig. 4:

Fig. 4.   Absorption of collaborator.

Again, the absorption of a collaborator is a frequent phenomenon along the Chew world-line. This interaction is a singularly strong one, with Geoff absorbing the ideas and talents of a succession of eminent collaborators and turning them to his own clearly-defined ends.

I observed with fascination the absorption of Stanley Mandelstam, and benefited immensely from the resulting Chew-Mandelstam collaboration. Geoff had been struggling for some months to derive partial-wave dispersion relations, and had realized that he needed some sort of thenunknown double dispersion relation. I remember him describing to Pauli, who was listening intently and nodding inscrutably, what important things one could accomplish with partial-wave dispersion relations, if only one understood more about the analytic structure of the S matrix in two variables. Thus, when Stanley gave a cryptic ten-minute paper at the Washington meeting of the American Physical Society in 1958, Geoff was uniquely appreciative of its significance. The folk tale has it that Geoff whisked Stanley out of the meeting room and directly onto a plane to Berkeley!

Our diagram rules are completed by the two types shown in Figs. 5 and 6, which I shall discuss later in my talk:

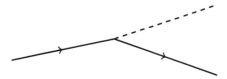

Fig. 5.   Radiation of an important new idea.

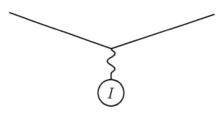

Fig. 6.   Interaction with an external field (Inspiration), with consequent change of direction.

In Fig. 7 I have represented some of the exciting events of the period 1958-61, in which the Chew-Mandelstam collaboration produced so many exciting new results.

Let me say a few words about the change in the direction of Geoff's world-line toward the right-hand side of Fig. 7. I was the first of the students emitted during the period shown, and went to the Institute for Advanced Study for a very exciting year in which I had the pleasure of preaching the new gospel to a very attentive and interested audience. I returned to Berkeley the following summer, to find to my shocked surprise that Geoff had made a major change of course in the meantime.

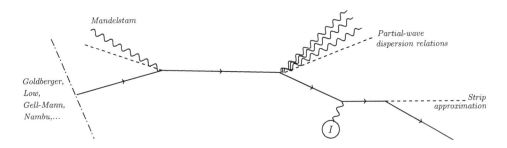

Fig. 7. The 1958–61 period: development of partial-wave dispersion relations.

He had seen the limitations of our old "nearby singularities" approach and was now intent on understanding the asymptotic behavior of scattering amplitudes. I was still attempting to follow the old course, and was shocked at how completely he had lost interest, and how completely he was convinced that the new direction was the one true pathway to truth! I have continued to marvel at Geoff's ability to adopt a working hypothesis with absolute conviction, and yet preserve the mental agility to change course as necessary — and to adopt the new working hypothesis with the same absolute conviction! In Fig. 8 I show a little more detail on the 1959–60 "shower" of students and ideas:

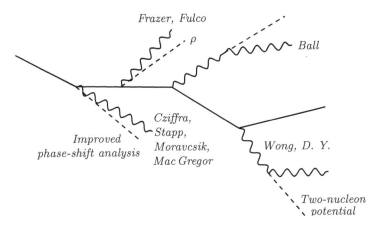

Fig. 8. The 1959–60 shower of students, ideas.[1–4]

Students and ideas are emitted, with students occasionally radiating their own ideas and students, to produce a veritable: shower. Fulco and I were so fortunate as to predict a resonance, which bore our names for a time, before we lost out to Sakurai's more euphonious "rho-meson". Geoff advised us every step of the way, but generously decided not to put his name on the paper. In addition to his evident desire to help two young physicists gain recognition, Geoff gave me a unique reason: He said that if he were to put his name on the paper, he would feel a continuing commitment to the problem of nucleon structure — he would be making it "his problem". He thought physicists should feel a commitment to a problem, rather than "skipping around" so much. Much as I admire Geoff's deep commitment, I'm afraid I proceeded to "skip around"!

Some of the developments in Fig. 8 led to quantitative results; or, to use the euphemism of the era which has, I hope, passed out of fashion, "semiquantitative" results. For example, the proton-proton phase shift analysis which explicitly included pion exchange was used to determine the pion mass, yielding[5] $m_{\pi^0} \approx 125$ MeV, against the actual $m_{\pi^0} \approx 135$ MeV. The same analysis determined the pion-nucleon coupling constant as $g^2 = 14$, whereas the Chew-Low[6] effective-range analysis of pion-nucleon scattering gave $g^2 = 15$. The analysis by Fulco and myself of electron-nucleon scattering predicted $m_\rho \approx 550$ MeV, *vs* the actual[7] 769 MeV. Coming as they did after decades of failure to formulate a quantitative theory of strong interactions, thest(results were very exciting.

From the perspective of two decades later, these accomplishments may seem quite modest compared to the present, when QCD provides a systematic theory of strong interactions. It is hard to find anyone who will admit to being a QCD skeptic today, but let me present an imaginary dialog between such a skeptic (S) and an orthodox QCD believer (Q):

> *Q:* QCD is a major advance, in that it provides a systematic theory of strong interactions. Its perturbation expansion is quantitatively successful.

> *S:* Yes, but that expansion is not valid in the region of the S-matrix results we have discussed.

> *Q:* Non-perturbative QCD, especially lattice gauge theory, promises to make systematic calculations which have never been possible before — such as, hadron masses.

> *S:* But in fact even the modest advances discussed above compare favorably with the accomplishments of lattice QCD to date!

> *Q:* But is the S-matrix approach used in those calculations fundamental, or does it just relate nearby singularities of the S matrix?

> *S:* No, it wasn't a fundamental theory, but is *QCD fundamental?*

In his contribution to the proceedings of this Jubilee, Weinberg discusses the question of what form the fundamental theory of matter might take, when examined on energy scales beyond those available at accelerators. I commend his paper to your attention, because it recounts the vicissitudes of S-matrix and field theory approaches, and because it points out eloquently the need to keep an open mind concerning the form of whatever may turn out to be the really fundamental theory which describes all particles, including gravity. Let me summarize Weinberg's argument briefly.

1. Our current "standard model" will break down before we reach some large mass scale, characteristic of gravity or grand unification. "Perhaps the theory of the new ultra-high energy scale will not be a quantum field theory at all. We don't know." But if the ultimate theory is not a quantum field theory, why do simple, beautiful field theories work so well today? In answer, Weinberg quotes two "folk theorems". He says, "I quote them partly because I think they can probably be formulated in precise terms and proved, and though I haven't done it, they are true."

2. Folk Theorem I: General quantum field theory, with the most general Lagrangian, is without content, it's just a way of implementing the axioms of S-matrix theory. Examples of quantum field theory used in this way are (i) the effective field theories used to derive soft-pion theorems, and (ii) Sakurai's vector meson theory of strong interactions. But if this theorem is true, why do we find that our successful field theories are so simple and beautiful? The answer lies in the second folk theorem:

3. Folk Theorem II (Harmonic Oscillator Theorem): At energies much lower than the natural energy scale of the fundamental theory, an effective field theory can be found, whose Lagrangian is dominated by terms with the fewest fields and/or derivatives. "Where possible, the interactions in this effective field theory will be so simple that they allow the cancellation of infinities to go through as they did in quantum electrodynamics."

   I call this a "harmonic oscillator theorem", because it points out that apparent simplicity and beauty can arise through nothing more fundamental than a power series expansion. Could it be that we who marvel at the profound beauty and simplicity of gauge field theories will appear to future generations to be as naive as a freshman who marvels at the simplicity of Hooke's law?

Weinberg concludes, "If our quantum field theories of which we're so proud are just the debris of some really fundamental field theory which describes all of physics including gravity, it may be that the really fundamental theory will have nothing to do with fields; it may not look like a quantum field theory at all. I think we have to leave this as an open possibility and maybe, in fact, that it will be something like an S-matrix theory."

Whatever the form of the fundamental theory which underlies our current efforts, experience tells us that much of what we have done will find a place in that theory. Let me attempt, at the peril of angering the reader by sins of omission and commission, to list some key ideas and concepts which have proved important in the development of S-matrix theory. Directly or indirectly, Geoff's influence has been felt in the development of all of them. Many are candidates for inclusion in a fundamental theory.

— ANALYTICITY: Dispersion relations, Mandelstam representation, partial-wave dispersion relations.
— NEAR-BY SINGULARITIES: Peripheral interactions, strip approximation, multi peripheral model.
— NUCLEAR DEMOCRACY: Bootstrap models.
— UNITARITY: Asymptotic behavior, Froissart bound.
— AXIOMATIC S-MATRIX THEORY.
— REGGE POLES.
— CROSSING: Duality, string models, topological expansion.

Let me conclude by an even lengthier list, the list of Geoff's Ph.D. students. My list (I hope it is complete!) numbers 60 students to date. If this is not unique in theoretical physics, it certainly has not been exceeded often!

Ph.D. Students of Geoffrey F. Chew

Noyes, Pierre[a]	Berkeley	1950
Blair, John[a]	Illinois	1951
Segall, Benjamin	Illinois	1951
Friedman, Marvin H.	Illinois	1952
Schey, Harry	Illinois	1953
Gartenhaus, Solomon	Illinois	1955
Lichtenberg, Don B.	Illinois	1955
Salzman, Freda	Illinois	1956
Franklin, Jerrold	Illinois	1956
Frazer, William R.	Berkeley	1959
Ball, James S.	Berkeley	1960
Cziffra, Peter	Berkeley	1960
Wong, How-Sen	Berkeley	1960
Desai, Bipin R.	Berkeley	1961
Young, James A.	Berkeley	1961
Kim, Yong Duk	Berkeley	1961
Fulco, Jose R.[a]	Buenos Aires	1962
Balazs, Louis A. P.	Berkeley	1962
Singh, Virendra	Berkeley	1962
Thiebaux, Martial L. Jr.	Berkeley	1962
Muzinich, Ivan J.	Berkeley	1963
Sakmar, Ismail A.	Berkeley	1963
Taylor, John R.	Berkeley	1963
Der Sarkissian, Michael[a]	Penn State	1963
Ahmadzadeh, Akbar	Berkeley	1964
Jones, C. Edward	Berkeley	1964
Arndt, Richard Allen	Berkeley	1964
deLany, Vincent M.	Berkeley	1965
Stack, John D.	Berkeley	1965
Chu, Shu-Yuan	Berkeley	1966
Gross, David J.	Berkeley	1966
Schwarz, John H.	Berkeley	1966
Rothe, Heinz J.	Berkeley	1966

Chau, Ling-Lie	Berkeley	1966
Finkelstein, Jerry L.	Berkeley	1967
Markley, Francis L.	Berkeley	1967
Scanio, Joseph J.	Berkeley	1967
Wang, Jiunn-Ming	Berkeley	1967
Arbab, Farzam	Berkeley	1968
Dash, Jan W.	Berkeley	1968
Lee, Huan	Berkeley	1968
Tan, Chung-I	Berkeley	1968
Brower, Richard C.	Berkeley	1969
Misheloff, Michael N.	Berkeley	1970
Ting, Peter Di-Hsian	Berkeley	1970
DeTar, Carleton E.	Berkeley	1970
Sivers, Dennis W.	Berkeley	1970
Tow, Don Mei	Berkeley	1970
Chan, Chun-Fai	Berkeley	1972
Chen, Chih Kwan	Berkeley	1972
Sorensen, Cristian	Berkeley	1972
Ghandour, Ghassan I.	Berkeley	1974
Shankar, R.	Berkeley	1974
Koplik, Joel I.	Berkeley	1974
Millan, Jaime	Berkeley	1977
Lucht, Philip H.	Berkeley	1978
Sursock, Jean-Pierre	Berkeley	1978
Weissmann, George	Berkeley	1978
Levinson, Mark A.	Berkeley	1983
Espinosa-Marty, Raul	Berkeley	1983

[a] *De facto*, but not *de jure* Ph.D. students

References

1. A good general reference for this period is Chew, G. F., *The Analytic S Matrix* (Benjamin, New York, 1966).
2. W. R. Frazer and J. R. Fulco, *Phys. Rev. Lett.* **2**, 365 (1969).
3. P. Cziffra, M. MacGregor, M. Moravcsik, and H. Stapp, *Phys. Rev.* **114**, 880 (1959).
4. D. Y. Wong, *Phys. Rev. Lett.* **2**, 406 (1959); H. P. Noyes and D. Y. Wong, *Phys. Rev. Lett.* **3**, 191 (1959); A. Scotti and D. Y. Wong, *Phys. Rev.* **138**, B145 (1965).
5. P. Signell, *Phys. Rev. Lett.* **5**, 474 (1960).
6. G. F. Chew and F. E. Low, *Phys. Rev.* **101**, 1570 (1956).
7. E. B. Hughes, T. A. Griffy, M. R. Yearian, and R. Hofstadter, *Phys. Rev.* **139**, B458 (1965).

How Geoff Got Started*

James S. Ball

19-YEAR OLD GEORGE WASHINGTON STUDENT EXPOSES U.S. ATOMIC BOMB PROJECT

Security Agent Interrogating Physics Student Geoff Chew
About Knowledge of Manhattan Project

While not all of the above actually happened something rather close to this scenario did and may have hastened Geoff's assignment to the Los Alamos project. The real story, as well as I have been able to reconstruct it, is as follows:

In 1943, with national price and wage controls in effect, all requests for salary increases were reviewed by the War Labor Board. One request in October, 1943 came from a physicist operating a cyclotron. Salary increases were usually justified by the importance of the worker's contribution to the war effort. This physicist's work was so far removed from the popular stereotype of the female riveter in the war plant that his request was treated as some kind of joke and the review board thought the case very amusing.

A young Washington Post reporter, Jean Craighead, was at that time assigned to write feature articles about the actions of the board. She didn't find the physicist's request particularly funny and she remembered that in a physics class taken several years earlier the professor had said that the next war would very likely be won by technological superiority. As a result she decided to write an article to illustrate how this physicist's work might lead more or less directly to winning the war.

Her first step was reviewing her college physics text, which, as it was published in 1939-40, contained some information on nuclear fission. I mention this because in texts published later, this material had been removed. To make sure that her facts were up to date she needed to consult someone currently studying physics. This problem was solved when she discovered that her friend, Ruth Chew had a brother who was a physics major at George Washington University.

After consulting with 19-yearold Geoff Chew, she wrote the following article

---

*Reprinted from *A Passion for Physics: Essays in Honour of Geoffrey Chew, including an Interview with Chew*, eds. C. DeTar, J. Finkelstein and C.-I. Tan (World Scientific, 1985).

which was published in the Washington Post on Sunday, October 31, 1943. While the reaction of the general reader is not known, the discussion of an atomic bomb fueled by uranium was very exciting reading for the U.S. security agencies guarding the secrecy of the Manhattan project. They moved swiftly, confiscating all copies of the article, including removing it from the Washington Post's archives. Miss Craighead was questioned intensively and forced to reveal that she had consulted with Geoffrey Chew. After a long session of interrogation of both Geoff and Miss Craighead, it became clear that this was not a case of a leak in security but simply an intuitive guess which happened to be uncomfortably close to the truth.

While no security records of this incident have been found, the following seems consistent with the security practices of that time: (1) Geoff knew too much to be allowed to be on his own; (2) he had done nothing wrong so he couldn't be legally locked up, therefore send him to Los Alamos where, to the first approximation, everyone was locked up and certainly where he could be watched. In any case, within 4 months of the publication of this article, Geoff had joined the scientific staff at Los Alamos.

Miss Craighead married in January 1944 and has since become a successful author under her married name of Jean George. As for Geoff Chew....

Just an Atom-Smasher

He Can't Get a Raise—Hasn't Accomplished Anything

By JEAN CRAIGHEAD

A young fellow who has been studying much of his life on the matter of blowing up nations with an atom would like to get a wage increase from the War Labor Board.

Preoccupied with discovering the formula for demolishing Berlin with a teaspoonful of dust,

An atom-smasher is just an atom-smasher

before the Berlin boys master this upsetting trick, he nevertheless needs a new pair of shoes and a winter overcoat.

In the laboratory where the young man lives, one seldom has such simple thoughts as the corner grocery and the Nation's wage stabilization policy.

The drift of the conversation usually tends toward bombarding isotopes, the effect of an alpha particle on the electric forces of an atom of uranium, and the explosive annihilation of Berlin.

So that it must have been with some bit of embarrassment that he found himself wondering about time and a half and incentive payments. Nevertheless, the very ordinary thought that he needed a wage increase for razors and cigarettes, outweighed the science of isotopes, and he took time to write to the War Labor Board to ask how it might be done in his case.

Why Not Another Job?

The letter stumped the WLB, and they looked through their directives and executive orders to see what could be done in the way of a pay adjustment for an atom smasher.

They suggested the Little Steel Formula, but the young man's natural inclinations run to bigger things, and 15 per cent of the amount he was receiving would hardly be enough to buy one new shirt for Wednesday evening forums, much less the latest textbook on extra-nuclear electrons.

The board then suggested that he be reclassified and given a raise on the basis of another job. But an atom smasher is an atom smasher, and after studying seven years to understand the trade, one would hardly want to make a shift.

On the other hand, the board stated, one can always obtain a merit increase.

In labor circles a merit increase means many things, but to a person occupied with the spontaneous desintegration of Axis cities and the Nazi war machine by bombarding an atom, a merit increase means only one thing: Finding the formula that will unlock uranium and with a dull roar of separating particles, buckle and rupture Berlin and Tokyo into fire and dust.

Don't Expect It Very Soon

Chances are slim that either side will have the formula before the next war, and that is a little long to wait for a merit increase.

Now this young man is no fictional character; he is very much alive, and very much kicking. But, because of the secret nature of his work, his name cannot be published.

In regards to the work he is doing, let me suggest that you keep on buying war bonds and turning out rifles. Although the dramatic effect of an atomic explosion in the Ruhr Valley is overwhelming, artillery is still a good thing in the absence of a formula.

Even the inspiration of a wage increase won't advance the research out of proportion, and we

Sketches by Jean Craighead.

A stadium around Germany, so we can see the show

can make use of the Flying Fortresses for some time to come.

When the young man with labor problems and his fellow genii do discover how to use the energy in an atom, it would be nice if they would throw up a stadium around Germany and invite the United Nations to the explosion of the Nazi war machine.

It would be a great spectacle—guns and war plants would spin apart under the violence of the impact, and erupt skyward in a fanfare of colors and electronic reactions.

Fig. 1.   Jean Craighead's article as it appeared in the Sunday, October 31, 1943 edition of the Washington Post.

© 2022 World Scientific Publishing Company
https://doi.org/10.1142/9789811219832_0003

Interactions with Geoff Chew*

Owen Chamberlain

Geoff Chew and I were both at Los Alamos during the last half of World War Il, though we had little contact most of that time. I came to notice Geoff as we were returning to graduate school after the interruption of the war.

We had all learned that Enrico Fermi and a number of the world's best physicists were going to the University of Chicago and that Sam Allison had a good number of University of Chicago fellowships to hand out. I applied for one of these fellowships but was turned down. I remember pounding Allison's desk at Los Alamos and saying I hoped to demonstrate that I deserved a fellowship and that my being refused was a mistake in judgment. (I was not one to pound desks, but on this occasion I felt very strongly.)

A month or so later Sam Allison called me back to his office. This time he reported that Geoff Chew had received a National Research Council fellowship, and that freed up a University fellowship which was awarded to me.

Well, that was only the beginning of Geoff's beneficial influence on me. When I arrived at Chicago, about March of 1946, I found one room of Eckart Hall in which there were the desks of Frank Yang (C. N. Yang), Murph Goldberger and Geoff Chew. Frank Yang sometimes helped us with problems, but for the most part I think he was going his own way, rather than interacting with the others. I felt I was befriended by Murph. Frequently I joined with Geoff and Murph to make a sort of working trio. Our object was to solve any interesting problem in sight. That was the real fun of physics! Murph seemed to know the most math, Geoff to provide the ideas, and I served to remind us of a method that had worked on a problem the previous week.

One of the high points in our graduate education was Clarence Zener's course: The Fundamentals of Solid State Physics. In the front row of seats there were, sitting side-by-side, Murph, Geoff, and I. I don't think we had any agreement among ourselves, but we worked out a pattern of behavior that we delighted in. Whenever Prof. Zener would say it can be proved that such-and-such is true, one of us would

---

*Reprinted from *A Passion for Physics: Essays in Honour of Geoffrey Chew, including an Interview with Chew*, eds. C. DeTar, J. Finkelstein and C.-I. Tan (World Scientific, 1985).

put up his hand and say: "Oh, Prof. Zener, how do you show that?" Well, this would take the lecture completely off the material he had planned on. It would take Zener into areas in which he was not fully prepared. But we found out a lot about how Zener thought about things.

We were afraid of Zener, I think. When we went to his office to complain that the problems were too hard we always went in groups of six or seven students. One day we complained that a problem as given didn't contain enough information to make a solution possible. He said, "Of course they don't have enough information. Do you think that when you have a problem in the laboratory, it comes with a list of the assumptions you must make in order to solve it? Make the necessary assumption!"

I thought that, apart from Enrico Fermi's private night course, the Zener course was the best course I ever had. When it was repeated the following year it was generally agreed that it fell flat by comparison. I think it lacked that extra stimulus that Geoff and Murph and I provided.

Those were great days in all our lives. We had had our education interrupted by war work, but we came back to the classroom with a more mature attitude, with a better idea of just what we wanted to learn, as well as a real determination to learn it. Several days a week we ate lunch with Enrico Fermi. If we had had a question Fermi couldn't answer we could always have taken it to Edward Teller or Gregor Wenzel or Maria Mayer.

In the academic year 47–48 I watched Murph and Geoff working together on various thesis topics. It was generally recognized that they were both ready to be awarded the degree, but still the right thesis topics had to be found. Several times it happened that they would turn out a nice piece of work that looked like good material, and, just as they were finishing up, there would appear in the *Physical Review* someone else's paper that covered pretty much the same ground. Then they would have to start all over again. In the end Geoff submitted a very respectable paper on proton-deuteron scattering, but Murph had to settle for a Monte Carlo calculation of a proton making successive collisions inside a carbon nucleus, which did not display the full capability of the man.

The proton-deuteron scattering became a steady topic for Geoff. He made successive improvements in his theory, then I think he made one improvement that was not an improvement. Then he made more real improvements. He invented the approximation known as the impulse approximation. Some said it should have been called the Chew approximation. It was the theoretical basis for a number of experiments I took part in at the 184-inch cyclotron.

During the earliest years in Berkeley, Geoff was the star at the annual Physics Department Picnic, for he dominated the baseball game between the faculty and students. (Anyone with a PhD played with the faculty.) Geoff could hit the ball the full length of the Meadows picnic area. I remember remarking to Geoff on how well he had hit the baseball. He answered that that was not a really squarely hit ball. That a well-hit ball would go a lot farther than that. In fact, I believe at one time Geoff had to make a choice between being a professional physicist and being a professional baseball player. It's a good thing for physics that he decided as he did. It was also a good thing for Geoff, for in later years it became clear that Geoff's back would never have stood the strain of professional baseball.

Then there came the year of the Oath. It took Geoff away from Berkeley to the University of Illinois. I think it was at that time that Geoff took on the chairmanship of the FAS (Federation of Atomic Scientists) committee on passports. This was in the period when many scientists were being denied passports if the Government felt that the U.S. would be better off if the scientist in question did not go abroad. Geoff very ably collected the information about who were being denied passports and made the information known to the public and to the Congress. His committee was an important element in the focus that caused the government to change its ways. We are now much better off.

Now it is well known that Emilio Segre is a great teacher. One of the things that I learned from him was that it was a good idea to stop in and see Fermi once in a while. He stopped in Chicago to see Fermi almost every time he went to the East Coast. I followed suit, stopping to see Fermi whenever I went East. But I did him one better. Not only did I spend a day in Chicago, but I spent the next day in Urbana seeing Geoff. These contacts were particularly valuable while Geoff was concentrating on the p-d scattering, for they directly affected our experimental program.

Later, when Geoff did return to Berkeley he became the work-horse of the theoreticians on whom everybody relied. There were several times that Geoff gave lecture series aimed at helping us experimentalists to understand current theory. I remember one series about the analyticity of the S matrix and another on Regge-pole theory.

Throughout his career he has been ready and able to explain in simple terms recent developments in theory.

I once asked Geoff how he decided which experimentalists were reliable and which unreliable in reporting their results. Geoff answered that he didn't judge experimentalists on the basis of their reliability, but on the interest of the problem they were working on, I think his answer is most appropriate and I think this attitude has served him well over the years.

And, over the years, we have all benefited from Geoff. He has at every turn been both a help and a stimulus. I am pleased to have this opportunity to express my pleasure in having him as a colleague. Thank you.

Interactions with Geoff Chew

Steven Weinberg

University of Texas at Austin, USA

I came to the Berkeley Radiation Laboratory in 1959, and a year later joined the University of California faculty on its Berkeley campus. In those days, Geoff Chew was the guiding spirit of the group of high energy theorists at Berkeley. In addition to the gracious hospitality that Geoff and Ruth Chew gave to me and my wife Louise at their house in Orinda, Geoff was remarkably kind to me professionally. It was remarkable, because I did not fall in with the spirit of the work pursued by his group.

The program pursued by Geoff and his collaborators at Berkeley is most often called S-matrix theory. The idea was that we theorists should not concern ourselves with things that cannot be observed, like the quantum fields of particles like baryons or mesons, but instead should deal only with observable probability amplitudes, assembled into the S-matrix. Instead of assuming field equations as fundamental principles, the work of particle theory should be based on assumptions about the S-matrix: its unitarity and various symmetry principles, such as Lorentz invariance, charge conservation, and for the strong interactions also some approximate symmetry principles, such as isotopic spin conservation. In addition it was necessary to make an assumption of maximum analyticity: Expressed as a function of Lorentz-invariant energies and momentum transfers, the S-matrix is assumed to be as analytic as possible, with only those poles and branch points required by unitarity.

I thought that the aims of this program were very attractive, if they could be achieved. It is always good to pare down to an essential minimum the assumptions on which our work is based. But I did not think that the S-matrix program could be made to work as a substitute for quantum field theory.

For one thing, in the late 1940s quantum field theory had scored great successes in understanding electromagnetic interactions. Also, by the end of the 1950s we had a quantum field theory of weak interactions that worked well as applied to beta decay in the lowest order of perturbation theory, although it gave unphysical results when extended to higher orders. I did not see how quantum field theory could be part of the fundamental structure of physical theory as applied to the electromagnetic and weak interactions but not the strong interactions.

Also, on a more practical level, I doubted the ability of theorists to deal with functions of several complex variables in implementing the principle of maximal analyticity. I certainly doubted my ability. It seemed to me that in practice we would simply have to assume that the analytic structure of S-matrix elements was whatever was provided by quantum field theory.

Instead of participating in S-matrix theory I spent my term at Berkeley working on an assortment of different topics, from weak interactions to cosmology. Fortunately for me, Geoff was no mandarin who demanded loyalty to his own scientific program. Instead he gave me his support in these years, endorsing my rapid promotion, and helping me to negotiate a long leave of absence so that my wife could go to law school at Harvard.

During the years at Berkeley I became fascinated with the idea of spontaneously broken symmetry, which as far as I could see could not be formulated in the context of S-matrix theory. Then shortly after moving to Cambridge in 1966 I figured out how to use a field theory of strong interactions with a spontaneously broken symmetry to calculate the S-matrix elements for pion-pion and pion-nucleon scattering, with a success that had not been possible in S-matrix theory.

Yet this work brought me some way back to the spirit of the program of S-matrix theory. The quantum field theory of pions and nucleons and other hadrons that we used to work out the implications of spontaneous symmetry breaking could not be taken seriously as fundamental principles of physics. At first this field theory was used only as a calculational tool that was guaranteed to give the same results as current algebra in the extreme soft pion limit. The theory was not renormalizable, and few physicists thought that both nucleons and pions were elementary particles whose fields would appear in the fundamental equations. Then, in trying to push beyond the leading soft-pion results, my point of view changed. I realized that if one does calculations using the most general possible quantum field theory involving fields for a set of particles like nucleons and pions, respecting all broken and unbroken symmetries and including terms in the Lagrangian with any numbers of fields and derivatives, then the result of these calculations would simply be the most general S-matrix for these particles consistent with these symmetries and with the general principles of unitarity and locality (that is, cluster decomposition). The field theory is not part of the fundamental laws of nature, but is an *effective* theory, used to implement general principles. Of course these calculations cannot yield exact results, because the Lagrangian will contain an infinite number of terms with unknown coefficients, which are needed anyway to cancel infinities, but the effects of almost all of these terms can be made as small as one likes by considering only energies sufficiently far below some limit. So it turns out that the S-matrix program was pursuing the right goals, but the tool for realizing them was quantum field theory.

In 1984 I was at a conference on cosmology at (of all places) the Vatican. At the close of the conference the participants were invited to Castel Gandolfo to meet the pope, but I had to forgo that pleasure, I had agreed to speak at a meeting in Berkeley a few days later in celebration of Geoff's 60th birthday, and I was determined to report my view that the present development of effective field theory was a realization of Geoff's earlier program of S-matrix theory. Delayed by a strike of air traffic controllers in Rome, I arrived in Berkeley with 15 minutes to spare before my talk. I may have been incoherent, and I certainly was tired, but I hope that Geoff was still able to take some pleasure from news of the conversion of a skeptic, making up in a small way for his kindnesses to me years before.

The Ultimate Structure of Matter[a]

This talk was not given at the celebration of Geoff Chew's birthday, but three years earlier, at the celebration of the 50th anniversary of the founding of the Lawrence Berkeley Laboratory. However, my connection with the "Rad Lab" from beginning to end was largely shaped by my connection with Geoff Chew, its leading theorist. During the time I was at Berkeley, from 1959 to 1966, Geoff led a movement that for some years dominated the theory of strong interactions, a movement away from quantum field theory and toward the so-called S-matrix approach. Almost inevitably, therefore, my talk at the Rad Lab's anniversary celebration centered on how I viewed the position of S-matrix theory in the history and the future of theoretical physics. I had rather hoped that Geoff would be pleased that I thought the time might be approaching for a swing back to S-matrix theory, but I later learned that he himself has been moving toward field theory. Nevertheless, it seemed to me that this talk might still make an appropriate contribution to Geoff's *festschrift*, that he will enjoy for old time's sake.

The article presented here is essentially the same as the talk given at Berkeley in October 1981, with only the lightest editing to put it into readable English. I have even left in one scientific mistake; in 1981 I expected that the first intermediate vector boson to be discovered would be the Z^0, because its decay provides a clearer signature, while in fact the available energy in the CERN experiment was so close to the Z^0 threshold that it was the $W^\pm$ that was first discovered.

*   *   *

I want to make it clear from the outset that the rather grandiose title of this talk, *The Ultimate Structure of Matter* was not chosen by me. It was chosen by the organizing committee. However, I did leave it, although I had a chance to change it. This was partly because that way I could blame it on the organizing committee. Also in fact it is precisely what I'm going to talk about. To be a little bit more

---

[a]Reprinted from *A Passion for Physics: Essays in Honour of Geoffrey Chew, including an Interview with Chew*, eds. C. DeTar, J. Finkelstein and C.-I. Tan (World Scientific, 1985).

specific, I want to talk about an old question in physics: What are the fundamental entities of which we regard our universe as being composed — particles or fields?

I don't mean this in the sense of how we should look at our existing theories. It really isn't terribly important, given a theory, whether you describe it in words having to do with particles or words having to do with fields. The important thing is whether it works. The question I'm asking is in what direction we will have to look in the future for more satisfactory theories of matter. In other words, my question is not philosophic but strategic. It is a question that quantum physicists have had to struggle with a great deal and it's appropriate to talk about it here at Berkeley, because many contributions to our enlightenment on this issue have come from here. I'm going to describe something of the history of this old question and then tell you where I think it stands now in today's physics.

In its modern form, the question of a universe of particles or a universe of fields is roughly about as old as the "Rad Lab". In the first decades of the 20th century the question didn't arise, or at least not in its modern form. Physicists then lived comfortably with a kind of dualism. There were undoubted particles like the electron of J. J. Thomson, and the atomic nucleus, discovered by Rutherford. And then there were fields. There was the electromagnetic field, and the gravitational field. True, it was worked out during the period from Einstein in 1905 to Dirac in 1927 that light has a particle nature, that electromagnetic waves can in some sense be thought of as consisting of particles, called (by the Berkeley chemist, G. N. Lewis) photons. Also, there was an effort early in the century by Abraham and Poincaré and others to understand the electron as a bundle of energy of the electromagnetic field. But no one at that time, I believe, dreamed of turning it around and thinking that such material particles as the electron or the proton might have anything to do with an electron field or a proton field. And this was not changed, despite what is sometimes said, by the advent of the quantum mechanics of the 1920's. In that quantum mechanics, as it developed in 1925–26, the description of nature was changed from a description in terms of the trajectories of particles (where a particle is at any one moment and how fast it's going) and fields (what are the values of the electric and magnetic fields at each position in space at any one moment) to a description in terms of wave functions, probability amplitudes, quantities that give you the probability of finding a certain configuration. But these probability amplitudes were still regarded as depending on the positions of particles and the values of fields. The position of every electron in the universe and the values of the electric and magnetic fields at every point in space were taken as the arguments, the independent variables, the things on which the wave function depended.

Dirac attempted in 1928 to formulate a relativistic quantum mechanics; he attempted to take this theory of particles and fields, this dualistic theory, and make it consistent with the principles of special relativity. Dirac's approach was strikingly successful as applied to electrons and electric and magnetic fields, but with the benefit of hindsight we can now say that it could not be extended to the rest of physics, and in particular not to the weak and strong nuclear forces. The fact that it was not generally successful has often been lost sight of because the mathematical machinery invented by Dirac for this purpose has become part of the standard stock in trade of all theorists that followed him. But in fact Dirac's effort to make a relativistic quantum mechanics of particles and fields was not the way of future physics.

Then in 1929, for the first time, there appeared a unified view of the constituents of the universe. I refer to the work of Heisenberg and Pauli in a pair of articles written in 1929, one of them published in 1930. In these articles, Heisenberg and Pauli constructed what we have come to call quantum field theory. The name tells exactly what it is. In this theory, the fundamental constituents of matter are taken to be fields. There is an electromagnetic field: that's no surprise. There's also an electron field; there is a proton field; there is a field for every fundamental particle. The particles emerge when quantum mechanics is applied to these fields, but the particles themselves are mere epiphenomena, just bundles of energy of the field. The energy of the fields are concentrated in little knots and the knots go zipping around and that's what we call particles. But the underlying reality is the field.

The quantum field theory of Heisenberg and Pauli led to a clearer view of what is meant by an elementary particle. An elementary particle is a bundle of energy of one of the fundamental fields that inhabit the universe. Everyone thought the electron, for instance, was an elementary particle, so they assumed that the fundamental field theory would have to involve something called the electron field. Likewise the photon was regarded as a fundamental particle, so the fundamental field theory would be regarded as also containing an electromagnetic field, of which the photon was the quantum or bundle of energy. Other particles, like, say, the nucleus of the iron atom or this blackboard eraser, were regarded as composites. And the fact that they were composites was given a precise meaning in the sense that the basic field equations, which govern the system of fields that makes up the universe, do not contain a field for the iron nucleus or a field for this eraser. These bodies are composed of more elementary particles whose fields do appear in the fundamental equations.

Quantum field theory scored an immediate triumph in 1933 when Fermi used it to develop a theory of the kind of radioactivity known as beta decay, which is the prototype of a whole class of elementary particle interactions which have become to be known as the weak interactions. Beta decay is a process in which a nucleus changes from one element to another element, emitting in the process a negative or a positive electron and a neutrino or an anti-neutrino. This was a process that involved the creation of new particles that could not have existed inside the nucleus. And it was a process that could not possibly be understood within the framework of the old particle quantum mechanics, either in its original form or in the relativistic version of Dirac.

I don't know why Fermi's achievement did not by itself convince physicists of the need for a quantum field theory. Perhaps part of the reason, or perhaps the whole reason, for a hesitancy about quantum field theory after its initial development by Heisenberg and Pauli, was the fact that it immediately ran into a terrible mathematical trouble. In the first few months after the second Heisenberg-Pauli paper was written in 1929, quantum field theory was found to be plagued by a terrible inconsistency, that also afflicted the dualistic particle-field theory of Dirac. One of the first to encounter this problem was a young professor of physics, then new on the faculty at Berkeley, J. Robert Oppenheimer.

Oppenheimer set out to calculate the contribution to the energy of atomic states from the interaction of the atomic electrons with the quantum field of the photon. And Oppenheimer found, to his surprise and chagrin, that the answer was infinite. It wasn't just that the whole zero-point of energy was being shifted by an infinite amount, which wouldn't be observable. Even the energy difference between two atomic energy levels, the quantity which is directly measured when you observe the frequency of light given off by an atom, came out to be infinite in Oppenheimer's calculation.

It was regarded as a disaster. Waller, in Sweden, discovered much the same thing (though he was considering free electrons rather than electrons in atoms), and he told the result to Pauli, and Pauli did not believe it, because it seemed to mean the end for the quantum field theory that he and Heisenberg had just developed. Then other infinites were discovered. Theoretical physicists would set out to calculate some perfectly sensible thing like the energy difference between two states of a hydrogen atom, and the answer they would get would be a perfectly nice number, and then they would carry the calculation to the next order of approximation, and in the next order of approximation they would get an infinite answer. The infinite answer takes the form of an infinite sum over all the ways that momentum and energy can flow between the electron and the radiation field. Other infinites were found in other physical processes. (In fact, more infinites were found than were actually there; there were errors made in some of these calculations which did much to confuse the issue.) There soon developed a general feeling of pessimism about the whole future of the field view of nature.

Many physicists retreated to a position that while the field concept might work as an approximation, there was something basically wrong with it, and if you carried your experiments to, say, one order of magnitude in energy higher than the energies which were then accessible in the kind of accelerator that Lawrence was working with here, that then you would find that the field concept just did not work. (That's always a very common sentiment among physicists, that if you carry experiments to one higher order of magnitude in energy than the energies that are now accessible, that then you'll find that existing ideas don't work. And it's sometimes true.) In particular, Oppenheimer was very much impressed by the fact that in the cosmic ray experiments which were being done at that time, there were discrepancies between the existing theory of electrons and photons and what was observed for cosmic ray showers. At the time no one realized that this was due to the production of particles called mesons, of whose existence physicists were then unaware. Oppenheimer interpreted the cosmic ray data as indicating a breakdown of quantum field theory itself.

Because of the problem of the infinities, there began a return to a view of nature as particles rather than fields. It started with John Wheeler in 1937 and then Heisenberg in 1944. Heisenberg took the point of view, following an ancient and honorable tradition in physics, that the laws of physics should not only make predictions solely about observables, but should not in the formulation of the laws refer to anything but observables. That is, every ingredient in physical law must be something that can be directly observed; physics has no business talking about things that are in principle outside the range of experiment. This satisfied a deep urge in physicists. This philosophic doctrine, which I believe is sometimes called logical positivism, is a recurrent theme in the physics of our century. It was, for instance, very useful to Einstein in his work on relativity. I think physicists often have a feeling that when their theories don't work it's because they've been naughty and introduced unobservable quantities, and if they would only purify themselves and return to things that are observable, then everything would work out.

For Wheeler, and then Heisenberg after him, the observables were the probabilities, or to be more technical, the probability amplitudes, for various collisions among particles. These give the probability for instance, that if you start with two particles coming toward each other at such and such an energy and angle, then you'll end up, say, with three particles going out at such and such energies and angles. All of these probability amplitudes were united by Wheeler and Heisenberg into a quantity called the S matrix, S for *Streung* or scattering. This matrix, an infinite array of complex numbers, would give you all the probabilities for all conceivable collision processes among particles. And the idea was that these were the only things in physics that were ever going to be observed. You would never ever look into a collision and see the local field theoretic processes that had been described by Heisenberg and Pauli so you shouldn't think about them, you should just think about the S matrix and make a theory in which all the laws were formulated in terms of the S matrix.

The issue was now squarely joined. On one side was a field theory of nature in which the underlying reality is a world of quantum fields and in which particles were merely bundles of the energy of the fields. In this view, the laws have to be formulated in terms of the equations that govern the fields — equations like Maxwell's equations, that govern the electromagnetic field. In opposition to this was a particle or S matrix theory, in which the underlying reality is a world of various kinds of particles, and in which when the field idea is useful at all (as everyone knows it is useful in dealing with electromagnetism), the fields are to be regarded as just some kind of collective state of huge numbers of particles, of the sort called a coherent state, but these fields are merely a convenient mathematical abstraction for describing large numbers of particles. In this view, the laws of nature have to be formulated not in terms of field equations but in terms of axioms that describe the S matrix, the array of all the probabilities for all different collision processes.

From the mid 1930's onwards, the mood of physics was going to swing back and forth several times, from S-matrix theory back to quantum field theory, then back again to S-matrix theory, then back again to quantum field theory. I want here to raise the question whether the time is approaching when we will have another swing, back to something like S-matrix theory.

Before I go into this history of these swings of opinion and my comments about where we're headed now, I think I must admit that I have been guilty of a historical oversimplification. Although we may now look back and see a clear opposition between a particle or S matrix view and a quantum field theory, nothing in life is ever that clear. It certainly wasn't that clear in the 1930's or 1940's. One of the great confusions in this story is that for certain problems (though by no means for all problems), Dirac's relativistic quantum mechanics, the dualistic quantum mechanics of particles and fields, of electrons and the electromagnetic field, were equivalent to the quantum field theory of Heisenberg and Pauli. Physicists often referred to them as if they were interchangeable. Weisskopf recently has written in some reminiscences that the paper that he and Pauli wrote in 1934 was designed specifically to demonstrate the need for a thorough-going field theory. (The issue hinged on the nature of anti-matter.) Dirac's view was that in addition to the electrons that we see normally, there is an infinite number of electrons with energies lower than the zero-energy of empty space, the so-called negative energy electrons. Every once in a while there's a hole in the sea of negative energy electrons and that hole we see in the laboratory as an electron but of opposite charge (because a hole in the sea of negatively charged electrons would appear as a positively charged particle). This "antielectron" or positron was discovered in 1932. Pauli and Weisskopf showed that this view of the nature of antiparticles which was built into Dirac's relativistic quantum mechanics was inadequate; they did this by showing that other sorts of particles also had antiparticles, particles that could not possibly form a sea of negative energy particles. These are the particles that physicists call bosons. They cannot form a stable sea of negative energy particles; if every negative energy state is occupied it becomes even easier for bosons to fall into these states. We now understand that Pauli and Weisskopf were right, and that in fact every non-neutral particle has an antiparticle, and these antiparticles are in no sense to be thought of as holes in a sea of negative energy particles. Among the non-neutral bosons are the $W^\pm$ particles which are exchanged in nuclear beta decay. The fact that every particle has an antiparticle became settled in most people's minds when the antiproton was discovered here in Berkeley in the 1950's. However even now the hole theory still appears in textbooks. I suppose it's an example of physicists not taking the trouble to rewrite their history.

Now back to the main line of my talk. I want to talk now of the swings of opinion between quantum field theory and S-matrix theory. The first revival of quantum field theory came in the late 1940's through the work of a number of people: Feynman, Schwinger, Tomonaga, Dyson, and others. It was found that the infinities that had been discovered by Oppenheimer and Waller and others in the 1930's were in fact due to a simple misinterpretation of the theory. (That is, to what we can now say with the benefit of hindsight was a simple misinterpretation. Nothing is simple when it actually happens.) The misinterpretation was the identification of the quantities e and m which appear in the field equations with the electric charge and mass of the electron as they're actually measured in the laboratory. It became clear that when we measure, say, the mass of the electron in the laboratory, we're not measuring the quantity m which appears in the field equations, but in fact we're measuring the mass m plus the effects of a huge number of so-called radiative processes, in which the electron emits and then reabsorbs photons many times. And all these processes are always going on all the time. You cannot ever measure the mass of the electron apart from these radiative processes. The Department of Energy may turn off accelerators but it can't turn off these processes. As a result, no one has seen a bare electron, an electron without its cloud of photons. And once you realize that the quantities appearing in your equations are not the ones that are measured and you reinterpret your equations to express them in terms of the measured quantities, all the infinities simply cancel. I'm making it sound really quite simple, although none of it was easy to see.

Quantum field theory now worked magnificently. A few years ago, just to give an example, I looked up the numbers for the comparison of theory and experiment for what's called the magnetic moment of the electron. (That's just its strength as a magnet, in natural atomic units.) The experimental value was 1.0011596541 and the theoretical value is 1.00115965234. The discrepancy is in the tenth decimal place and is easily accounted for by both experimental and calculational uncertainties.

There is an interesting historical point here which I suspect not too many people know because probably not many people have read Oppenheimer's 1930 paper. (Incidentally, that paper appeared in a journal which had not, until then, been the scene of many important publications in fundamental theoretical physics, the *Physical Review*.) One of the triumphs of the revival of quantum field theory in the late 1940's was Bethe's calculation of what's called the Lamb shift, which is just the splitting in energy (due to emission and absorption of photons) of two otherwise equal energy levels of the hydrogen atom. He calculated it and it agreed with the experimental value that had just been determined by Willis Lamb. If you look back at Oppenheimer's paper you will find that he had everything there needed for the calculation of the Lamb shift. All he needed to do was to give the crank one more quarter turn and he would have had Bethe's formula for the Lamb shift, and could have calculated the numerical value. Really, all Oppenheimer had to do was put in the numbers and throw away the infinite effects of intermediate states involving relativistic electrons, and he would have gotten the right answer, or at any rate the answer to as good an approximation as Bethe's. He didn't do it because he didn't have confidence in quantum field theory. What happened in the late 1940's was precisely a restoration of confidence in quantum field theory.

But that confidence didn't last very long. More problems were found and there was another revival of S-matrix theory in the late 1950's and early 1960's. One of the problems was that although these infinities beautifully cancelled in the theory of electricity and magnetism and electrons, which is known as quantum electrodynamics, the infinities would not cancel in that way in the theory that Fermi had developed to describe the radioactive process of beta decay. (The generic name for the force that produces these processes, as I said before, is the weak interaction, because it has an intrinsically weak strength, which makes processes like beta decay go very slowly.) For the weak interactions (of which many more examples were known at that time) this lovely trick of cancelling infinities just didn't work. That was, of course, an old story but it was hoped that the new idea of absorbing the infinities into a redefinition or, in other words, a renormalization of the electron's mass and charge, would also work when you applied the idea to the weak as well as the electromagnetic interactions. But it just didn't work.

The second problem was the apparent hopelessness of calculations involving strong nuclear forces. (These are the forces that hold the nucleus of the atom together. It's the strength of these forces that makes the nucleus so much smaller than the atom and that gives rise to such enormous energies when you disrupt the nucleus, typically a million times larger than the energies that are released when you disrupt an atom in an ordinary chemical reaction.) In quantum electrodynamics one has a lovely situation in which the next to lowest order approximation to any calculated quantity is about $1/137$ of the lowest order quantity (in rough magnitude). The next to next to lowest order is another factor of 137 smaller, so if you want values to a certain accuracy you just take a certain number of terms in your perturbation expansion and you get pretty good results. In the strong interactions, the quantity which in quantum electrodynamics is $1/137$, the number that defines how strong the force is, is more like 1, and so the first term is of order 1, the second term is of order 1-squared, the third term is of order 1-cubed, and you just don't make any money that way.

Finally, during this period from the late 50's to the early 60's, there was a profusion of new particles being discovered, very largely here at Berkeley at the Bevatron and very largely through the capabilities opened up by the bubble chamber. These particles seemed every bit as elementary as the proton and the neutron which are the constituents of the nuclei of ordinary atoms. In fact some of them even form families with the proton and neutron. And others look just as good. And there were so many of them. Physicists even got to the point where they had to carry around a booklet, which is published now here at Berkeley, listing all the particles. It was clear that anything you need a telephone directory for can't be elementary.

Remember, this idea of elementarity was tied to the idea of a field theory. That is, the elementary particles were those that were associated with the fundamental fields described by the field equations that were seen as the basic laws of nature, but it didn't look like there were any particles that were any more elementary than any others.

Here at Berkeley, building on earlier work of Chew, Gell-Mann, Goldberger, Low, Nambu, and others on dispersion relations, Chew and Mandelstam and Stapp and their co-workers set about reviving S-matrix theories, but now with a much more specific and mathematically powerful set of axioms than in Heisenberg's time. I hope the nonphysicists in the audience will forgive me, but I'll just name these by the code words, the buzz words that become common among physicists: unitarity, analyticity, Lorentz invariance and clustering. (I won't tell you what any one of them is.) These were the basic properties which it was argued, a theory would have to have to be physically sensible at all. And the hope was that if you demanded these properties, they would provide so many equations relating different elements of the S matrix that the whole theory would become uniquely defined and you could actually solve the equations and come up with numbers for physical quantities. And on top of this you would have the happy feeling in the back of your mind that you were doing what a scientist should do: You were dealing at every point with physical observables, and you were not getting involved with the mythical quantities, the quantum fields, that Heisenberg and Pauli had used in 1929.

Now the S-matrix theory as developed by Chew and his co-workers did not in fact prove in any sense a failure. It did not lead to results on which we now turn our backs. It was however, to a certain extent, bypassed by the main stream of the history of physics in the following years. This first revival of S-matrix theory was in the 1950's and the early 1960's. But it was followed by a second revival of quantum field theory (the first one was in the 1940's), extending from the late 1960's to the present.

The reasons for the second revival of quantum field theory as I said had nothing to do with any failure of S-matrix theory, but with successes in other directions. First of all, theories of the weak interactions and of the strong interactions were developed during this period which were just as good from the point of view of infinities as the older theory of quantum electrodynamics. When I say they were just as good, I mean again that if you were just careful to properly identify the physical quantities that you're talking about, then all the infinities would cancel. Theories were developed for both the weak and strong interactions in which that was true. In fact, it was not just accidentally true; these theories were built on an analogy with the theory of electromagnetism. In the case of the weak interactions it was more than an analogy; there was actually a unification with electromagnetism, so that increasingly, physicists no longer refer to weak and electromagnetic interactions, but just call them the electroweak interactions. The theory of strong interactions was also constructed in close analogy with quantum electrodynamics; it is called quantum chromodynamics. In fact if you look at a page of equations, until you're told exactly how many values the varying parameters take and what the indices mean, etc., the equations look the same. There are reasons why it took so long to develop these theories, having to do with gauge symmetries and broken symmetries and things like that that I will not have time to go into here.

Another reason for the second revival of quantum field theory was the fact that it was found that the strong interactions, although very strong at distances typical of the size of the particles in the nucleus, get progressively weaker as we go to very, very short distances or, equivalently, to very, very high energies. This means that it is now possible to use our theory of the strong interactions to do actual calculations in the same way that we use quantum electrodynamics to calculate what happens inside an atom. The calculations tell us what happens to elementary particles at very high energy, much higher than is typically found in the nuclei of atoms. And the successes of these calculations in comparison with experiments are sufficient to convince us (or will be in a short time) that this theory is correct. I can't claim that this theory of strong interactions is entirely verified experimentally but there's not much question that it's correct.

Oddly though, the old questions of how to calculate the nuclear forces that hold the nucleus of the atom together, or the kind of thing that was worried about at Berkeley when I was here in the late 50's, like what is the cross section for scattering a pion on a nucleon at 700 MeV, cannot be answered today any more than they could have been 20 years ago. This is an evolution that I think often happens in physics. You don't solve all the problems that concern one generation of physicists; instead, the next generation finds that there are more urgent problems. The important thing in the end is not to solve every problem, but to solve enough problems so that you know you have the right theory. And that is what we're in the process of doing with the theory of the strong forces.

The third thing that happened during the second revival of quantum field theory was the realization that there are particles that seem pretty elementary after all. There's the good old electron, and it comes with a family of siblings, the muon, the tauon, and so on, and related particles called neutrinos. No one has ever found any structure inside them, and there aren't an enormous number of them, just the electron, mu, tau and their corresponding neutrinos. In addition, in place of the proton and the neutron and other particles that were being discovered in Berkeley in the 1950's and early 60's, we now have a much smaller set of strongly interacting fundamental particles called quarks. The proton and the neutron and all those other particles discovered here at Berkeley seem to be composites, made up of quarks. In addition, we have one other class of elementary particles, containing the photon, the good old quantum of light, and its siblings, particles called gluons and other particles too heavy to have been produced yet, called W and Z particles. (The Z particle is due to be discovered in Geneva pretty soon now.) All these particles are regarded as elementary, in the sense that they are manifestations of the quantum fields that appear in the underlying field equations, and they are what they are because of these field equations.

In 1975 1 gave a talk about all this at Harvard, and although I don't now remember much of what I said, I remember the title I used for the talk, and it gives a pretty good idea of what I think had been happening in the preceding decade. The title was *The Renaissance of Quantum Field Theory*.

That might seem the end of the story. These have been exciting times. Quantum field theory is riding very high and one might be forgiven for a certain amount of complacency with it. But perhaps we will see another swing away from quantum field theory. Perhaps that swing will be back in the direction of something like S-matrix theory, back to a view of particles as fundamental. There are several reasons that I can point to for this. (By the way, in case you didn't notice, I'm now finished with the historical part of my talk, and up to 1981.) One of the reasons is the continued failure to make a mathematically satisfactory quantum field theory of gravity. The problem again is these damned infinities that Oppenheimer and Waller discovered in 1930. There's no quantum theory of gravity which is free of these infinities and we don't have any good idea of how to make one. From all indications the existing quantum field theory, at least of gravity, and perhaps quantum field theory in general, needs some kind of modification at any energy at or below the very high energy of 10^{19} proton masses. (That's a one with 19 zeros.) Something new has to happen in physics because our existing theories simply break down at these energies.

Another hint of a new energy scale in physics comes from the fact that I've mentioned before, that the strong interaction strength decreases as you go to high energies. If the strength of the strong interactions decreases as you go to high energy, then perhaps it's merely the accident that we are doing experiments at relatively low energy that makes the strong interactions look so much stronger than the other interactions. Perhaps the strong and electroweak interactions really all have the same strength at a fundamental level. The decrease in the strength of the strong interactions is only logarithmic with energy, so the energy at which the strong and electroweak forces become comparable has to be enormously high; in a very wide variety of theories, it is found that that energy is about 10^{15} proton masses. So here again we're led to contemplate enormously high scales of energy.

These hints suggest that there is a fundamental scale of energies in physics far beyond anything that is accessible or will ever be accessible to our accelerators, until someone finds some way of putting a macroscopic amount of energy, like the energy in an automobile tank full of gasoline, on one elementary particle. Perhaps the theory of this new ultra-high energy scale in physics will not be a quantum field theory at all. We don't know. We can't do experiments at these energies.

But if it's not going to be a quantum field theory, the question naturally arises, what are these beautiful theories that we've delighted so much in developing? If the underlying truth is not a quantum field theory, then how come the quantum field theories that we have developed, quantum electrodynamics and then the generalization of quantum electrodynamics to include the Weak interactions, and the theory known as quantum chromodynamics that describes the strong interactions, why do these beautiful field theories work so well? What are they if they're not fundamental?

The answer may perhaps be provided by two theorems. These are what I believe Wightman calls "folk theorems", that is, things that have never been proved but are well known to be true. I think these theorems can probably be formulated in precise terms, and proved, though I haven't done it.

The first folk theorem is that if we write down the most general quantum field theory (for the physicists here I'll say the most general Lagrangian) including all possible terms in the theory that satisfy the appropriate symmetries, and if we calculate processes to any given order of perturbation theory, that is to any order of approximation, then what we get, provided we are talking about the most general possible quantum field theory and not some specific theory, is simply the most general possible S-matrix element which to that degree of approximation satisfies the S-matrix axioms of Chew, Mandelstam, Stapp *et al.* Another, briefer, way of saying this is that field theory is in itself without content. Quantum field theory, divorced from specific theories, but just the general idea of quantum field theory, is just a convenient way of implementing the axioms of S-matrix theory.

In fact quantum field theory has been used in precisely this way ever since 1967 in studying the interactions of low energy π-mesons, particles which no one today regards as elementary and yet which are described by a quantum field theory which is widely used to calculate their various reactions. The reason that this field theory works is precisely because all that quantum field theory does for you is to reproduce the most general S matrix consistent with the symmetries that you're assuming and consistent with the axioms of S-matrix theory.

But if that's all quantum field theory is, if quantum field theory is just a clever mathematical trick for implementing the axioms of S-matrix theory, then why are the detailed working theories that we've developed, quantum electrodynamics and quantum chromodynamics and quantum electroweak dynamics and so on, why are these theories so beautifully simple? You expect to find simplicity in physics if you deal with physics at a fundamental level. If you're dealing with something that's just a lot of mathematical trickery, then why should it look so simple? Why, for example, are the field equations of quantum electrodynamics, or Maxwell's equations for that matter, so simple?

The answer to that question may be found in a second folk theorem. The second folk theorem says that theories with a natural energy scale (and I'm thinking here of energies like 10^{19} proton masses or 10^{15} proton masses or whatever), if studied at much lower energies than the natural energy scale, will always be found to be described to a good approximation by an effective field theory which is as simple as possible. (In technical language, the effective Lagrangian is dominated by terms with the fewest fields and/or derivatives.) Where possible, the interactions in this effective field theory will be so simple that they allow the cancellation of infinities to go through as they did in quantum electrodynamics. I suggest that this is why quantum electrodynamics and quantum chromodynamics and quantum electroweak-dynamics are as simple as they are. Where this is not possible, where the symmetries simply don't allow interactions that are that simple, then the physics will be dominated by interactions which do not allow the cancellation of infinities, and which are also very weak, being suppressed by powers of the natural mass scale, whether it's 10^{15} or 10^{19} proton masses or whatever.

We see at least one example of such very weak forces in the world we study today: gravitation. The fact that gravitation seems outside the scope of quantum field theory is from this point of view just due to the fact that the symmetries that it has to satisfy are too stringent to allow field equations which are simple enough for this cancellation of infinities to occur.

Another possible class of extremely weak interactions which has been much discussed lately and which we may discover in the laboratory are the interactions which could lead to the decay of the proton or to the mass of the neutrino. Elaborate experiments are underway now looking for these things, and they may be found.

If our quantum field theories of which we're so proud are just the debris of some really fundamental theory which describes all of physics including gravity, it may be that the really fundamental theory will have nothing to do with fields; it may not look like a quantum field theory at all. I think we have to leave it as an open possibility that maybe, in fact, it will be something like an S-matrix theory.

On the Uniqueness of Physical Theories*

David J. Gross

It is a pleasure and an honor to participate in the celebration of the sixtieth birthday of Geoff Chew. Over the years many people have expressed surprise upon hearing that I was a student of Chew. To paraphrase their comments: "Funny — you don't look Chewish". I suppose this has something to do with my role in the development of QCD, a quantum field theory of the strong interactions. I shall take advantage of this special occasion to address the astonishment of all of these people, to acknowledge my debt for the education that I received from Geoff and to explain how my personal journey to QCD was a continuous one.

Although this road has led me in a direction different from that of Chew, the goal was common — the search for a unique theory of hadronic structure, a goal I was inspired to pursue as a student of Chew at Berkeley. The uniqueness of physical theories is the central motif in Chew's approach to physics. It is therefore appropriate to address this issue from the perspective of the modern gauge theory of elementary particles and in light of current attempts to construct unified theories.

1. The Road from N/D to QCD

Let us start in Berkeley in the middle of the 1960's. High energy physics was in a period of rapid change, on both an experimental and theoretical level, and Berkeley was in the middle of much of the action. I was one of a large group of Chew's students. This was a very enjoyable experience. In large part, this was due to Geoff's generous and supportive attitude towards his students, No matter how many students he had he would always find time for each, sharing with us his ideas and insights and treating us as full partners in a common effort.

Most of all it was an exciting time. Geoff transmitted to us his unique passion for physics. We were not merely doing phenomenology of the strong interactions, but were embarked on a great adventure to find a unique theory of hadrons. There is

*Supported in part by NSF Grant No. PHY80-19754.
Reprinted from *A Passion for Physics: Essays in Honour of Geoffrey Chew, including an Interview with Chew*, eds. C. DeTar, J. Finkelstein and C.-I. Tan (World Scientific, 1985).

nothing as important to convey to students as this sense of adventure. Formulas and theorems can be learned from books; attitudes and goals one acquires by personal example. Geoff inspired us to think big, to attempt to achieve ambitious goals and in particular to search for uniqueness in physical theories.

On a more practical level one of the most important things I learned in Berkeley, in large part by observing Geoff, was the correct attitude towards experimental physics. Not just the obvious dictum that one must follow closely and critically experimental developments, but rather that elusive, intuitive sense of which experiments are truly important and must be reckoned with, and which are not and may be ignored. Some of this rubbed off on me at Berkeley and was to serve me well.

From Berkeley I went to Harvard. This was the heyday of current algebra. Like many others I was very impressed with the predictive success of the assumed structure of current commutators. Clearly the properties of these observable hadronic currents placed strong restrictions on hadronic dynamics. By this time most of the easy stuff had been done, and the implications of global current algebra were well understood as consequences of broken chiral symmetry. I therefore studied the less understood properties of the algebra of local currents. These were more model dependent — but that was fine, they thus contained dynamical information which went beyond global symmetry. Furthermore, as it was soon realized, they could be tested in deep inelastic lepton-hadron scattering experiments.

In 1967, Callan and I proposed a sum rule to test the then popular Sugawara model, a dynamical model of local currents. Bjorken noted that this sum rule, as well as dimensional arguments, implies the scaling of deep inelastic electron-proton scattering cross sections. This was shortly confirmed by the new experiments at SLAC. In 1968, Callan and I noted that by measuring σ_L/σ_T one could determine whether the constituents of the hadron had spin zero ($\sigma_T = 0$) or spin 1/2 ($\sigma_L = 0$). The experiments indicated spin 1/2. These experiments had a profound impact on me. They showed that the proton behaved, over short times, as if it were made up of pointlike objects with spin 1/2 and (as later neutrino-proton scattering indicated) baryon number 1/3. From then on I was convinced of the reality of quarks, not just as mathematical objects useful as mnemonic devices for summarizing hadronic symmetries, but as physical pointlike constituents of the nucleon. But how could that be? Surely strong interactions must exist between the quarks which would smear out their pointlike behavior. It soon became clear that in a field theoretic context only free field theory could produce exact scaling, Once interactions were included, scaling and the sum rules went down the tube. Yet the experiments indicated otherwise. This paradox and the search for an explanation of scaling was to preoccupy me for the following four years.

My trust in experiment and my distrust of field theory served me well. In 1970, I met S. Polyakov at the Kiev conference, an uninvited but already impressive participant. We had long discussions about deep inelastic scattering. Polyakov knew a lot about the renormalization group and explained to me that in field theory one expects anomalous dimensions. I retorted that the experiments show otherwise. He responded that this contradicts field theory. We departed, he convinced that experiments at higher energies would change, I that the theory would change.

By the end of 1972 1 had learned enough field theory, in particular renormalization group methods from Ken Wilson, to tackle the problem head on. I decided to prove that field theory could not explain the experimental fact of scaling and thus that field theory was not appropriate to describe the strong interactions. The plan of attack was twofold: first to show that ultraviolet stability, later called asymptotic freedom, was necessary to explain scaling and second to prove that no field theory was asymptotically free. In the spring of 1973 Callan and I had completed a proof of the first argument and Coleman and I were close to a proof of the second, There was one hole in the arguments, non-Abelian gauge theories, which for technical reasons were hard to treat. With Frank Wilczek, who had started his graduate work with me that year, we tried to close that hole. The discovery that non-Abelian gauge theories were asymptotically free (made at the same time by Politzer who, working under Coleman, was trying to generalize the Coleman-Weinberg mechanism to Yang-Mills theories) came to me as a total surprise. Wilczek and I quickly realized that color gauge theories of quarks could easily explain the deep-inelastic scattering (albeit with logarithmic modifications of scaling — a bonus since that provided an experimental test of the theory) and that the infrared growth of the coupling (infrared slavery) might provide a mechanism for quark confinement. Rather than killing field theory I was led to a rather unique field theory of the Strong interactions. QCD might be wrong, though few would argue that today, but since 1973 it has never had a serious competition.

2. Arbitrary Parameters

One of the features of QCD that has always appealed to me is its uniqueness. I refer not to the fact that it seems to be uniquely singled out by experiment but rather to the fact that it contains essentially no adjustable dimensionless parameters.

Arbitrary, adjustable parameters in a fundamental theory are embarrassing. The search for uniqueness is often motivated by a desire to explain and fix such arbitrary parameters. Many of us strongly believe that eventually all such parameters will be calculable. Einstein expressed this view eloquently in his autobiographical notes: "I would like to state a principle, which cannot be based upon anything more than a faith in the simplicity, i.e. intelligibility, of nature; that is to say, nature is so constituted that it is possible logically to lay down such strongly determined laws that within these laws only rationally, completely determined constants occur (not constants, therefore, whose numerical value could be changed without destroying the theory)."

This is a marvelous statement of the ultimate goal of fundamental theoretical physics — the motivating force behind the bootstrap and unified field theory alike. It is a rather arrogant goal and a recent one in the history of physics. This is understandable — until one begins to understand the "how" of things one is in no position to ask for the "why".

The first reference I have found where the issue of the number of fundamental constants of nature is seriously addressed is a book of Eddington, published in 1934. Eddington made a list of the seven *primitive* constants of nature. This was a remarkably good list, considering subsequent developments. It consisted of the following: Planck's constant h, the velocity of light c, Newton's gravitational constant G, the electron charge e, the electron mass m, the proton mass M and the cosmological constant Λ.

Of these seven fundamental constants of nature three are, of course, arbitrary; they simply provide units of length, time and mass. It is customary to pick h, c and G to provide natural units, in which length, time and mass are measured in units of the Planck length $\sqrt{hG/c^3}$, time $\sqrt{hG/c^5}$ and mass $\sqrt{hc/G}$. That leaves four dimensionless ratios, pure numbers, whose precise values are amenable to rational understanding. These four numbers can be thought of as representing four categories of physical parameters. They are:

1. The fine structure constant, $\alpha = e^2/hc \sim 1/137$. In this category we would also include the values of other gauge couplings, such as α_{strong}. The values of these couplings are typically of order $1 - 10^{-3}$.

2. The ratios of the electron and proton masses, $m_e/M \sim 1/2000$. In this category we would, today, include the multitude of mass ratios that particle theories attempt to explain: the ratios of hadronic, quark, lepton and weak boson masses. These ratios range from 1 to 10^{-9}.

3. The ratio of the proton mass to the Planck mass, $M\sqrt{G/hc} \sim 10^{-19}$. Our ignorance of the origin of this exceedingly small parameter is referred to nowadays as the "hierarchy problem". In this category we also have the ratios of mass scale of gravity and/or unified gauge theories to that of the weak or strong interactions. These ratios are all believed to be of order $10^{-15} - 10^{-19}$.

4. The ratio of the cosmological constant to the Planck energy density, $\Lambda(G/hc)^2 < 10^{-120}$.

The incredibly small bound on this ratio suggests that $\Lambda = 0$, although we do not understand why this must be so.

Thus Eddington's list is quite up-to-date. It includes four of the primary goals in the contemporary search for uniqueness. In the two first categories much progress has been made; the last two are still clouded in mystery. That is not to say that we can predict the precise value of gauge couplings, quark and lepton masses, or even calculate hadronic mass ratios precisely, but we have understood many of the principles that will determine these. In the case of the hierarchy constant, and especially the cosmological constant, we are far from constructing a mechanism that would uniquely determine them.

3. Progress Towards Uniqueness

The fundamental constant that has attracted most attention has been a. Who has not dreamt of calculating $\alpha^{-1} = 137.03604\ldots$? In fact Eddington and many others have tried, with no success. Today we realize that these attempts were premature, that QED is not a complete or internally consistent theory and that the effective a at the unification scale, where its value is of order $1/10$, will most likely be determined by a solution to the hierarchy problem.

However QCD provides us with a gauge theory of the strong interactions which is internally consistent and realistic, in which the strong α_s and hadronic mass ratios are completely determined and calculable. It is instructive to examine how this occurs and to draw some lessons from this, albeit partial, solution to the problem of uniqueness.

Before considering QCD let us first examine a toy "model of the world". Consider the following non-relativistic Hamiltonian, that describes a world consisting of charged particles of equal mass and charges $\pm e$.

$$H = \sum_i \frac{\mathbf{p}_i^2}{2m} + \sum_{i \neq j} \frac{\pm e^2}{|\mathbf{x}_i - \mathbf{x}_j|}. \tag{1}$$

Although there appear to be two free parameters in H this "theory" actually has no adjustable parameters. That is because all physical observables will be expressible in terms of the dimensionful parameters e, m and h: the numerical values of these is therefore a matter of convention. Another way of seeing why the theory contains no adjustable coupling is to note that under a change of scale: $x \to \lambda x, t \to \lambda^2 t, h \to h, m \to m$, we have

$$H(\mathbf{p}/\lambda; \lambda\mathbf{x}; e^2/\lambda) = \frac{1}{\lambda^2} H(\mathbf{p}, \mathbf{x}; e^2). \tag{2}$$

Therefore a change of the value of the coupling e^2 is equivalent to a change in the scale of length, time and energy.

However this theory does not quite satisfy Einstein's criterion, since there is no reason why the potential must have the form $V = e^2/x$. It could just as well be of the form $V = e^2/x + g/x^2 + fx + \ldots$ Therefore the "theory" has an infinite number of free dimensionless parameters: $g, f/e^2, \ldots$ which we have arbitrarily set to zero. What principle could eliminate the nonuniqueness of this theory? One possibility is to demand extra symmetry of the theory. Thus if we demand scale invariance (namely that under a change of scale the potential energy scales like the kinetic energy) then $V(x)$ must equal g/x^2, and $ef, \ldots$ will vanish as a consequence of the symmetry.

Imposing scale invariance leaves us with a theory with three parameters, h, m and g. However the coupling is now dimensionless. Consequently g is a free parameter. Under a change of scale we now have

$$H\left(\frac{\mathbf{p}}{\lambda}, \lambda\mathbf{x}, g\right) = \frac{1}{\lambda^2}H(\mathbf{p}, \mathbf{x}; g), \tag{3}$$

so that g is unchanged. Thus although we have achieved our goal of explaining why the theory has only three parameters we are left with an arbitrary adjustable dimensionless coupling constant g. Furthermore, this "theory" has a fatal disease. Due to the absence of a length scale in the theory, all energies must be continuous and there can be no discrete states. Such a theory is highly unphysical.

The formulation of QCD, the quantum gauge theory of the strong interactions, also involves three parameters, h, c and g. This is a consequence of a variety of symmetries. The Lagrangian is

$$\mathscr{L} = \frac{1}{2}\operatorname{Tr} F_{\mu\nu}F^{\mu\nu} + \sum_i \bar{\psi}_i(i\partial\!\!\!/ + g A\!\!\!/)\psi_i. \tag{4}$$

The symmetries that eliminate all but these parameters are the following. Gauge symmetry requires that the gluon-gluon coupling, g_{GGG}, equal the quark-gluon coupling g_{Gqq}. Chiral invariance precludes a quark mass term. Finally the requirement of renormalizability, together with chiral invariance, leads to scale invariance which limits the parameters to three.

A distinction must be made between the status of local color gauge symmetry and global chiral invariance, If we alter the equality of g_{GGG} and g_{Gqq}, breaking gauge invariance, we totally change (if not destroy) the structure of the theory; however the addition of a small quark mass, which breaks chiral invariance, has but a small effect on the theory. This would not please Einstein. In the real world the vanishing of (bare) quark and lepton masses is a consequence of local chiral invariance which cannot be disturbed without totally destroying the full (SU(3) × SU(2) × U(1)) theory. For the moment, however, we shall ignore the electroweak gauge interactions.

So far QCD is similar to the previous scale invariant toy model, and would appear to suffer from the same difficulties. Although containing only three parameters (h, c, g), one combination is dimensionless and adjustable, $\alpha_s = \frac{g^2}{hc}$ and scale invariance should imply continuous energies. How does QCD avoid this conundrum?

QCD avoids the pitfalls of scale invariant theories with a finite number of degrees of freedom due to the phenomenon of renormalization. The QCD vacuum behaves as a physical medium containing virtual charged particles (gluons and quarks). The polarization of this vacuum produces a distance-dependent coupling. In QCD, the effects of the gluons dominate. They behave as permanent magnetic dipoles, producing a paramagnetic medium with a permeability $\mu > 1$. In a relativistic theory, this yields a dielectric constant $\epsilon = \frac{1}{\mu} < 1$, as if the vacuum anti-screens color charge. Thus the gauge charge g depends on the distance r from the source, decreasing as $1/\ln(\frac{1}{r})$ for small r and, presumably, increasing without bound for large r. This behavior is responsible for the characteristic dynamical features of

QCD; asymptotic freedom, which leads to free pointlike behavior at short distances and infrared slavery, which leads to color confinement at large distances.

Of concern to us here is the fact that renormalization provides a dynamical mechanism for the breaking of scale invariance, without destroying the symmetries of the theory that restrict the number of parameters to three. Consequently g depends on the distance scale. The precise dependence is quite complicated. It is described by the renormalization group which determines how g varies as the scale of lengths is varied. This breaking of scale invariance means that g can be traded off for a choice of a unit of length — dimensional transmutation. For example a unit of length can be chosen to be the distance at which the effective g is equal to one, and this length can be replaced by a more physical choice, say the Compton wave length of the proton. We are therefore in the same situation as in the first toy model considered above — the theory contains three *dimensionful* parameters; h, c and say, m_p = mass of the lowest bound state with quark number equal to 3. All physical observables can be, in principle, calculated in terms of these. All dimensionless parameters can be determined and there are no arbitrary, adjustable parameters.

4. Lessons and Future Directions

Great progress towards uniqueness has also been achieved in the modern theories of the weak and electromagnetic interactions. As a consequence of unified gauge symmetry the number of independent couplings is greatly reduced; as a consequence of chiral gauge symmetries arbitrary fermionic mass parameters are eliminated. Here too, a crucial ingredient of these theories is the existence of dynamical mechanisms that lead to symmetry breaking but which does not decrease the predictive power of the theory by introducing new arbitrary parameters. Indeed our present frustration with the many (~ 20) arbitrary parameters in the standard model $SU(3) \times SU(2) \times U(1)$ is due to our incomplete understanding of the mechanism of dynamical chiral gauge symmetry breaking.

It is fair to say that we have gone a long way towards determining the parameters of first two entries in Eddington's list (symbolized by α and m_e/m_p). Little progress has been achieved, however, in determining the second pair of constants. This is not surprising since these appear to require an unprecedented extrapolation of present knowledge to distances of $10^{-29} - 10^{-33}$ cm, wherein lie the secrets of gauge theory unification and quantum gravity. What lessons can be extracted from the success of the past decades that could serve as a guide in the ongoing search for uniqueness?

The primary lesson is that the elimination of superfluous constants is a consequence of the discovery of new *symmetries* of nature. However it is not sufficient to simply dream up new symmetries. One must also explain why these symmetries are not apparent. This often requires both the discovery of *new hidden degrees of freedom and a mechanism for dynamical symmetry breaking*. Thus the establishment of color gauge symmetry required the discovery of the carriers of color charge (quarks and gluons) as well as the elucidation of the dynamical mechanism of confinement by which these remain hidden. The establishment of the electroweak gauge symmetry required discovery of the dynamical mechanism of spontaneous gauge symmetry breaking that produces the apparent differences in the structure of the weak and electromagnetic interactions. Future progress towards uniqueness will require both new symmetries and new dynamics.

It is no surprise therefore that much of exploratory particle theory is devoted to a search for new symmetries. Some of this effort is based on straightforward extrapolations of established symmetries and dynamics, as in the search for unified gauge theories (SU(5), SO(10), etc.), or in the development of a predictive theory of dynamical chiral gauge symmetry breaking (technicolor, preons, etc.). Ultimately more promising, however, are the suggestions for radically new symmetries and degrees of freedom.

First there is supersymmetry, a radical and beautiful extension of spacetime symmetry to include fermionic charges, which has the potential to drastically reduce the number of free parameters. Most of all it offers an explanation for the existence of fermionic matter, quarks and leptons, as compelling as the argument that the existence of gauge mesons follows from gauge symmetry. Supersymmetry will undoubtedly play an important role in a future unified theory. At this point, however, we have not developed an understanding of the dynamics of supersymmetry breaking which is adequate to allow for a direct confrontation with experiment.

An even greater enlargement of symmetry, and of hidden degrees of freedom, is envisaged in the current efforts to revive the ideas of Kaluza and Klein, wherein space itself contains new, hidden, dimensions. These new degrees of freedom are hidden due to the spontaneous compactification of the new spatial dimensions, which partially breaks many of the space-time symmetries of the large manifold. The unbroken isometries of the hidden, compact, dimensions can yield a gravitational explanation of the emergence of gauge symmetries (and, in supergravity theories, fermionic matter). A combination of supersymmetry and Kaluza-Klein has the potential to yield truly unique theories, and could shed light on the hierarchy and cosmological constant problems. The most radical of these attempts to construct unified theories are based on the quantum theory of strings. Here we are contemplating an enormous increase in both degrees of freedom (the basic dynamical entity is not a pointlike object but rather a one-dimensional extended object) and in symmetry. One of the most exciting features of these string theories, which have the possibility of containing all of known low energy physics, is their large degree of uniqueness. If a unified string theory turns out to be correct, it could not only allow us to calculate all of Eddington's fundamental constants but could even determine the number of spatial dimensions.

Finally, recent developments have taught us that gauge symmetries lead to strong global constraints that are not deducible from local considerations. Theories which are apparently symmetric may contain anomalies when one considers the global structure of configuration space. Consistency requires the absence or cancellation of such anomalies and leads to strong constraints on the freedom to adjust parameters. Indeed it appears that as we contemplate even greater extensions of local gauge symmetry, consistency requirements lead to even stronger constraints, thus bolstering our hopes of ultimate uniqueness.

It is unlikely, however, that the ultimate goal will ever be achieved. The more we understand, the more questions we learn to ask. So while we might eventually calculate all of the arbitrary parameters that can now be enumerated, success will most likely produce a new set of questions and parameters whose relevance is not yet appreciated. This is just as well, since the most exciting and enjoyable element of the search for uniqueness is the search itself.

My Postdoctoral Years at Berkeley with Geoff Chew

Steven Frautschi

Walter Burke Institute of Physics,
California Institute of Technology,
Pasadena, CA 91125, USA

After graduating from Stanford in 1958 and spending a year at the Yukawa Institute for Fundamental Physics in Kyoto, Japan, I had the good fortune to be accepted as a postdoc at Berkeley for 1959–61. Upon hearing this, one of my Japanese colleagues said, "Chew is there!"

What was so special about Chew? First of all, he was a superb group leader, articulating a deep program (understanding strong interactions via the tools of analyticity, crossing, and unitarity) and providing each student with a niche where they could contribute to the cause. When Chew spoke at meetings, he always outlined his general program, and then took care to indicate how each and every graduate student, as well as postdoc, fit into this program.

When I arrived in 1959, his research group comprised several graduate students, plus postdocs including Stanley Mandelstam, myself, and Marcel Froissart. Most weekdays he would stop by our postdoc office at noon and walk over to lunch at the nearby faculty club with Mandelstam and whoever else was available. Once a week, lunch was devoted to a full group meeting including graduate students where Chew or others would report on current developments or their own research projects. Mostly, the subject was Chew's bootstrap and S-matrix program but sometimes it was a broader journal club. I remember one occasion when I was asked to study up and report on the then — unfamiliar topic of the Renormalization Group.

There was even an annual excursion to Candlestick Park where he would introduce his international group to major league baseball and the San Francisco Giants. As a young man, he had been a standout baseball player and had tried out for a pro team until a bad back ended that effort. At the annual Physics department picnic in Berkeley, his hitting and throwing were on a level far above everyone else, until his back acted up and forced him to stop.

Upon meeting him for the first time, people were often surprised to find that Chew was not Chinese (similarly, his collaborator Francis Low had not a drop of Chinese blood). Chew's father was English, and he told us his middle initial F. stood for Fouquar, an old Norman name. However, he explained that his great grandfather on his mother's side had managed a teak plantation in Burma and married a local woman, presumably accounting for Chew's high cheekbones and permanent suntan.

But of course what really made Chew's group exciting was his research program. Up to the late 1950s, most of the precision particle physics scattering data was limited to two-body scattering in the energy region below one BeV. Several resonances had been identified in pion-nucleon scattering, most notably the $J = 3/2, I = 3/2$ ("3-3") resonance. It quickly became apparent that the pion-nucleon coupling (known from Yukawa's pion exchange in nucleon-nucleon scattering) was too strong to allow perturbation theory calculations to come close to matching experiment. Theorists had shifted to the firmer ground of fixed momentum transfer dispersion relations derived from analyticity properties of scattering amplitudes plus crossing and unitarity. Chew had been a leader in this work, especially in the landmark Chew-Low paper[1] which showed for the static model of the pion-nucleon interaction that the scattering amplitude is an analytic function of the energy, that the "forces" corresponding, for example, to nucleon exchange between pion and nucleon could be associated with calculable singularities of the scattering amplitudes in unphysical regions, and that this could approximately explain a number of experimentally observed features of the low energy pion-nucleon system including the 3-3 resonance.

The one-dimensional dispersion relations used through 1957 did not describe all unphysical singularities and were insufficient to determine the S-Matrix. The inclusion of all the forces requires a knowledge of singularities in momentum transfer as well as energy; this information was provided by the double dispersion relations proposed in 1958 by Mandelstam.[2] Chew happened upon a presentation of Mandelstam's novel ideas at the 1958 spring American Physical Society meeting in Washington D.C. Initially unable to understand what Mandelstam was talking about, Chew took him aside for two hours of intensive discussion. Becoming convinced that Mandelstam had special insight, Chew invited him to Berkeley on the spot.[3]

During my years at Berkeley, Chew's program of S-Matrix studies was centered on understanding and utilizing Mandelstam's contributions. The first year the focus was on the double dispersion relations and s, t, u variables, where any one of s, t, u could serve as the energy variable and the other two as forward and backward exchange variables. This suggested all sorts of approaches, models, and approximations. Personally, I cut my teeth on a study with my college classmate Dirk Walecka (then at Stanford) of single and two particle exchanges in low energy pion-nucleon scattering using the Mandelstam representation. That suggested extending the analysis to high energy to describe forward peaks (high s, low t) and backward peaks (high s, low u) in terms of the nearest singularities in the t and u channels (the "strip approximation"). Here I had the excitement of working directly with Chew on ideas that were relevant to early data from the new high energy accelerators that were coming on line.[4]

A focus of my second year at Berkeley was the struggle to make scattering amplitudes at asymptotically high energies behave when hadrons with spin greater than one were exchanged. Already in the early 1940s Werner Heisenberg had noted that such exchanges would cause the scattering cross section to grow as a power of the energy at energies much greater than mc^2, approaching infinity in the high energy limit. That contradicted what we knew of scattering cross sections for cosmic rays, it contradicted physical expectations for a short-range force, and in 1961 Marcel Froissart proved a landmark theorem that such growth was impossible. So, either no hadron existed with spin greater than one, or it seemed that the rapid growth of the one-particle high spin exchange cross section would have to be cancelled by messy, hard-to-compute higher order corrections.

By 1961 experimental developments were making it imperative that this theoretical crisis be resolved. New facilities capable of accelerating protons to extreme high energies of 30 mc^2 were coming into operation at the CERN Lab in Geneva, Switzerland, and Brookhaven Lab on Long Island, New York. And strongly interacting particles with higher spin were beginning to be discovered, presaging a flood of such discoveries during the 1960s.

In a seemingly unrelated 1959 paper[5], Tulio Regge had introduced new ideas into the theory of low energy (kinetic energy much less than mc^2) scattering of a particle by a potential. Regge considered angular momentum J as a complex variable. He showed that Schrödinger's equation allowed for a continuous set of solutions $J(E)$ as energy E increased, even though these mathematical solutions could represent physical bound states or resonances of the particle in the potential only at energies where J reached the quantum-allowed angular momentum values $J = 0, 1, 2, \ldots$. If the potential was attractive enough, one would get a sequence of physical states of progressively higher angular momentum and energy, which we dubbed a "Regge trajectory."

Remarkably, Mandelstam noticed the low energy Regge paper in the journal Nuovo Cimento and suggested that it might solve our high energy problems! Following up on his suggestion, Chew and I proposed[6] that all hadrons lie on Regge trajectories. This meant that the appropriate mathematical object to exchange was not the individual J spin particle (with it's dangerous high energy behavior if J exceeded one) but the whole moving Regge trajectory which, even if it represented one or more high spin particles, could have the property of falling below one in the kinematic region where the exchange scattering took place. It followed that the high energy behavior was automatically well-behaved without the necessity of making high energy corrections.

Our proposal had the following consequences:

(1) The conceptual problem of reconciling the existence of high spin particles with finite high energy cross sections was resolved.

(2) We related the energy dependence of various high energy reactions such as $\pi^- + p \to \pi^0 + n$ to exchanged Regge trajectories, and predicted that high energy forward scattering peaks should shrink as the energy increases. These features were quickly verified experimentally as the CERN and Brookhaven accelerators came into action.

(3) Since we had at most two experimental points per Regge trajectory when we wrote our paper, we drew straight lines through them to represent the interpolating Regge trajectories. Over the next couple of years, I attended several meetings where a speaker, unlike us, had taken the trouble to actually calculate the Regge trajectory for scattering in a Yukawa (or other hopefully realistic) potential. The displayed trajectory would rise with energy, then fall back along a curve; the speaker would say with a flourish, "See how straight the trajectory is" and the audience would laugh. But those single channel potential calculations missed a crucial feature of the high energy world: every time the energy rises by $m_\pi c^2$, a new channel opens up. In fact, our straight line choice proved serendipitous — as more data points became available it turned out that the Regge trajectories really were approximately straight lines, with a quasi-universal slope. Thus, there was equal spacing in the energy variable s between successive states such as $J = 0, 2,$ and 4. Since energy is proportional to frequency in wave mechanics, that meant we were observing equal spacing in frequency. And a familiar feature of waves such as those on a violin string is equal spacing between the frequencies of successive overtones. So, though we were not explicitly aware of it at the time, we had stumbled upon evidence that hadrons are not points but strings (mesons, for example, are now thought of as gluon strings with a quark at one end and an antiquark at the other end, and the universal slope of the Regge trajectories is related to the string tension). In this way our 1961–62 papers became part of the twisting road that led to string theory.

Another implication of Regge trajectories was their impact on the question: which of the observed hadrons could be elementary particles? As we've already stated, any hadron with $J > 1$ must lie on a moving Regge trajectory, and moving trajectories are the result of attractive potentials acting dynamically, so such hadrons cannot be elementary. Our proposal that all hadrons, even with $J = 0, 1/2,$ or 1, lie on Regge trajectories was a speculative leap which implied that none of the observed hadrons are fundamental.

This proposal was related to Chew's bootstrap idea. The bootstrap idea was roughly that hadrons arise from potentials associated with exchange of other hadrons in a self-consistent way, without starting from an interaction in a Lagrangian field theory. The term "bootstrap" had been introduced in 1959 by Chew and Mandelstam[7] in connection with $\pi\pi$ scattering, where exchange of the ρ meson provides an attractive potential for production of the ρ resonance. Another example was provided by the "reciprocal bootstrap": an expanded version of the 1956 Chew–Low model for πN scattering wherein exchange of the nucleon and P 3-3 resonance produced an attractive "potential" leading to a nucleon bound state and the P 3-3 resonance.[8] These crude models provided a hint of how prominent $J = 1$ and $J = 1/2$ hadrons — even the nucleon — could arise through self-consistent dynamics. The Chew–Frautschi proposal that all hadrons lie on Regge trajectories implied that all hadron poles might be determined by a collective bootstrap — even though we only had the explicit examples of low energy $\pi\pi$ and πN scattering.

I spent the year after leaving Berkeley as an Assistant Professor at Cornell. There Hans Bethe asked me to give a series of seminars on Regge poles and related subjects. I poured into these lectures what I'd learned at Berkeley in countless discussions with Chew and his group. When Chew was invited to give an extensive series of lectures [15 one-hour talks over three weeks!] at a summer school in Bangalore, India, he demurred, but recommended me as a possible substitute. Thanks to the work I'd already prepared for Bethe, I was able to accept and enjoy a whirlwind trip to India (weekend visits to ancient temples, a tiger preserve, and the Nilgiri Hills, as well as a lecture every weekday.) The Bangalore lectures took printed form in my book, "Regge Poles and S-Matrix Theory."[9]

After I left Berkeley in 1961 Chew's "democratic" hadron bootstrap (all hadrons are made of each other) was never falsified. But it lost out to the much more predictive quark model associated with a gauged quantum field theory. It became generally accepted that no hadrons are elementary, but for the completely different ("aristocratic") reason that they are all made of quarks and gluons. Regge trajectories proved useful in describing high energy scattering phenomenologically, but their predictive power was compromised by Regge cuts. High energy experimentalists developed means of detecting multiparticle final states, which are harder to analyze using only analyticity, unitarity, and crossing.

And yet "dual resonance" models[10] were developed which contain in one elegant analytic function straight Regge trajectories (with associated "daughter trajectories") in the direct and exchange channels. These models provided impetus for the development of string theory. Thus, Regge poles are alive and well. And just as small dinosaurs live on as birds even though their great grandparents might not recognize them, Chew's vision of the bootstrap with Regge trajectories dynamically interwoven with Regge exchanges lives on in string theory.

References

1. G. F. Chew and F. Low, *Phys. Rev.* **101**, 1570 (1956).
2. S. Mandelstam, *Phys. Rev.* **112**, 1344 (1958).
3. G. F. Chew, *Memorial Volume for Stanley Mandelstam* (World Scientific, 2017), p. 1.

4. G. F. Chew and S. C. Frautschi, *Phys. Rev.* **124**, 264 (1961).

5. T. Regge, *Nuovo Cimento* **14**, 951 (1959); *ibid.* **18**, 948, (1960).

6. G. F. Chew and S. C. Frautschi, *Phys. Rev. Lett.* **7**, 394 (1961); *ibid.* **8**, 41 (1962).

7. G. F. Chew, in *Proceedings of the Kiev Conference on High Energy Physics*, Plenary Session III, 332 (1959); G. F. Chew and S. Mandelstam, *Nuovo Cimento* **19**, 752 (1961); F. Zachariasen, *Phys. Rev. Lett.* **7**, 112, (1961).

8. G. F. Chew, *Phys. Rev. Lett.* **9**, 233 (1962).

9. S. Frautschi, *Regge Poles and S-Matrix Theory* (W.A. Benjamin Inc., 1963). This graduate level monograph sold 7500 copies at a time when the publisher's breakeven point was 2200 copies.

10. G. Veneziano, *Nuovo Cimento* **57A**, 190 (1968).

Salesman of Ideas*

John Polkinghorne

We used to call Geoff Chew "the handsomest man in high energy physics." I know of at least one senior secretary in a British physics department who kept a photograph of him near her desk, That frank and open face, with just a hint of his one-eighth Burmese ancestry, and his tall commanding figure, made him one of the few theorists in the pin-up class. Allied to this was considerable personal charm and an ingenuous manner. Geoff was definitely a man from whom one would be happy to buy a used car.

Indeed it was as a salesman that Geoff exerted some of his most important influence on high energy physics, peddling, of course, not automobiles but ideas. They could be his own or they could be those of others, for Geoff was remarkably quick to get onto the scent of a new discovery and generous in propagating other men's theories. I remember a "Rochester" Conference at CERN in the late fifties at which Geoff expounded with characteristic vigour and conviction the recent work of the then young and comparatively unknown Stanley Mandelstam. Stanley had just written a paper in which he introduced his conjectured two-variable representation of the scattering amplitude, incorporating analyticity and crossing properties. At the time the paper was not easy reading but Geoff rightly persuaded us all of its importance and the need to get to grips with it. There was a saying at the time that "there is no God but Mandelstam and Chew is his prophet." They were a striking pair together — the long and the short of it, one talkative, the other quiet. The combination was powerful indeed.

There was a sequence of "Rochester" Conferences at which Geoff fulfilled this salesman role. His talks were always eagerly awaited, both because of their inspirational and encouraging tone which helped to sustain one's possibly flagging spirits, and also because of his ability to put his finger on whatever was most promising in the year's crop of ideas. Prompted by Mandelstam, Geoff was among the first to appreciate and attempt to exploit Regge's ideas on high energy behaviour. He also recognized the significance of the bound on high energy behaviour obtained from

---

*Reprinted from *A Passion for Physics: Essays in Honour of Geoffrey Chew, including an Interview with Chew*, eds. C. DeTar, J. Finkelstein and C.-I. Tan (World Scientific, 1985).

very general principles by Marcel Froissart. For a while the message was that one should "saturate Froissart", a name, I remember, which Geoff pronounced at the time in a manner more American than Gallic.

But of all the ideas that Geoff presented to the physics community, none was sold with greater fervour than that of the S matrix and the bootstrap. It was a grand notion. When a few years ago I wrote a little book, *The Particle Play*, about the development of high energy physics I was rebuked by some reviewers for having included a chapter on these ideas, now so out of fashion. I remain unrepentant. Bootstrappery was a significant episode in the subject and the concept is one that deserves intellectual appreciation, whatever its eventual fate as a physical principle may prove to be. Its attraction lies in its audacity. The exchange of particles, according to our understanding, creates forces. May not these forces then prove sufficient in turn to create the particles exchanged? In this way the equation of the world would become a gigantic self-consistency condition expressing the possibility that everything is made out of everything else. The universe would, in an act of breathtaking legerdemain, have lifted itself into being by its own bootstraps. It is difficult to think of a more grandiose or exciting proposition.

One of the consequences of bootstrappery would be the abolition of the idea of elementary particles. If everything is made of everything then the supposition of special basic constituents becomes redundant. Geoff coined a slogan for selling this: "nuclear democracy". He proclaimed this antielitist physics with enthusiasm. (Oddly enough, in Regge theory he permitted himself a more hiererchical terminology, calling leading trajectories the "Queen" and the "King" and gallantly placing the Queen above the King. This idiosyncratic terminology never caught on.)

All good sales campaigns depend for their success upon timing. In this respect things were perfect. The S-matrix idea cashed in on the disillusionment then current with quantum field theory, so successful perturbatively in dealing with the comparatively weak electromagnetic interactions but apparently powerless to tackle the non-perturbative problem of the strong interactions. Bootstrappery seemed particularly appealing at a time when so-called "elementary" objects were proliferating, as experimentalists discovered resonance after resonance and before the quark model brought recognized order to the chaos of high energy physics.

Geoff proclaimed these ideas with a fervour that went beyond that of the salesman and approximated to that of the impassioned evangelist. There seemed to be a moral edge to the endeavour. It was not so much that it was expedient to be on the mass-shell of the S matrix as that it would have been sinful to be anywhere else.

The intensity of this conviction is summed up for me by the recollection of a small conference at La Jolla in the sixties. It was made clear to us that it was time to decide what our positions were. Speaking with directness and simplicity Geoff said something to the effect that we should all abandon the fruitless pursuit of out-moded field theory and supply our efforts to the elucidation of the S matrix. He had run up his colours and nailed them to the mast. One thought of Martin

Luther, "Here I stand, I can do no other." Arthur Wightman, sitting in the front row, turned a bright red but behaved with characteristic courtesy at this labelling of his life's work as vanity.

Geoff's unflagging commitment to the S matrix maintained the momentum of a massive Berkeley programme toiling with ideas of bootstrappery through successive versions of the strip approximation. A great deal of valuable hadronic physics was learnt in the process but in the end it collapsed under the weight of its own complexity. After all, if everything is made of everything that is going to be a rather involved notion to express and manipulate. Moreover it became plain that the times were not after all favourable to egalitarian ideas of nuclear democracy. It became clearer and clearer that quarks were cast for an elitist elementary role.

Intellectual history is full of ups and downs and the years since those heroic times have also seen an astonishing reflowering of quantum field theory. Like Mark Twain, the report of its demise has proved exaggerated.

Theoretical physicists of my generation owe a great debt of gratitude to Geoff Chew. He was our inspirer and encourager, guiding and leading us by his enthusiasm and example through an era of great activity in the subject. He spoke always with integrity and intensity, so that you readily bought his ideas. We also owe him a great debt of affection, for he was and is one of the nicest men you could wish to meet.

The Bootstrap: Still a Relevant and Prolific Idea

Gabriele Veneziano

Theory Department, CERN, CH-1211 Geneva 23, Switzerland
Collége de France, 11 place M. Berthelot, 75005 Paris, France
gabriele.veneziano@cern.ch

Geoff Chew's bootstrap idea fascinated me as a graduate student and strongly influenced the direction of my early research activity for about 10 years. Although with the advent and triumph of QCD the bootstrap eventually lost some of its appeal for strong interactions, its philosophy of getting insight — as well as important results — from mere self-consistency is still very actively pursued in many areas of theoretical physics. This modest contribution will try to trace back the evolution of Geoff's original idea, as I felt it, through half a century of theoretical particle physics.

1. Fascinated as a Graduate Student

As a graduate student, first in Florence in the group led by Raoul Gatto, then at the Weizmann Institute under the supervision of Hector Rubinstein, Geoff Chew's bootstrap idea strongly appealed to me and influenced my research. The group of Gatto was mainly working in the theory and phenomenology of weak interactions and made important contributions in that area (e.g. the Ademollo–Gatto theorem). In fact my (unpublished) master thesis had to do with non-leptonic weak decays of hadrons using some then popular group-theoretic framework. But already my first publication, in 1965, had to do with strong interactions where, together with my advisor, I attempted to get the mass of the Δ_{33} baryon from an approach vaguely resembling a bootstrap. Even the following publications from my short post-doctoral Florence period dealt with strong interactions, particularly through the use of Current Algebra.

Chew's bootstrap was a very ambitious one: solving the strong-interaction problem by using only some very general requirements like analyticity and unitarity of the S-matrix. Of course Geoff knew very well that something specific to strong interactions had to be added to his program: otherwise how could one distinguish them from, say, QED, a theory that was also supposed to obey those same general requirements?

What was typical of the strong interaction was its strength, of course, its short range (i.e. absence of massless particles) and, especially, the extreme richness of its spectrum of (mostly unstable) particles. This led Geoff to base his "Regge bootstrap,"[1,2] on what he called "Nuclear Democracy,"[a] the assumption that *all* strongly interacting particles (hadrons) lie on Regge trajectories (for positive values of their argument) and that *all* asymptotic behaviors are determined by *the same* Regge trajectories (for negative values of their argument). His hope was that this basic assumption, together with the very general properties mentioned above, would completely fix the strong-interaction S-matrix: indeed a fascinating idea!

However, it was not until I went to the Weizmann Institute as a Ph.D. student that I fully embarked on a bona-fide research program centered on a variant of Chew's bootstrap idea. The turning point took place at Erice's summer school during its 1967 edition. There were many exciting courses, but my mind was particularly struck by a side remark made by Murray Gell-Mann. He advertised a very recent paper by Dolen, Horn and Schmid[5] (DHS) that had just come out of the Caltech theory group. The paper had to do with pion-nucleon charge exchange ($\pi^- p \to \pi^0 n$) where DHS observed that the contribution to the (imaginary part of the) scattering amplitude coming from s-channel baryonic resonances was well represented, on average, by the contribution coming from the t-channel bosonic Regge trajectories. The two contributions were "Dual" to each other (this property became thus known as DHS duality) so that adding them would give the wrong scattering amplitude (by about a factor 2).

Gell-Mann stressed that this observation could lead to a new and easier bootstrap than the one advocated by Chew. I remember Murray calling it a "cheap bootstrap" as opposed to the one of Geoff, that he considered (technically-speaking) "expensive." The main difference, qualitatively, was that Chew's bootstrap involved the non-linear unitarity constraint while the duality bootstrap, at least superficially, did not.

2. Solving the Duality Bootstrap

Back at the Weizmann Institute we formed a very determined collaboration to tackle the "easy bootstrap" program. Our gang of four consisted of M. Ademollo, then at Harvard, H. Rubinstein, M. Virasoro and myself, all three at the WIS. We were later joined for parts of the project by two graduate students: M. Bishari and A. Schwimmer.

DHS had shown experimental evidence for their duality for the reaction $\pi^- p \to \pi^0 n$ where there is a net exchange of a unit of charge in the t-channel. The Regge poles relevant to the process included, for instance, the one containing the ρ-meson, while the s-channel resonances included the already mentioned spin-3/2 Δ_{33} baryon. For implementing the bootstrap it was much better to choose a process in which s

---

[a] A similar idea is at the basis of R. Hagedorn's "Statistical Bootstrap"[3] which also made contact later with Dual-Resonance-Models and String Theory, see Ref. 4 for an account of that story.

and t-channel shared the same resonances/Regge trajectories even at the expense of being experimentally unaccessible (the advantage of being theorists!). Therefore, instead of looking at pion-nucleon scattering, we picked up the process: $\pi\pi \to \pi\omega$ which allows the same Regge trajectories (with the very restrictive quantum numbers of the ρ-meson and excitations thereof) in all three channels. The first results[6] were quite encouraging and looked compatible with the approximate linearity of Regge trajectories (meaning a linear relation between spin and squared mass: $J \sim \alpha'M^2$ with an almost constant Regge-slope parameter α'). Experimental evidence for such a behavior had been stressed for sometime by S. Mandelstam.[7]

I will skip the rest of that story: it led, eventually, to my 1968 paper[8] describing that reaction in terms of an Euler Beta-function, and marking the beginning of what became known as the "Dual Resonance Model" (DRM). In turn the DRM evolved within a few years into String Theory. From that perspective one can argue that the postulates of Chew's original bootstrap had to be augmented by a dynamical assumption: that all hadrons are strings! The nuclear-democracy assumption was thus confirmed as being a necessary ingredient but not a sufficient one without an extra qualification. Actually, the idea that hadrons must have a string-like structure can be seen as a direct consequence of $J \sim \alpha'M^2$ in which $T \equiv (2\pi\alpha')^{-1}$ has the meaning of a (classical) string tension (energy per unit length). In the words of Hagedorn (see Ref. 4), string theory gives a microscopic interpretation of a basic consequence of his statistical bootstrap, the exponential growth (with energy) of the number of hadrons and a limiting temperature, in analogy with what statistical mechanics does with respect to thermodynamics.

Is it possible that with such an extra assumption, strings, one has to end up with a unique theory? In the modern reinterpretation of string theory[9] in which it is elevated[10] to a candidate quantum theory for all interactions, including gravity, this does not look to be the case since we know of several distinct, yet consistent, superstring theories. One can argue that M-theory, by unifying the known consistent superstrings, provides an affirmative answer to that question. Nevertheless, its pretended uniqueness is undermined by the fact that M-theory has many solutions ("vacua") each one with its physical realization. It also pretends to describe all interactions, including those to which the bootstrap is not supposed to apply. Does this mean that the bootstrap had died with the birth of string theory? I will now explain why it did not, at least immediately.

3. Topological Unitarization and More Bootstrap

In the summer of 1970, while visiting CERN, I started to discuss with S. Fubini, A. Di Giacomo and L. Sertorio the issue of how to implement unitarity in the DRM (which was already evolving into becoming string theory). People had developed techniques to compute loop corrections to the tree-level dual amplitudes. Although this was conceptually sufficient it looked to us like a long shot towards efficiently

adding unitarity to the model. Furthermore, there was a problem intrinsically related to DHS duality. In QFT one is instructed to resum non-perturbatively infinite
sets of diagrams, e.g. bubble diagrams corresponding to mass renormalization and
decay widths of unstable resonances (and there was an infinite number of them in
the model!). But, because of duality, it was not possible — or at least awkward —
to separate diagrams with a given Feynman-diagram-like structure, from all other
diagrams of a given duality-preserving topology.

   We then proposed[11] to consider a new unitarization procedure which would
avoid the above (non duality preserving) separation. To begin with the whole class
of planar dual diagrams would provide resonances with a width so as to avoid an
obvious clash with unitarity while maintaining the good features of the narrow-
resonance approximation. Then, full unitarity would force upon dual diagrams of
more complicated (i.e. non-planar) topology to be added perturbatively.

   This was, as far as I know, the first example of a resummation of diagrams
according to a duality-preserving topological classification. It became known later
as "Dual Topological Unitarization" and was followed successfully by several groups
(see, e.g. Refs. 12 and 13) for describing high-energy soft hadronic physics. I found
the whole idea very appealing and worked on various aspects of it (defining the
bare Pomeron, classifying different contributions to inclusive cross sections, etc.)
for a number of years. But concerning our subject here, Chew and the bootstrap,
the most interesting aspect was the so-called planar bootstrap.[14] This was the
statement that the first term of this topological unitarization had to satisfy *exactly*
and non-perturbatively a version of unitarity called planar unitarity.

   This planar approximation to the full theory represented a simplified version
of Chew's grand bootstrap (somehow midway between the "cheap bootstrap" of
Gell-Mann's 1967 lecture and the really "expensive" one). I will give an example of
it. Consider the s-channel discontinuity of a 4-point function in the planar approximation: this is still given by the well-known Cutkowsky cutting rules whose result
is:

$$\frac{1}{2i}\text{Disc}_s A_{2\to2}^{\text{planar}} = \sum_{n=2}^{\infty} A_{2\to n}^{\text{planar}} \otimes_{\text{planar}} A_{2\to n}^{*\text{planar}} = \int \frac{1}{2i}\text{Disc}_{M^2} A_{3\to3}^{\text{planar}} \qquad (1)$$

where $\otimes_{\text{planar}}$ means that the n particles in $A_{2\to n}$ and $A_{2\to n}^*$ have to be suitably
paired in order to preserve the planarity of the outcome. Also, in the last equation
we have used exact unitarity sum rules[15] as well as A. Mueller's generalization of
the optical theorem to inclusive cross sections.[16] Finally, one can argue that in
the planar limit both the four and the six-point amplitude are dominated, at high
energy, by $(q-\bar{q})$ Regge-pole exchanges. Since (1) relates amplitudes with a different
number of legs — and consequently different powers of the strong coupling — the
end result is a self-consistency (i.e. bootstrap) non-linear relation to be satisfied by
the mesonic Regge trajectories and their couplings. Making some reasonable ansatz
on these, one was able to obtain quite accurate predictions, e.g. for the ρ-trajectory.

   Geoff liked very much this reformulation of its old bootstrap program since it could justify, in the planar approximation, assumptions that were not so obvious in his original program. It allowed, for instance, to work with Regge poles only, since, as S. Mandelstam had shown,[17] Regge-cuts (branch points in the complex-J plane) could only originate from non-planar diagrams. Regge pole couplings also obey simple factorization properties, just like ordinary resonances. Geoff worked intensively on the subject, first in collaboration with C. Rosenzweig with whom he published several papers as well as an interesting review article.[18] One idea he pushed[19] went under the name of "asymptotic planarity" according to which in some kinematical limit there was a dynamical reason to justify the dominance of planar diagrams and therefore an expansion in topological complexity. Later on he collaborated with Poenaru[20] on further extensions of the topological bootstrap idea incorporating in it also multi-sheeted surfaces of the kind suggested by QCD.[21]

Meanwhile QCD was knocking at the door and slowly started to impose itself as the correct theory of strong interactions. Its celebrated Asymptotic Freedom (AF) was looking interestingly related to asymptotic planarity. Both would justify, for instance, the so-called Okubo–Zweig–Iizuka (OZI) rule which disfavors non-planar (hairpin-like) diagrams.

Another aspect of QCD which, if I remember well, struck a cord in Geoff's heart was a statement, first made (I believe) by R. Dashen, that QCD predicts the couplings among hadrons to be fixed, instead of being functions of the strong coupling appearing in the Lagrangian. This sounded very much like what I described earlier as a result of the planar bootstrap. Yet the explanation within QCD was quite different: a consequence of "dimensional transmutation" by which the original dimensionless coupling would be converted, through its renormalization-group running, into a dimensionful energy scale, the typical mass of hadrons arising from confinement (even in the absence of any input quark-mass parameter). Conversely, hadronic couplings (in those same units) would be fixed ... well almost as we shall now discuss.

4. Large-N QCD: A Bootstrap Without Bootstrap?

At first sight QCD looks quite in line with one of Chew's basic assumption of nuclear democracy. He was advocating that all hadrons are, at a fundamental level, equal. According to the Nuclear democracy idea, they all lie on Regge trajectories in the time-like region whereas the same Regge trajectories in the space-like region would dominate the high-energy behavior of scattering amplitudes. As already mentioned, such an assumption looks very well motivated in the planar approximation. In QCD hadrons are also all equal albeit in a different sense: they are all bound states of the same constituents, quarks and gluons. Their stringy nature also comes out as a result of color confinement, whereby a thin tube of chromo-electric field pops up as we try to separate the quarks inside the hadrons by injecting energy.

We mentioned that hadronic couplings and masses will all be fixed in terms of a single mass/energy parameter usually denoted by $\Lambda_{\rm QCD}$. This is believed to be true in the real world modulo effects due to quark masses. Does then QCD make the same claim as the bootstrap about fixing couplings and masses? The bootstrap would say that mere self-consistency would fix those quantities. Instead, QCD would say that an extra information is needed, the fact that the gauge group underlying QCD is $SU(3)$. One can imagine different versions of QCD (with the same number of quarks) but with a gauge group $SU(N_c)$ where N_c represents the number of "colors" a quark of a given flavor can have.

In 1974 't Hooft came up with the brilliant idea[22] of considering such an infinite set of theories and, in particular, the limit in which $N_c \to \infty$ while keeping the so-called 't Hooft coupling $g^2 N_c$, as well as the number of quark flavors N_f, fixed. He then showed that, in this limit, planar diagrams without quark loops dominate over all other diagrams. An extension of that limit,[23] in which N_f/N_c is also kept fixed, actually gives precisely dominance of the diagrams of the planar bootstrap. What could not be better for the bootstrap then? The problem is that the hadronic couplings are no longer fixed once and for all but depend on N_c (and also on N_f/N_c in the limit of Ref. 23). In particular, hadronic couplings become arbitrarily small as $N_f/N_c \to 0$. That meant, however, that the planar hadronic bootstrap itself could not possibly be sufficient for determining everything: something hidden (like the color degrees of freedom) was also necessary.

On can then go back to the planar bootstrap and ask how it could possibly be satisfied in the limit $N_f/N_c \to 0$. A (qualitative) answer was given in Ref. 23: in that limit resonances become more and more long-lived so that, the effective phase space at a given energy grows like the inverse of the resonance width. This increases the r.h.s. of the planar-unitarity equation making it possible for it to be satisfied in spite of the higher power of the tiny hadronic couplings.

I don't know how Geoff reacted to all these developments within QCD. For me, 't Hooft's paper was a turning point since everything I had found new and appealing in the DRM and hadronic String Theory was potentially already build-in in QCD provided one accepted to organize diagrams according to topology rather than loop-number. And although analyticity, unitarity and crossing represented very powerful and useful constraints on different topological structures, they could not replace the particular microscopic physics underlying a given theory.

5. The Bootstrap is Still Alive!

Even if the original bootstrap idea has slowly faded away, the power of self-consistency is undeniable and keeps feeding modern theoretical physics: and it is still called a bootstrap! It's most fashionable version today is represented by the so-called conformal bootstrap which bears an amusing resemblance to the "easy bootstrap" we discussed above.

The simplest example of a conformal bootstrap consists in taking a 4-point function of some operators in a conformal field theory (CFT) and in performing the operator-product expansion (OPE) of any two pairs of them. The result should be the same independently of how we make the pairings. The three possibilities correspond to performing the OPE in the s, t or u channels. The advantage of working with a CFT is that the OPE is very much constrained in terms of the (generally anomalous) dimensions of the operators. In this way one has succeeded[24] in obtaining powerful results (often in the form of constraints) on the possible structure of generic CFTs.

The analogy with the "easy bootstrap" based on imposing DHS duality is quite striking and is related to the existence, in the latter case, of a two-dimensional CFT underlying the dual-resonance model. Indeed Mack[25] has recently managed to make a clean connection between the old and the new bootstrap after reformulating the latter in the space of Mellin–Barnes transforms.

Much work is still going on in this and other areas, both analytically and numerically, showing how bootstrap methods, if not in their most extreme form, are still alive and well in the particle theory community.

References

1. G. F. Chew, "Nuclear democracy and bootstrap dynamics", University of California Radiation Laboratory preprint, UCRL-11163,12 (1963).
2. G. F. Chew and S. C. Frautschi, "Regge Trajectories and the Principle of Maximum Strength for Strong Interactions", *Phys. Rev. Lett.* **8**, 41 (1962).
3. R. Hagedorn, "Statistical thermodynamics of strong interactions at high-energies," *Nuovo Cim. Suppl.* **3**, 147–186 (1965)
4. G. Veneziano, "The Hagedorn Spectrum and the Dual Resonance Model: An Old Love Affair," in *Melting Hadrons, Boiling Quarks? From Hagedorn Temperature to Ultra-Relativistic Heavy-Ion Collisions at CERN*, ed. J. Rafelski, Nato ASI series Vol. 346, p. 72 (Springer, 1995).
5. R. Dolen, D. Horn and C. Schmid, "Prediction of Regge Parameters of rho Poles from Low-Energy pi N Data", *Phys. Rev. Lett.* **19**, 402 (1967); "Finite energy sum rules and their application to pi N charge exchange", *Phys. Rev.* **166**, 1768 (1968).
6. M. Ademollo, H. R. Rubinstein, G. Veneziano and M. A. Virasoro, "Bootstrap of meson trajectories from superconvergence", *Phys. Rev.* **176**, 1904 (1968).
7. S. Mandelstam, "Dynamics based on indefinitely rising regge trajectories," *Conf. Proc.* C **670828**, 605–620 (1967).
8. G. Veneziano, "Construction of a crossing - symmetric, Regge behaved amplitude for linearly rising trajectories", *Nuovo Cim. A* **57**, 190 (1968).
9. J. Scherk and J. H. Schwarz, "Dual Models for non-hadrons", *Nucl. Phys. B* **81**, 118 (1974).
10. M. B. Green and J. H. Schwarz, "Anomaly Cancellation in Supersymmetric D=10 Gauge Theory and Superstring Theory", *Phys. Lett. B* **149**, 117 (1984).
11. A. Di Giacomo, S. Fubini, L. Sertorio and G. Veneziano, "Unitarity in dual resonance models," *Phys. Lett. B* **33**, 171–174 (1970).
12. A. Capella, U. Sukhatme, C. I. Tan and J. Tran Thanh Van, "Dual parton model," *Phys. Rept.* **236**, 225–329 (1994).
13. A. B. Kaidalov and K. A. Ter-Martirosian, "Pomeron as Quark-Gluon Strings and Multiple Hadron Production at SPS Collider Energies," *Phys. Lett. B* **117**, 247–251 (1982)

14. G. Veneziano, "Duality and the bootstrap," *Phys. Lett. B* **34**, 59–62 (1971).
15. G. Veneziano, "Inclusive approach to unitarity," *Phys. Rev. Lett.* **28**, 578–580 (1972).
16. A. H. Mueller, "O(2,1) Analysis of Single Particle Spectra at High-energy," *Phys. Rev. D* **2**, 2963–2968 (1970)
17. S. Mandelstam, "Cuts in the Angular Momentum Plane. 2," *Nuovo Cim.* **30**, 1148–1162 (1963).
18. G. F. Chew and C. Rosenzweig, "Dual Topological Unitarization: An Ordered Approach to Hadron Theory," *Phys. Rept.* **41**, 263–327 (1978).
19. G. F. Chew and C. Rosenzweig, "Asymptotic Planarity: An S Matrix Basis for the Okubo-Zweig-Iizuka Rule," *Nucl. Phys. B* **104**, 290–306 (1976).
20. G. F. Chew and V. Poenaru, "Topological Bootstrap Theory of Hadrons," *Z. Phys. C* **11**, 59 (1981).
21. G. C. Rossi and G. Veneziano, "A Possible Description of Baryon Dynamics in Dual and Gauge Theories," *Nucl. Phys. B* **123**, 507–545 (1977).
22. G. 't Hooft, "A Planar Diagram Theory for Strong Interactions," *Nucl. Phys. B* **72**, 461 (1974).
23. G. Veneziano, "Some Aspects of a Unified Approach to Gauge, Dual and Gribov Theories," *Nucl. Phys. B* **117**, 519–545 (1976).
24. R. Rattazzi, V. S. Rychkov, E. Tonni and A. Vichi, "Bounding scalar operator dimensions in 4D CFT," *JHEP* **12**, 031 (2008).
25. G. Mack, "D-independent representation of Conformal Field Theories in D dimensions via transformation to auxiliary Dual Resonance Models. Scalar amplitudes," arXiv:0907.2407 [hep-th].

© 2022 World Scientific Publishing Company
https://doi.org/10.1142/9789811219832_0009

Geoffrey Chew and Regge Poles

A. Donnachie

University of Manchester, UK

P. V. Landshoff

University of Cambridge, UK

We review Chew's contribution to our understanding of the long-range strong interaction in high-energy scattering processes, with particular reference to his thinking on the pomeron.

Our understanding of high-energy hadron scattering is firmly based on the work of Geoffrey Chew. In the 1960s a talk by him at a physics conference was like a religious meeting: he preached that the analytic S-matrix approach to understanding high-energy scattering was the right one, and that field theory was unhelpful or even wrong. Indeed, in his book[1] published in 1966 he said "It is entirely possible that hadron scattering amplitudes may have the analytic properties commonly ascribed to them, while at the same time the association of fields with these particles is meaningless." In a sense, we now know that he was right: while we can be confident that QCD is the field theory that describes strong interactions, we can use it to calculate only processes dominated by short-range forces and extending this calculation to those that make up much the largest part of the total cross section may always remain beyond our reach.

Our knowledge of the analytic properties of the S-matrix is described in detail in a book[2] with a title almost identical with that[1] of Chew's and published in the same year.

In 1959 Tullio Regge[3] applied to nonrelativistic particle scattering a complex-angular-momentum formalism that had been developed for optics. Building on work by Froissart and Gribov,[4] Stanley Mandelstam very soon extended this to the relativistic theory with some clever mathematics.[5,6] The formalism yields an integral that converts the amplitude A from a function of the squared centre-of-mass energy s and the momentum transfer t to one that is a function of s and the angular momentum j, which is then treated as a complex variable. The integral converges for sufficiently large values of Re j, so that $A(s,j)$ is analytic for those values. With Frautschi, Chew introduced[7] an assumption which he called "maximal analyticity of the second degree", that $A(s,j)$ can be continued to physical values of j and that all physical particles correspond to "Regge poles"

$$A(s,j) \sim \frac{F(t)}{j - \alpha(t)}. \tag{1}$$

Chew and Frautschi noted[7] that mesons fall into families of straight-line trajectories $\alpha(t)$ on a plot in which particle spins are plotted against their squared mass. Figure 1 shows the four nearly-degenerate leading trajectories.

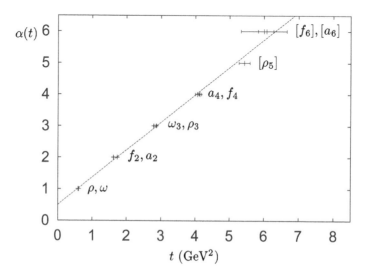

Fig. 1.   Plot from Ref. 6 of four degenerate Regge trajectories: particle spins plotted against their squared masses t. The particles in square brackets are listed in the data tables,[8] but there is some doubt about them. The straight line is $\alpha(t) = 0.5 + 0.9t$.

Regge poles control the high-energy behaviour of scattering amplitudes at fixed momentum transfer. The exchange of the families of particles corresponding to the various trajectories $\alpha_i(t)$ contributes to the amplitude

$$T_{ab}(s,t) = \sum_i \gamma_{ai}(t)\gamma_{bi}(t)(\alpha_i' s)^{\alpha_i(t)}\xi_i(t) \tag{2}$$

where the $\gamma_{ai}(t)$ and $\gamma_{bi}(t)$ are real and are the coupling functions of the ith exchange to the particles a and b and the "signature factors"

$$\xi_i(t) = 1 \pm e^{-i\pi\alpha_i(t)} \tag{3}$$

according to whether the C-parity of the exchange is ± 1. So, because of the optical theorem, at large s the contribution to the total cross section of the ith trajectory is proportional to $s^{\alpha_i(0)-1}$.

Chew maintained[1] that the leading trajectory, the one that contributes the highest power to the total cross section, should have the quantum numbers of the vacuum. This means, for example, that it contributes equally to pp and $p\bar{p}$ scattering. This is in agreement with the Pomeranchuk theorem[9] which states that, $\sigma^{\text{Tot}}(pp)/\sigma^{\text{Tot}}(\bar{p}p) \longrightarrow 1$ as $s \longrightarrow \infty$. Thus the trajectory is known as the Pomeranchuk trajectory, or the pomeron for short. It seems to be in agreement with data to assume that it too is linear in t:

$$\alpha_{I\!P}(t) = \alpha_{I\!P}(0) + \alpha'_{I\!P} t \, . \tag{4}$$

Chew originally believed[13] that the pomeron is not an aristocratic trajectory distinct from all others but merely the highest reggeon. He proposed that the intercept $\alpha_{I\!P}(0)$ of the pomeron trajectory should be 1, but we now know that total cross sections are not constant at high energy, but rise slowly with energy. So nowadays[10,11] the intercept is known to be close to 1.1. He also believed that the $f_2(1270)$ lay on the trajectory.[14] But, as Fig. 1 shows, the intercept of the f_2 trajectory is close to $\frac{1}{2}$ and we now believe that pomeron exchange corresponds to gluon exchange, so that any particles along the pomeron trajectory are glueballs. Meyer and Teper[12] have performed lattice calculation of the masses of the leading $2^{++}, 4^{++}$ and 6^{++} glueballs and find that the trajectory is approximately linear, with an intercept close to 1.

In 1976, with Stephens and Rosenzweig, Chew used the Regge-pole formalism to give an excellent description of the total cross sections for $\bar{p}p$, pp, $\pi^- p$, $\pi^+ p$, $K^- p$ and $K^+ p$ for $p_{lab} < 30$ GeV, but their pomeron trajectory had an intercept that was even a little below 1. In 1992 we did a similar analysis[15] of the data up to the very much higher energies available then, with $\alpha_{I\!P}(0) = 1.08$, and included also the γp cross section.

Chew stressed[16] that when the Regge-pole contribution (2) is applied to elastic scattering, it results in the low-t "diffractive peak" shrinking as the energy increases. This is shown in Fig. 2, where the fit to the data is that described in Ref. 11.

As Chew was well aware, Regge poles are not the only important singularities in the complex-j plane. There are also cuts, corresponding to the simultaneous exchange of two or more reggeons. For double pomeron exchange the trajectory is

$$\alpha_{I\!P I\!P}(t) = 1 + 2\alpha_{I\!P}(0) + \tfrac{1}{2}\alpha'_{I\!P} t \tag{5}$$

so, although the cut contributions seem to be small at low values of s and t, this is not the case at larger values. They do not simply contribute powers of s to the amplitude, there are also unknown factors involving log s. And, as Mandelstam showed,[19] the coupling of the cuts to hadrons is complicated and so also it cannot be calculated. So the best one can do is to resort to models, such as in Ref. 11.

Chew realised[21] that pomeron physics is applicable to a wide range of processes, not just elastic scattering and total cross sections. An example is[20] high-energy single-particle inclusive processes $A(p_1) + B(p_2) \rightarrow A(p'_1) + X$. In these processes particle A "radiates" a pomeron which then interacts with particle B to produce the system X.

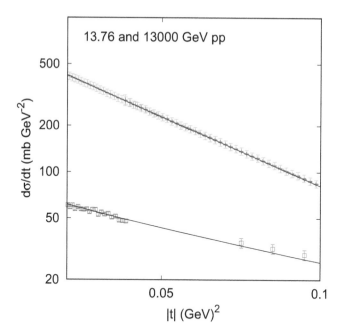

Fig. 2. pp elastic scattering data at 13 TeV (upper[17]) and 13.76 GeV (lower[18]) with fits from Ref. 11.

Data on two-particle inclusive cross sections $p_1 + p_2 \rightarrow p_1' + p_2' + \pi\pi$ were beginning to be available. Chew was interested[21] in "double-pomeron exchange", that is the region where both $(p_1' + p_{\pi\pi})^2$ and $(p_2' + p_{\pi\pi})^2$ are large. Each initial hadron radiates a pomeron and the two pomerons interact to produce the $\pi\pi$ system. The results of the calculation were in qualitative agreement with data. The relevance of this is that it paved the way for the modern industry in central exclusive production in hadron-hadron collisions.[22–24]

References

1. G. F. Chew, *The Analytic S Matrix* (Benjamin, 1966).
2. R. J. Eden, P. V. Landshoff, D. I. Olive and J. C. Polkinghorne, *The Analytic S-Matrix* (Cambridge University Press, 1966).
3. T. Regge, *Il Nuovo Cimento* **14**, 951 (1959).
4. M. Froissart, *Phys. Rev.* **123**, 2047 (1961); V. N. Gribov, *Sov. Phys. JETP* **41**, 1962 (1961).
5. S. Mandelstam, *Annals of Physics* **19**, 254 (1962).
6. A. Donnachie, H. G. Dosch, P. V. Landshoff and O. Nachtmann, *Pomeron Physics and QCD* (Cambridge University Press, 2002).
7. G. F. Chew and S. C. Frautschi, *Phys. Rev. Lett.* **8**, 49 (1962).

8. Particle Data Group, *Eur. Phys. J. C* **15**, 1 (2000).

9. I. Y. Pomeranchuk, *Sov. Phys. JETP* **7**, 499 (1958).

10. J. R. Cudell, K. Kang and S. K. Kim, *Phys. Lett. B* **395**, 31 (1997).

11. A. Donnachie and P. V. Landshoff, *Phys. Lett. B* **798**, 135008 (2019).

12. H. D. Meyer and M. J. Teper, *Phys. Lett. B* **605**, 344 (2005).

13. G. F. Chew and C. Rosenzweig, *Phys. Lett. B* **58** (1975) 93; *Phys. Rev. D* **12**, 3907 (1975).

14. S. R. Stevens, G. F. Chew and C. Rosenzweig, *Nucl. Phys. B* **110**, 355 (1976).

15. A. Donnachie and P. V. Landshoff, *Phys. Lett. B* **296**, 227 (1992).

16. G. F. Chew, *Comments Nuclear and Particle Physics* **1**, 4 (1967).

17. G. Antchev *et al.*, *Eur. Phys. J. C* **79**, 9 (2019).

18. J. P. Burq *et al.*, *Nucl. Phys. B* **217**, 285 (1983); C Akerlof *et al.*, *Phys. Rev. D* **14**, 2864 (1976).

19. S. Mandelstam, *Nuovo Cimento* **30**, 1127 (1963).

20. R. B. Appleby *et al.*, *Eur. Phys. J. C* **76**, 10 (2016).

21. D. M. Chew and G. F. Chew, *Phys. Lett. B* **53**, 191 (1974).

22. A. Donnachie and P. V. Landshoff, *Int. J. Mod. Phys. A* **29**, 1446007 (2014).

23. M. Albrow, V. Khose and C. Royon (eds.), Central exclusive production in hadron–hadron collisions, *Int. J. Mod. Phys. A* **29**, Issue 28 (2014).

24. J. Adam *et al.*, *JHEP* **07**, 07 (2020), arXiv:2004.11078.

Memories:
Geoff Chew, S-matrix, Bootstraps and Dual Topology

Carl Rosenzweig

Department of Physics, Syracuse University,
NY 13244-1130, USA

From 1974 until 1977 I had the privilege of working closely with, and learning from, Geoff Chew. It was an exciting time in physics. The paradigm shift to gauge theories as the fundamental tool for understanding interactions of elementary particles was rapidly gaining steam but the S Matrix approach was still viable and popular. Experimental results were plentiful and the November revolution (discovery of J/ψ) occurred. My PhD, while ostensibly from Harvard, was supervised by S. Fubini at MIT. He and Veneziano were exploring the implications of Veneziano's newly discovered model for strong interactions. Both Fubini and Veneziano were, at that time, in the S-Matrix camp, not seeing the current state of field theory as capable of explaining the abundance of data involving strong interactions, particularly at high energy. Even at Harvard there was little optimism about the future of field theory. (Remember I was a student before 't Hooft!) The pendulum began to strongly swing back to field theory especially for Weak interactions after 't Hooft's brilliant work in 1972[1] and then in strong interactions after the discovery of asymptotic freedom (1973) by Geoff's former student David Gross and collaborators.[2]

I spent the academic year 1973–74 as a post-doc at The Weizmann Institute of Science in Israel, arriving just a few days before the Arab invasion which set off the Yom Kippur war. Needless to say in those first weeks physics was not at the top of anybody's agenda. A large fraction of the staff and students were in the army or patrolling the grounds of the Institute in their military uniforms. It was not until January '74 that a more normal academic life began. We were all trying to learn the rudiments of gauge theories while some of us were proceeding with our decidedly non-field theoretical investigations. Veneziano arrived in Israel from CERN after the beginning of the new year to assume his professorship at Weizmann and he and I resumed our collaboration. He was very excited about his new approach to strong dynamics based on an expansion in the number of flavors,[3] modeled on 't Hooft's expansion of QCD in the number of colors.[4] His enthusiasm was contagious, and we began work on a formal development of the theory. This resulted in our paper on the implications of the zeroth order (planar) theory for Regge intercepts and couplings.[5]

Returning to Berkeley, September 1974, I was disappointed with my meager output for my post-doc year at Weizmann, even if being in the middle of a war is a good excuse. Chew was extremely happy to see me as he had met Veneziano over the summer and was quite taken with the new approach. His excitement immediately raised me from my funk as he corralled me into many discussions of this new expansion. He was intrigued by the viewpoint of a topological expansion as a way to organize a theory of strong interactions. I also think, that for his entire career, Chew was captivated by the presence of small parameters in physics which allowed physicists to isolate and explore in depth some limited sector of the physical world. The smallness of alpha, the electromagnetic coupling, is the best example of this. The number of flavors effectively, (accounting for the heavier strange quark mass, although Chew would not have phrased it this way), a number between 2 and 3, gave a new small parameter $(1/N_{\text{flavor}})$ to be exploited in a systematic way through this Dual Topological expansion. Chew had no interest in 't Hooft's original expansion which had an even smaller number $1/N_{\text{color}}$ since this was just a theoretical construct whereas flavors were a real physical quantity.

A major attraction of the new topological approach was that at the zeroth order level, that of the planar theory, it already incorporated a good deal of unitarity and also what was then known as the duality properties of strong interactions. (Harari-Rosner diagrams.[6]) We immediately began studying the major non planar contribution (down by $1/N_{\text{flavor}}$) which we called the cylinder correction. This led us to a Pomeron with special couplings.[7] I remember this being very exciting and a lot of fun. In the midst of this work the J/ψ exploded into physics. Those of us active at this time will never forget the unique excitement and challenge this presented. Ironically for me, the unraveling of the nature of these particles led to the triumph of field theory and the shuttling off the main stage for S-Matrix approaches. Nevertheless, the dramatic properties of these mesons, especially the extreme narrowness of the peaks, forced theorists to come up with a good explanation of the Okubo–Zweig–Iizuku (OZI) rule.[8] Geoff and I worried about this too but were not prepared to ascribe the OZI rule to asymptotic freedom. At the planar level the OZI rule was exact (a good start!) and was violated at the first non-planar or cylinder level. The suppression was orders of magnitude beyond $1/N_{\text{flavor}}$ and we struggled to find additional suppression. We eventually realized that the interfering phases of the intermediate states making up the cylinder would do just this and led us to conjecture the asymptotic planarity limit of the planar plus cylinder level of the expansion.[9] For me this was the most satisfying result we found. It was an important property of the theory and at the same time it made contact with exciting new experimental results and provided insight to an old puzzle.

Although $q\bar{q}$ bound states were well known in the quark model the fact that if the quarks were massive quarks, like the $c\bar{c}$, the state was an almost molecular object got us thinking about a new possibility. Geoff was more acquainted with the nuclear community and the possible existence of $N\bar{N}$ bound states and started the ball rolling on baryonium.[10] We also had a lot of fun with that. It was somewhat surprising and satisfying to see a recent revival of these ideas in the $cq\,\overline{cq}$ sector.[11]

Despite the fact that Geoff was also chair at this time he took advantage of LBL to disappear there for the afternoon to concentrate on physics. Mornings were for department business. This schedule worked perfectly for a young post-doc! Once, apologizing for spending time on administration, he said to me, in wonder, that many years ago he, Goldberger, Low and Nambu wrote some influential papers together.[12] Nobody would have believed then that all four would go on to be department chairs. Goldberger no surprise, but the other three? (Low had a distinguished administrative career at MIT.)

Sometimes our discussions were halted for an extended interval as Geoff stared at the board lost in thought and then came up with a great insight as if out of nowhere. His knack for knowing that something was important is well illustrated by the well-known story about the first time he heard Mandelstam give a short talk about the double dispersion relation. Geoff didn't understand Mandelstam's talk but immediately knew that it was important. I also remember once Geoff showed me a paper by a well-regarded physicist he knew. He shook his head and said he knows the person doesn't believe in this theory but wrote the article any way. Geoff couldn't understand how anyone could write about something that he didn't believe in!

As I moved to Pittsburgh and then Syracuse we continued to work together on dual topology. Geoff was better than I was at keeping in touch via mail and phone. We produced several interesting papers[13] but Geoff was moving in a more formal direction than I was comfortable with. Our joint work culminated with a large review in Physics Reports.[14] By then I was joining the exodus from the S-Matrix movement toward QCD. One of the virtues of teaching was that by signing up to teach a field theory course my transformation was hastened.

A prime, but indirect, lesson I learned from our collaboration was the importance of keeping in touch with the data. At the beginning we were deep in the phenomena of Regge theory, the duality between the Regge picture of high energy scattering and the low energy dominance by particle states and the manifestation of the Pomeron in high energy cross sections. The discovery of the J psi unleashed a torrent of new and confusing data that forced us to think in new directions. I have always tried to keep this in mind. Of course, it was much easier in the 1970's when every time an accelerator turned on there was tons of new data and puzzles. I do not envy students of today with the paucity of truly new data and the abundance of experimental articles entitled "Search for..." It was sad for me to see that Geoff, who had immersed himself in the nitty gritty of explaining experiments for over 30 years, gradually got seduced by some sophisticated mathematics which didn't come close to replacing his amazing intuition about what was actually going on.

A previous volume of tribute to Geoff was aptly entitled "A Passion for Physics."[15] The passion continued well past retirement. In his golden years he devoted himself to a study of Cosmology (is this a common feature of retired S-Matrix physicists?[16]) writing at least 14 (by my count) papers[17] in his own quest of the holy grail of a complete theory of everything. He had no more success than Einstein but his enthusiasm shines through.

Legacy

My students today are unfamiliar with Geoff Chew and his major contributions to particle physics. They find it hard to imagine that a colloquium by Chew would fill the largest auditorium at MIT (standing room only). Yet his effect on the field is still with us as many of his ideas are now part of physics folklore (without references to his fundamental contributions). It is inconceivable that our knowledge of the interactions of hadrons would have progressed to where we are today without his long list of achievements and insights. But more importantly his ideas are still influencing modern research. While Geoff was not a fan of string theory there is no doubt that the road to superstrings went through the S-matrix and Regge trajectories (and another of Geoff's students, J. Schwartz).

Currently a search on the arXiv for particle physics papers including the word "S-Matrix" in the abstract returns well over 100 papers in the past year alone. Geoff did not invent the S-matrix, but he popularized it and showed how it can be an important tool to study the mysterious interactions of hadrons. Even more, he eventually advocated that the S-Matrix be the fundamental concept in unraveling the nature of the subatomic world.[18] Today this idea has seen a revival sometime associated with strange words such as amplituhedron[19] but never with a reference to Chew. I find it always easy to understand the first ten to fifteen minutes of a seminar on this topic if you remember what Geoff taught.

Similarly, Geoff's usage of the word bootstrap to explain his program of building up a theory (preferably an S-Matrix theory) piece by piece from basic principles applied to an initially restricted regime, is now commonplace. Again, a search on the arXiv for particle physics related articles employing the term bootstrap returns well over 100 references in the past year alone. The terminology and the idea were Geoff's.[20] Geoff, I am sure, would have viewed with both pleasure and consternation the currently popular program of the cosmological bootstrap![21]

Geoff Chew was a wonderful man and a great physicist. I was privileged to know him.

References

1. G. 't Hooft, *Nucl. Phys. B* **33**, 173 (1971).
2. D. Gross and F. Wilczek, *Phys. Rev. Lett.* **30**, 1343 (1973); H. David Politzer, *Phys. Rev. Lett.* **30**, 1346 (1973).

3. G. Veneziano, *Nucl. Phys. B* **74**, 365 (1974).
4. G. 't Hooft, *Nucl. Phys. B* **72**, 461 (1974).
5. C. Rosenzweig and G. Veneziano, *Phys. Lett. B* **52**, 335 (1974).
6. H. Harari, *Phys. Rev. Lett.* **22**, 562 (1969); J. Rosner, *Phys. Rev. Lett.* **22**, 689 (1969).
7. G. Chew and C. Rosenzweig, *Phys. Lett. B* **58**, 93 (1975); *Phys. Rev. D* **12**, 3907 (1975).
8. S. Okubo, *Phys. Lett.* **5**, 165 (1963); G. Zweig, CERN report TH 412 (1964); J. Iizuka, *Suppl. Prog. Theor. Phys.* **37-38**, 21 (1966).
9. G. Chew and C. Rosenzweig, *Nucl. Phys. B* **104**, 290 (1976); C. Rosenzweig, *Phys. Rev. D* **13**, 3080 (1976).
10. G. Chew, in *Antinucleon-Nucleon Interactions*, eds. G. Ekspong and S. Nilsson (Pergamon Press, 1977), p. 515; C. Rosenzweig, *Phys. Rev. Lett.* **36**, 697 (1976).
11. A. Esposito *et al.*, *Phys. Rep.* **668**, 1 (2017).
12. G. Chew, M. Goldberger, F. Low and Y. Nambu, *Phys. Rev.* **106**, 1337 (1957); *Phys. Rev.* **106**, 1345 (1957).
13. G. Chew and C. Rosenzweig, *Phys. Lett. B* **63**, 429 (1976); *Annals Phys.* **105**, 212 (1977); *Phys. Rev. D* **15**, 3433 (1977).
14. G. Chew and C. Rosenzweig, *Phys. Rep.* **41C**, 263 (1978).
15. C. DeTar, J. Finkelstein and C. Tan, *Passion for Physics* (World Scientific Publishing, 1985).
16. J. George and C. Rosenzweig, *JCAP* **01**, 028 (2020).
17. G. Chew, LBL report, Feb 11, 2019, Chiral-Electromagnetic-Gravitational Theory of Every 'Thing' and references therein.
18. G. Chew, *Analytic S Matrix; A Basis for Nuclear Democracy* (W.A. Benjamin Inc., 1966).
19. See N. Arkani-Hamad *et al.*, *Grassmannian Geometry of Scattering Amplitudes* (Cambridge University Press, 2016) and citations therein.
20. G. Chew, *Science* **161**, 762 (1968).
21. D. Baumann *et al.*, arXiv:2005.04234 and references therein.

© 2022 World Scientific Publishing Company
https://doi.org/10.1142/9789811219832_0011

Geoff Chew and the S-Matrix

Alan White

Argonne National Laboratory

When I knew Geoff most closely he was very negative about field theory and passionate about the physical significance of the S-Matrix. He had concluded that it is the essential tool for comparing theory with experiment. As a consequence, conventional field theory had to be thrown out. He would emphasize "There is no experimental evidence for the presence of a field interaction in strong interaction physics".

I believe he was correct in saying this about a field interaction of the kind that he was familiar with and that history will judge him positively as a result. Unfortunately, perhaps, it became a major cornerstone of his immediate reputation and produced some antagonism amongst his contemporaries, setting him up as separate from the mainstream. Nevertheless, he had a major impact on my own thinking, since his arguments coincided so well with mine.

I interacted with all the Berkeley stars of the time. First visited Stanley Mandelstam, while he was on sabbatical in London, to discuss the use of t-channel unitarity. I then visited Berkeley, as a postdoc, where I began my discussions with Geoff and also started my interaction with Henry Stapp that produced our work on multiparticle dispersion theory and Regge theory.

Geoff always hammered me about the failures of Feynman diagrams. At first I simply just listened with interest. But as my own research proceeded and I understood the significance of multi-Regge infra-red divergences, I discovered a fundamental result that would have delighted Geoff. After incorporating the multi-Regge infra-red divergences in all the diagrams, the bound-state physical S-Matrix that reproduces the Standard Model is the only surviving non-perturbative element. This S-Matrix also contains additional new physics that is crucially relevant to explain much-needed beyond the Standard Model physics.

I never had a chance to describe this result to Geoff but I am sure that he would have been very happy with it.

Geoff Chew of Illinois

Jerrold Franklin

*Department of Physics, Temple University,
Philadelphia, PA 19122-1801, USA*

I was lucky to get to Illinois during the Chew-Low era. I had gone to an engineering school and didn't know which schools were good, or even the names of many physicists. Fortunately, Aaron Yalow, a physics professor recommended Illinois where he had gone.

The first time I heard the name Chew was several weeks after I got to Illinois. One of the students asked "Who is that graduate student who always asks the first question at colloquium?" Someone answered, "Oh, that's Professor Chew, he's the colloquium chairman." I could see, then, that he felt it was his job to start the speaker off with a helpful beginning question.

Chew always looked younger than he was. The last time I saw him was in 2002 when he gave a talk on his version of quantum mechanics at a quantum conference. He still looked like a young, but confident man.

One episode early in my knowledge of Chew occurred on November 29, 1954. Chew entered our quantum classroom without his usual smile. Then he spoke for about ten minutes about Enrico Fermi, describing with deep feeling, Fermi, the man and the physicist. Then, he went into his lecture. I learned a lot that day about Fermi, and about Chew, the man and the physicist.

I was last in a line of graduate students mentored by Chew at Illinois. 'Mentored' is the right word because he taught as much an attitude toward physics as he did the details. As I recall, the batting order was Sol Gartenhaus, Harry Schey, Don Lichtenberg, and me. I was 'clean up' because then Chew, Low, and I headed back East.

Unfortunately, Harry Schey can no longer speak for himself, so I will say a few words for him. Harry was well known for his acerbic, often self-deprecating wit. He had a love (real)- hate (put on) relationship with Chew.

Harry regularly related 'Chewy' episodes to me. One good one was, "I worked for two weeks, and the damned thing wouldn't work out. I finally brought it to Chew. He looked at it and said 'See Harry, if you put this in there and take the other out, it works out.' I asked him, 'How did you do that?' He said 'I don't know Harry. Sometimes things just work out for me.' Damn Chew!" Praise like that from Harry convinced me that I, too, had to work for Chew.

After taking Chew's quantum course, I got up the nerve to go to ask him to be my thesis advisor. He gave me a simple scattering problem to work out. After learning some field thory from Bethe-deHoffman, I came back to Chew with my answer. As I handed it to him he said "It's wrong." I blurted back, "You haven't even looked at it!" He replied calmly, "A cross section has to go like one over the energy squared." Of course he was right. I still have trouble with when to use the $\sqrt{2E}$ normalization, and B-dH hadn't been of much help. Anyway, I had learned my first valuable lesson from Chew. Harry was right, I had the right advisor.

Things went smoothly after that, Chew was helpful without leading. I could just walk into his office if any question came up. Then we discussed it like equals even though we weren't. I think I was his first 'Chew-Low' student.

One encouragement came when we invited the Chews to our 'student' apartment shortly after we were married. As he entered, he remarked "What a beautiful apartment. I hope you're not planning to stay too long."

One more anecdote will show Chew's power over nature. The seminars at Illinois were like Chew's 'secret' seminars at Berkeley, more discussions than formal talks. Chew was talking at one seminar, and Gilberto Bernandini, at Illinois on an extended visit (It did not remind him of Italy), asked a question. Chew responded with, "I think I can say something that may ring a bell." Just then, the classroom bell rang, on cue.

I have been using the name 'Chew'. There was a tradition at Illinois. You entered your PhD final with Professor Chew. When (and if) he congratulated you at the end, you replied, "Thank you Geoff."

THANK YOU GEOFF!

From the S Matrix to String Theory

John H. Schwarz

*California Institute of Technology,
Pasadena, CA 91125, USA*

This article presents a brief overview of the early history of string theory and supersymmetry. It describes how the S-matrix theory program for understanding the strong nuclear force, championed by Geoffrey Chew, evolved into superstring theory, which is a promising framework for constructing a unified quantum theory of all forces including gravity.

1. Introduction

I am pleased to have this opportunity to contribute to this volume, which is dedicated to the memory of Geoffrey Chew. I had the privilege of being one of his graduate student advisees in Berkeley. Chew's generation of physicists came of age during the second world war and the postwar period. During the war Chew was a junior participant in the Manhattan Project. Following the war he and Marvin Goldberger were Enrico Fermi's first two theoretical physics graduate students at the University of Chicago. Fermi inherited them from Edward Teller after Teller decided that he had more important things to work on. After graduating from Chicago, Chew spent his entire academic career at Berkeley except for a few years in the 1950s when he joined the University of Illinois. This move was prompted by a loyalty oath controversy at the University of California during the McCarthy period. I don't recall ever hearing Chew speak about his personal experiences, and I was too shy to ask. Our conversations always focused on research.

Chew was at his creative peak in the decade of the 1960s. I had the good fortune to be one of his ten or so students in the middle of that decade. Chew and Stanley Mandelstam were focused on constructing a theory of the strong nuclear force, *i.e.*, a theory of hadrons. Chew's approach to understanding the strong nuclear force was based on *S-matrix theory*. He argued that quantum field theory, which was so successful in describing QED, was inappropriate for describing a strongly interacting theory, where a weak-coupling perturbation expansion would not be useful. One reason for holding this view was that none of the hadrons seemed more fundamental than any of the others. Therefore a field theory that singled out some subset of the hadrons did not seem sensible. Also, it seemed impossible to formulate a quantum field theory with a fundamental field for every hadron.

Chew spoke of *nuclear democracy* and the *bootstrap principle* to describe his preferred approaches. Specifically, Chew advocated studying physical quantities, especially the S Matrix, which describes on-mass-shell scattering amplitudes. The goal was to develop a theory that would determine the hadron spectrum and the hadronic S matrix. Chew was a powerful spokesman for this approach. He was very influential in the theoretical physics community throughout the 1960s.

The quark concept also arose during the same period, but the prevailing opinion in the mid 1960s was that quarks are mathematical constructs, rather than physical entities, whose main use is as a mnemonic for understanding symmetries and quantum numbers. The SLAC deep inelastic scattering experiments in the late 1960s made it clear that quarks and gluons are physical (confined) particles. It was then natural to try to base a quantum field theory on them, and QCD was developed a few years later. Thus, with the wisdom of hindsight, it is clear that Chew *et al.* were wrong to reject quantum field theory. Nonetheless, their insights were very influential, perhaps even crucial, for the discovery of string theory, which can be regarded as the ultimate realization of the S-Matrix program.

2. S-Matrix Theory

Two of the key ingredients that went into the S-matrix theory program wee unitarity and maximal analyticity of the S matrix. These properties, inferred from quantum field theory, encode the requirements of causality and nonnegative probabilities. Another important ingredient was analyticity in angular momentum. The basic idea is that partial wave amplitudes $a_l(s)$, which are defined in the first instance for angular momenta $l = 0, 1, \ldots$, can be extended to an analytic function of l, $a(l, s)$. The uniqueness of this extension results from imposing suitable asymptotic behavior in l. The Mandelstam invariant s is the square of the center-of-mass energy of the scattering reaction. The analytic function $a(l, s)$ can have isolated poles called *Regge poles*. Branch points are also possible, but they were usually ignored. The position of a Regge pole is given by a Regge trajectory, $l = \alpha(s)$. A value of s for which $l = \alpha(s)$ takes a physical value corresponds to a physical hadron of spin l and squared mass equal to s.

Theoretical work in this period was strongly influenced by experimental results. Many new hadrons were discovered in experiments at the Bevatron in Berkeley, the AGS in Brookhaven, and the PS at CERN. Plotting masses squared versus angular momentum (for fixed values of other quantum numbers), it was noticed that the Regge trajectories are approximately linear with a common slope

$$\alpha(s) = \alpha(0) + \alpha' s \qquad \alpha' \sim 1.0\,(\text{GeV})^{-2}. \tag{1}$$

Using the crossing-symmetry properties of analytically continued scattering amplitudes, one argued that exchange of Regge poles (in the t channel) controlled the high-energy, fixed momentum transfer, asymptotic behavior of physical amplitudes:

$$A(s,t) \sim \beta(t)(s/s_0)^{\alpha(t)} \qquad s \to \infty, \, t < 0. \tag{2}$$

In this way one deduced from data that the intercept of the ρ trajectory, for example, was $\alpha_\rho(0) \sim .5$. This is consistent with the measured mass $m_\rho = .76\,\text{GeV}$ and the Regge slope $\alpha' \sim 1.0\,(\text{GeV})^{-2}$.

The approximation of linear Regge trajectories describes long-lived resonances, whose widths are negligible compared to their masses. This approximation is called the *narrow resonance approximation*. In this approximation branch cuts in scattering amplitudes, whose branch points correspond to multiparticle thresholds, are approximated by a sequence of resonance poles. This is what one would expect in the tree approximation to a quantum field theory in which all the resonances appear as fundamental fields. However, there was also another discovery, called *duality*, which clashed with the usual notions of quantum field theory. In this context duality means that a scattering amplitude $A(s,t)$ can be expanded in an infinite series of s-channel poles, and this gives the same result as its expansion in an infinite series of t-channel poles. To include both sets of poles, as usual Feynman diagram techniques might suggest, would amount to double counting.

3. The Discovery of String Theory

In 1968 Veneziano[1] discovered a simple analytic formula that exhibits duality with linear Regge trajectories. It is given by a sum of ratios of Euler gamma functions:

$$T = A(s,t) + A(s,u) + A(t,u), \quad \text{where} \quad A(s,t) = g^2 \frac{\Gamma(-\alpha(s))\Gamma(-\alpha(t))}{\Gamma(-\alpha(s) - \alpha(t))}, \tag{3}$$

g is a coupling constant, and α is a linear Regge trajectory

$$\alpha(s) = \alpha(0) + \alpha' s. \tag{4}$$

The Veneziano formula gives an explicit realization of duality and Regge behavior in the narrow resonance approximation. The function $A(s,t)$ can be expanded as an infinite series of s-channel poles or of t-channel poles, and the two expansions have overlapping regions of convergence. The motivation for considering formulas of this type was largely phenomenological. Incredibly, it turned out that formulas like this describe scattering amplitudes in the tree approximation of a consistent quantum theory!

A method of incorporating adjoint $SU(N)$ quantum numbers was formulated by Paton and Chan.[2] Chan–Paton symmetry was initially envisaged to be a global (flavor) symmetry, but it was shown later to be a local gauge symmetry. It is easily generalized to orthogonal and symplectic groups, a fact that became important later.

Very soon after the appearance of the Veneziano amplitude, Virasoro[3] proposed an alternative formula

$$T = g^2 \frac{\Gamma(-\tfrac{1}{2}\alpha(s))\Gamma(-\tfrac{1}{2}\alpha(t))\Gamma(-\tfrac{1}{2}\alpha(u))}{\Gamma(-\tfrac{1}{2}\alpha(t) - \tfrac{1}{2}\alpha(u))\Gamma(-\tfrac{1}{2}\alpha(s) - \tfrac{1}{2}\alpha(u))\Gamma(-\tfrac{1}{2}\alpha(s) - \tfrac{1}{2}\alpha(t))}, \tag{5}$$

which has similar virtues. Since this formula has total s, t, u symmetry, it describes particles that are singlets of the Chan–Paton symmetry group.

Over the course of the next year or so, *dual models* or *dual-resonance models*, the names used for this subject at that time, underwent a sudden surge of popularity, marked by several remarkable discoveries. One was the discovery (by several different groups) of an N-particle generalization of the Veneziano formula

$$A_N(k_1, k_2, \ldots, k_N) = g_{\text{open}}^{N-2} \int d\mu_N(y) \prod_{i<j} (y_i - y_j)^{\alpha' k_i \cdot k_j}, \tag{6}$$

where $y_1, y_2, \ldots, y_N$ are real coordinates, associated to the N scattered particles. Because of an $SL(2, \mathbb{R})$ symmetry, the integration is only over $N-3$ of them. I will omit the description of the measure $d\mu_N(y)$, which can be found in Ref. 4. This formula has cyclic symmetry in the N external particles. It should be multiplied by a Chan–Paton factor, with the same cyclic symmetry, and then summed over all cyclically inequivalent permutations.

Soon thereafter Shapiro[5] formulated an N-particle generalization of the Virasoro formula:

$$A_N(k_1, k_2, \ldots, k_N) = g_{\text{closed}}^{N-2} \int d\mu_N(z) \prod_{i<j} |z_i - z_j|^{\alpha' k_i \cdot k_j}, \tag{7}$$

where $z_1, z_2, \ldots, z_N$ are complex coordinates. Because of an $SL(2, \mathbb{C})$ symmetry, the integration is only over $N-3$ of them. This amplitude has total symmetry in the N external particles.

Both of these formulas for multiparticle amplitudes were shown to have poles whose residues factorize in a consistent manner on an infinite spectrum of single-particle states. This spectrum is described by a Fock space associated to an infinite number of harmonic oscillators, described by raising and lowering operators that satisfy

$$[a_m^\mu, a_n^{\nu\dagger}] = \delta_{mn} \eta^{\mu\nu} \qquad \mu, \nu = 0, 1, \ldots, d-1 \qquad m, n = 1, 2, \ldots \tag{8}$$

where d is the dimension of Minkowski spacetime, which was initially assumed to be four. There is one set of such oscillators in the Veneziano case and two sets in the Shapiro–Virasoro case. These spectra were interpreted as describing the normal modes of a relativistic string: an open string (with ends) in the first case and a closed string (loop) in the second case. Amazingly, the formulas were discovered before this interpretation was proposed. In the above formulas, the y coordinates parametrize points on the boundary of a string world sheet, where particles that are open-string states are emitted or absorbed, whereas the z coordinates parametrize points on the interior of a string world sheet, where particles that are closed-string states are emitted or absorbed. It is also possible to construct amplitudes in which both types of particles participate.

Having found the factorization, it became possible to compute radiative corrections (loop amplitudes). Gross *et al.*[6] discovered unanticipated singularities in a particular one-loop diagram for which the world sheet is a cylinder with two external particles attached to each of the two boundaries. The computations showed that this diagram gives branch points that violate unitarity. This was a very disturbing conclusion, since it seemed to imply that the classical theory does not have a consistent quantum extension. However, soon thereafter it was pointed out by Lovelace[7] that these branch points become poles provided that

$$\alpha(0) = 1 \quad \text{and} \quad d = 26. \tag{9}$$

Prior to this discovery, everyone assumed that the spacetime dimension should be $d = 4$. We had no other motivation to consider extra dimensions. It was the mathematics that forced us in that direction. Later, these poles were interpreted as closed-string states in a one-loop open-string amplitude. Nowadays this is referred to as *open-string/closed-string duality*. This is closely related to *gauge/gravity duality*, which was discovered 27 years later.

The analysis also required there to be an infinite number of decoupling conditions, which turned out to coincide with the constraints proposed by Virasoro[8] and further elaborated upon by Fubini and Veneziano.[9] Since the string has an infinite spectrum of higher-spin states, there are corresponding gauge invariances that eliminate unphysical degrees of freedom. The operators that describe the constraints that arise for a particular covariant gauge choice satisfy the Virasoro algebra

$$[L_m, L_n] = (m - n)L_{m+n} + \frac{c}{12}(m^3 - m)\delta_{m,-n}, \tag{10}$$

where m, n are arbitrary integers. These operators can also be interpreted as generators of conformal symmetry for the two-dimensional (2d) string world sheet. The central charge (or conformal anomaly) c is equal to the spacetime dimension d. This anomaly cancels for $d = 26$ when the contribution of Faddeev–Popov ghosts is included.

4. The RNS Model and the Discovery of Supersymmetry

In a very inspired and important development, Ramond[10] constructed a stringy analog of the Dirac equation, which describes a fermionic string. Just as the string momentum p^μ is the zero mode of a density $P^\mu(\sigma)$, where the coordinate σ parametrizes the string, he proposed that the Dirac matrices γ^μ should be the zero modes of densities $\Gamma^\mu(\sigma)$. Then he considered the Fourier modes of the dot product:

$$F_n = \int_0^{2\pi} e^{-in\sigma}\Gamma(\sigma) \cdot P(\sigma)d\sigma \qquad n \in \mathbb{Z}. \tag{11}$$

In particular,

$$F_0 = \gamma \cdot p + \text{additional terms.} \tag{12}$$

He proposed that physical states of a fermionic string should satisfy the following analog of the Dirac equation

$$(F_0 + M)|\psi\rangle = 0. \tag{13}$$

He also observed that in the case of the fermionic string the Virasoro algebra generalizes to a super-Virasoro algebra

$$\{F_m, F_n\} = 2L_{m+n} + \frac{c}{3}m^2\delta_{m,-n} \tag{14}$$

$$[L_m, F_n] = \left(\frac{m}{2} - n\right)F_{m+n} \tag{15}$$

$$[L_m, L_n] = (m-n)L_{m+n} + \frac{c}{12}m^3\delta_{m,-n}. \tag{16}$$

Ramond's paper does not include the central extension, which turns out to be $c = 3d/2$, where d is the spacetime dimension. Later, it was realized that consistency of the interacting theory requires $d = 10$ and $M = 0$. These conditions are the analogs of $d = 26$ and $\alpha(0) = 1$ for the bosonic Veneziano string theory.

Neveu and I[11] constructed a new interacting bosonic string theory, which we called the *dual pion model*. Our paper appeared a couple of months after Ramond's. It has a similar structure to the fermionic string, but the periodic density $\Gamma^\mu(\sigma)$ is replaced by an antiperiodic one $H^\mu(\sigma + 2\pi) = -H^\mu(\sigma)$. Then the Fourier modes

$$G_r = \int_0^{2\pi} e^{-ir\sigma} H \cdot P d\sigma \qquad r \in \mathbb{Z} + 1/2 \tag{17}$$

satisfy a similar super-Virasoro algebra. Neveu and I referred to this algebra as a *supergauge algebra*, a terminology that was sensible in the context at hand. The NS bosons and R fermions were combined in a unified interacting theory of bosons and fermions by Neveu and me[12] and by Thorn.[13] This theory (the RNS model) was an early version of superstring theory. As will be explained shortly, a few crucial issues were not yet understood.

After a few more months, Gervais and Sakita[14] showed that the RNS model is described by the string world-sheet action

$$S = T \int d\sigma d\tau \left(\partial_\alpha X^\mu \partial^\alpha X_\mu - i\bar{\psi}^\mu \gamma^\alpha \partial_\alpha \psi_\mu\right), \tag{18}$$

where the coefficient T is the string tension. They also explained that it has 2d supersymmetry, though that terminology was not used yet, by showing that it is invariant under the transformations

$$\delta X^\mu = \bar{\varepsilon}\psi^\mu, \qquad \delta\psi^\mu = -i\gamma^\alpha \varepsilon \partial_\alpha X^\mu, \tag{19}$$

where ε is an infinitesimal constant spinor. To the best of my knowledge, this is the first supersymmetric theory identified in the literature! There are two possibilities for the world-sheet fermi fields ψ^μ. When it is antiperiodic $\psi^\mu = H^\mu$, which gives the boson spectrum (NS sector), and when it is periodic $\psi^\mu = \Gamma^\mu$, which gives the fermion spectrum (R sector).

At about the same time as Ramond's paper, the 4d super-Poincaré algebra was introduced in a paper by Golfand and Likhtman,[15] who proposed constructing 4d field theories with this symmetry. This paper went unnoticed in the West for several more years. In fact, the celebrated paper of Wess and Zumino,[16] which formulated a class of 4d supersymmetric theories, was motivated by the search for 4d interacting analogs of the 2d Gervais–Sakita world-sheet action. The Wess–Zumino paper launched the study of supersymmetric field theories, which proceeded in parallel with the development of supersymmetric string theory. Wess and Zumino used the word *supergauge*, following the terminology of,[11] but in their subsequent papers they switched to *supersymmetry*. As they noted, this was more appropriate for what they were doing.

In 1976, Brink, Di Vecchia, and Howe[17] and Deser and Zumino[18] constructed a more fundamental world-sheet action with local supersymmetry. This formulation of the world-sheet theory improves upon the Gervais–Sakita action by also accounting for the super-Virasoro constraints. This made it clear that the significance of the super-Virasoro algebra is that the world-sheet theory, when properly gauge fixed and quantized, has *superconformal symmetry*. Again, the anomaly cancels for $d = 10$ when the Faddeev–Popov ghosts are included.

5. The Temporary Demise of String Theory

String theory is formulated as an on-shell S-matrix theory in keeping with its origins discussed earlier. However, the SLAC deep inelastic scattering experiments in the late 1960s made it clear that the hadronic component of the electromagnetic current is a physical off-shell quantity, and that its asymptotic properties imply that hadrons have hard pointlike constituents. Moreover, all indications (at that time) were that strings are too soft to describe hadrons containing pointlike constituents.

By 1973–74 there were many good reasons to stop working on string theory: a successful and convincing theory of hadrons (QCD) was discovered, and string theory had severe problems as a theory of hadrons. These included an unrealistic spacetime dimension ($d = 10$ or $d = 26$), an unrealistic spectrum (including a tachyon and massless particles), and the absence of pointlike constituents. A few years of attempts to do better had been unsuccessful. Moreover, convincing theoretical and experimental evidence for the Standard Model was rapidly falling into place. That was where the action was. Even for those seeking to pursue speculative theoretical ideas there were options other than string theory that most people found more appealing, such as grand unification and supersymmetric field theory. Understandably, string theory fell out of favor. What had been a booming enterprise involving several hundred theorists rapidly came to a grinding halt. Only a few diehards continued to pursue it.

6. Gravity and Unification

Among the problems of the known string theories, as a theory of hadrons, was the fact that the spectrum of open strings contains massless spin 1 particles, and the spectrum of closed strings contains a massless spin 2 particle (as well as other massless particles), but there are no massless hadrons. In 1974, Joël Scherk and I decided to take string theory seriously as it stood, rather than forcing it to conform to our preconceptions. This meant abandoning the original program of describing hadron physics and interpreting the massless spin 2 state in the closed-string spectrum as a graviton. Also, the massless spin 1 states in the open-string spectrum could be interpreted as particles associated to Yang–Mills gauge fields. Specifically, Scherk and I[19] proposed trying to interpret string theory as a unified quantum theory of all forces including gravity. Neveu and Scherk[20] had shown that string theory incorporates the correct gauge invariances to ensure agreement at low energies (compared to the scale given by the string tension) with Yang–Mills theory. Yoneya,[21,22] as well as Scherk and I, showed that it also contains gauge invariances that ensure agreement at low energies with general relativity.

To account for Newton's constant, the most natural choice for the fundamental string length scale was $l_s \sim 10^{-33}$ cm (the Planck length) instead of $l_s \sim 10^{-13}$ cm (the typical size of a hadron). Thus the strings suddenly shrank by 20 orders of magnitude, but the mathematics was essentially unchanged. The string tension is proportional to l_s^{-2}, so it increased by 40 orders of magnitude.

The proposed new interpretation had several advantages:
- Gravity and Yang–Mills forces are required by string theory.
- String theory has no UV divergences.
- Extra spatial dimensions could be a good thing.

Let me say a few words about the last point. In a nongravitational theory, the spacetime geometry is a rigid background on which the dynamics takes place. In that setup, the fact that we observe 4d Minkowski spacetime is a compelling argument to formulate the theory in that background geometry. On the other hand, in a gravitational theory that abides by the general principles laid out by Einstein, the spacetime geometry is determined by the dynamical equations. In such a setup extra dimensions can make sense provided that the equations of the theory have a solution for which the geometry is the product of 4d Minkowski spacetime and a compact manifold that is sufficiently small to have eluded detection. It turns out that there are many such solutions. Moreover, the details of the compact manifold play a crucial role in determining the symmetries and particle content of the effective low-energy theory in four dimensions, even when the compact dimensions are much too small to observe directly.

7. Superstrings

Gliozzi, Scherk, and Olive[23] proposed a truncation of the RNS string theory spectrum — *the GSO Projection* — that removes half of the fermion states and the "odd G-parity" bosons. In particular, the latter projection eliminates the tachyon. (This is the particle that we had originally wished to identify as a somewhat misplaced pion.) They showed that after the projection the number of physical bosonic degrees of freedom is equal to the number of physical fermionic degrees of freedom at every mass level. This was compelling evidence for *10d spacetime supersymmetry* of the GSO-projected theory. Prior to this, we knew about the supersymmetry of the 2d string world-sheet theory, discussed earlier, but we had not considered the possibility of spacetime supersymmetry (despite the work of Wess and Zumino). In fact, the GSO projection is not just an option; it is required for consistency of the theory, since the closed-string spectrum contains a massless gravitino.

In 1979 Michael Green and I began a collaboration, which had the initial goal of understanding and proving the 10d spacetime supersymmetry of the GSO-projected version of the RNS theory. The highlights of our work included Refs. 24 and 25, which developed a new formalism in which the spacetime supersymmetry of the GSO-projected RNS string is manifest, and Ref. 26, which classified the consistent 10d superstring theories and named them Type I, Type IIA, and Type IIB. We were excited about these (and other) developments, but they did not arouse much interest in the theory community at that time. String theory was still in the doldrums.

If a unified theory is to make contact with the Standard Model, and have a chance of being realistic, parity violation is an essential ingredient. However, parity-violating classical theories generically have *gauge anomalies*, which means that they cannot be used to define consistent quantum theories. The gauge symmetry is broken by one-loop quantum corrections, rendering the would-be quantum theory inconsistent. In the case of the Standard Model, if one were to modify the theory to have an unequal number of quark and lepton families, the theory would become inconsistent. When the numbers are equal all gauge anomalies beautifully cancel, and so the Standard Model is a well-defined quantum theory. These considerations raise the question whether the potential gauge anomalies in chiral superstring theories also cancel, so that they give consistent quantum theories.

The Type I superstring theory is a well-defined 10d theory at tree level for any $SO(n)$ or $USp(2n)$ gauge group, and for every such group it is chiral (*i.e.*, parity violating). However, evaluation of a one-loop hexagon diagram in 10d super Yang–Mills theory, which describes the massless open-string states, exhibits explicit nonconservation of gauge currents, signalling a gauge anomaly. The only hope for consistency is that inclusion of the closed-string (gravitational) sector cancels this gauge anomaly without introducing new ones.

Type IIB superstring theory, which only has a closed-string gravitational sector, is also chiral and therefore potentially anomalous. Alvarez-Gaumé and Witten[27] derived general formulas for gauge, gravitational, and mixed anomalies in an arbitrary spacetime dimension. Using their results, they discovered that the gravitational anomalies, which would imply nonconservation of the stress tensor, cancel in Type IIB superstring theory. In their calculation this cancellation appears quite miraculous, though the UV finiteness of the Type IIB loop amplitudes implies that it had to work. Thus, Type IIB is a consistent chiral superstring theory. On the other hand, it did not look promising for describing the real world, since it does not contain any Yang–Mills gauge fields. (More than a decade later, nonperturbative Type IIB solutions that do contain Yang–Mills fields were discovered.)

At that time, the last hope for constructing a realistic model seemed to reside with the Type I superstring theories, which are chiral and do contain Yang–Mills fields. Following some failed attempts, Green and I finally managed to compute the one-loop hexagon diagrams in Type I superstring theory. We found that both the cylinder and the Möbius-strip world-sheet diagrams contribute to the gauge anomaly, and we realized that there might be a gauge group for which the two contributions cancel. Green and I[28] showed that $SO(32)$ is the unique choice for which the cancellation occurs. Since this computation only demonstrated the cancellation of the pure gauge part of the anomaly, we decided to explore the low-energy effective field theory to see whether the gravitational and mixed anomalies also cancel. Using the results of Alvarez-Gaumé and Witten, we[29] verified that all gauge, gravitational, and mixed anomalies do in fact cancel for the gauge group $SO(32)$.

The effective field theory analysis showed that $E_8 \times E_8$ is a second (and the only other) gauge group for which the anomalies could cancel for a theory with $\mathcal{N} = 1$ supersymmetry in 10d. In both cases, it is crucial for the result that the coupling to supergravity is included. The $SO(32)$ case could be accommodated by Type I superstring theory, but we didn't know a superstring theory with gauge group $E_8 \times E_8$. We were aware of the article by Goddard and Olive[30] that pointed out (among other things) that there are exactly two even self-dual Euclidean lattices in 16 dimensions, and these are associated with precisely these two gauge groups. However, we did not figure out how to exploit this fact before the problem was solved by Gross *et al.*[31] Following these discoveries, as well as the work of Candelas *et al.*,[32] concerning promising candidates for the geometry of the compact dimensions, there was a sudden surge of interest in superstring theory.

After more than a decade, string theory had emerged from the doldrums. Much has happened since then. Suffice it say that many surprising and profound discoveries have been made in the subsequent 35 years. Moreover, insights derived from these studies have had a profound impact on fundamental mathematics and are also inspiring new approaches to understanding topics in other areas of physics.

8. Conclusion

Geoffrey Chew's focus on S-matrix theory led the theoretical particle physics community into directions that nobody could have anticipated, especially the development of string theory. Superstring theory is currently the leading framework for constructing a consistent unification of all forces including gravity. Following the discovery of AdS/CFT (or gauge-gravity duality) string theory, M-theory, and quantum field theory have become a largely unified subject. Whatever one chooses to call this unified subject, it is clear that there is much that remains to be understood. This challenge continues to attract many of the brightest young theorists.

References

1. G. Veneziano, Construction of a crossing-symmetric, Regge behaved amplitude for linearly rising trajectories, *Nuovo Cim. A* **57**, 190–197 (1968), doi:10.1007/BF02824451
2. J. E. Paton and H. M. Chan, Generalized Veneziano model with isospin, *Nucl. Phys. B* **10**, 516–520 (1969), doi:10.1016/0550-3213(69)90038-8
3. M. Virasoro, Alternative constructions of crossing-symmetric amplitudes with Regge behavior, *Phys. Rev.* **177**, 2309–2311 (1969) doi:10.1103/PhysRev.177.2309
4. J. H. Schwarz, The early years of string theory: A personal perspective, [arXiv:0708.1917 [hep-th]].
5. J. A. Shapiro, Electrostatic analog for the Virasoro model, *Phys. Lett. B* **33**, 361–362 (1970), doi:10.1016/0370-2693(70)90255-8
6. D. Gross, A. Neveu, J. Scherk and J. Schwarz, Renormalization and unitary in the dual-resonance model, *Phys. Rev. D* **2**, 697–710 (1970), doi:10.1103/PhysRevD.2.697
7. C. Lovelace, Pomeron form-factors and dual Regge cuts, *Phys. Lett. B* **34**, 500–506 (1971), doi:10.1016/0370-2693(71)90665-4
8. M. Virasoro, Subsidiary conditions and ghosts in dual resonance models, *Phys. Rev. D* **1**, 2933–2936 (1970), doi:10.1103/PhysRevD.1.2933
9. S. Fubini and G. Veneziano, Algebraic treatment of subsidiary conditions in dual resonance models, *Annals Phys.* **63**, 12–27 (1971), doi:10.1016/0003-4916(71)90295-8
10. P. Ramond, Dual theory for free fermions, *Phys. Rev. D* **3**, 2415–2418 (1971), doi:10.1103/PhysRevD.3.2415
11. A. Neveu and J. Schwarz, Factorizable dual model of pions, *Nucl. Phys. B* **31**, 86–112 (1971), doi:10.1016/0550-3213(71)90448-2
12. A. Neveu and J. Schwarz, Quark model of dual pions, *Phys. Rev. D* **4**, 1109–1111 (1971), doi:10.1103/PhysRevD.4.1109
13. C. B. Thorn, Embryonic dual model for pions and fermions, *Phys. Rev. D* **4**, 1112–1116 (1971), doi:10.1103/PhysRevD.4.1112
14. J. L. Gervais and B. Sakita, Field theory interpretation of supergauges in dual models, *Nucl. Phys. B* **34**, 632–639 (1971), doi:10.1016/0550-3213(71)90351-8
15. Y. Golfand and E. Likhtman, Extension of the algebra of Poincare group generators and violation of p invariance, *JETP Lett.* **13**, 323–326 (1971),
16. J. Wess and B. Zumino, Supergauge transformations in four dimensions, *Nucl. Phys. B* **70**, 39–50 (1974), doi:10.1016/0550-3213(74)90355-1
17. L. Brink, P. Di Vecchia and P. S. Howe, A locally supersymmetric and reparametrization invariant action for the spinning string, *Phys. Lett. B* **65**, 471–474 (1976), doi:10.1016/0370-2693(76)90445-7
18. S. Deser and B. Zumino, A complete action for the spinning string, *Phys. Lett. B* **65**, 369–373 (1976), doi:10.1016/0370-2693(76)90245-8
19. J. Scherk and J. H. Schwarz, Dual models for nonhadrons, *Nucl. Phys. B* **81**, 118–144 (1974), doi:10.1016/0550-3213(74)90010-8
20. A. Neveu and J. Scherk, Connection between Yang–Mills fields and dual models, *Nucl. Phys. B* **36**, 155–161 (1972), doi:10.1016/0550-3213(72)90301-X
21. T. Yoneya, Quantum gravity and the zero slope limit of the generalized Virasoro model, *Lett. Nuovo Cim.* **8**, 951–955 (1973), doi:10.1007/BF02727806
22. T. Yoneya, Connection of dual models to electrodynamics and gravidynamics, *Prog. Theor. Phys.* **51**, 1907–1920 (1974), doi:10.1143/PTP.51.1907
23. F. Gliozzi, J. Scherk and D. I. Olive, Supersymmetry, supergravity theories and the dual spinor model, *Nucl. Phys. B* **122**, 253–290 (1977), doi:10.1016/0550-3213(77)90206-1

24. M. B. Green and J. H. Schwarz, Supersymmetrical dual string theory, *Nucl. Phys. B* **181**, 502–530 (1981), doi:10.1016/0550-3213(81)90538-1

25. M. B. Green and J. H. Schwarz, Covariant description of superstrings, *Phys. Lett. B* **136**, 367–370 (1984), doi:10.1016/0370-2693(84)92021-5

26. M. B. Green and J. H. Schwarz, Supersymmetrical string theories, *Phys. Lett. B* **109**, 444–448 (1982), doi:10.1016/0370-2693(82)91110-8

27. L. Alvarez-Gaume and E. Witten, Gravitational anomalies, *Nucl. Phys. B* **234**, 269 (1984), doi:10.1016/0550-3213(84)90066-X

28. M. B. Green and J. H. Schwarz, The hexagon gauge anomaly in type I superstring theory, *Nucl. Phys. B* **255**, 93–114 (1985), doi:10.1016/0550-3213(85)90130-0

29. M. B. Green and J. H. Schwarz, Anomaly cancellation in supersymmetric $D = 10$ gauge theory and superstring theory, *Phys. Lett. B* **149**, 117–122 (1984), doi:10.1016/0370-2693(84)91565-X

30. P. Goddard and D. I. Olive, Algebras, lattices and strings, DAMTP-83/22.

31. D. J. Gross, J. A. Harvey, E. J. Martinec and R. Rohm, The heterotic string, *Phys. Rev. Lett.* **54**, 502–505 (1985), doi:10.1103/PhysRevLett.54.502

32. P. Candelas, G. T. Horowitz, A. Strominger and E. Witten, Vacuum configurations for superstrings, *Nucl. Phys. B* **258**, 46–74 (1985), doi:10.1016/0550-3213(85)90602-9

© 2022 World Scientific Publishing Company
https://doi.org/10.1142/9789811219832_0014

Geoffrey F. Chew (1924–2019): A Passion for Physics and the Ph.D. Thesis Professor for Me and Seventy-plus Others

Ling-Lie Chau (喬玲麗)[*]

University of California, Davis, CA 95616, USA
chau@physics.ucdavis.edu

No words are adequate to express my deep sadness of having Geoff no more as a friend in this world and no words are adequate to express my deep gratitude to him for having been my Ph.D. thesis professor that launched my career as a theoretical physicist. Here I give a brief description of my experience with him as my Ph.D. professor, some other anecdotes, and an Attachment of three photos at the end. For the detailed accounts of his biography, accomplishments, publications, the list of seventy-plus Ph.D.s he had graduated, etc., please see other contributions to the Memorial Volume for Geoff, to be published by the World Scientific.

I could not be happier when studying physics as a graduate student at Berkeley in 1961–66.[a] Geoff was then the most popular Ph.D. thesis professor in theoretical physics. I was happy that he accepted me as one of his Ph.D. candidate students in 1963. At one point there were thirteen of us! However, I felt that he was always available. I shared an office with another student of his at the Lawrence Lab on the hill. The office was right next to the door that led to the big balcony that looked west over the beautiful campus toward the San Francisco Bay, with the Golden Gate Bridge in view. Standing and walking around there, one can see the fogs flowing over the Golden Bridge or the sunsets! Such was the incredible place for me to do my graduate research, and being paid too! All these were arranged by Geoff.

---

[*]Professor Emerita in Physics and GGAM (Graduate Group of Applied Mathematics), UC Davis.
[a]The acceptance of me by UC Berkeley as a graduate student in physics was heavenly to me. Furthermore, the generous full year IBM Fellowship for my first year 1961–62 afforded me to live in the International House with room and meals, to have no teaching assistant duties, and to fully concentrate in studying! My streak of good luck of "being always at the right place at the right time" was brought to a new height — as told in the profile of me by my high school (1951–57), Tainan Girls School (TNGS) which celebrated its Centennial in 2017 and where the excellent science teacher cultivated my interest in physics and started my streak of good luck, https://tngs100.blogspot.com/2017/07/blog-post_70.html (The text is in Chinese, however there are English web links — in colored characters — that help to cover some of the contents in English.)

At the time Geoff's research interests were to formulate a theory for strong interactions using the analyticity properties of the scattering amplitudes. He said to me that there was a need to formulate analytic scattering amplitudes for two-to-two particles with higher spins, beyond 0 and $\frac{1}{2}$. I began to look into the matter. To make the long story short, suffice it to say that the then known method of constructing analytic amplitudes for two-to-two particles with spins 0 or $\frac{1}{2}$ could not be generalized to higher spins.

Every morning when I got to my office, I would find papers that Geoff left on my desk for me to look into. Months went by. One day, I read a paper and suddenly I got an idea how to formulate a general method to construct analytic amplitudes for two-to-two particles with arbitrary spins. I told Geoff about it. He immediately liked the idea and encouraged me to go on, despite of doubts expressed by other physicists. Geoff's steady guidance (more anecdotes later), feeding me with relevant papers, quick insightful understanding of and strong stand for my idea were very important to me. Looking back now, I am still so impressed by and so appreciative of him.

I worked my idea out for arbitrary spins, and importantly checked that my results for spins 0 and $\frac{1}{2}$ are consistent with the previously known results but through different methods that could not be generalized to arbitrary spins. I wrote up the paper and published it in Physical Review (1966, Vol. 142, pp. 1187–1194). By then, the initially doubting physicists were also convinced. Afterwards I quickly wrote another paper applying the results to analyze scattering properties of particles with higher spins published in Physical Review Letters (1966, Vol. 16, pp. 756–760). Nicely, I did not need to type up my papers. They were typed up by the staff members (then called secretaries) at the Lab! All I had to do for my thesis was to have a covering page with acknowledgement to Geoff, stapled together with the two published papers, so no typing was needed for my thesis!

Then I took the oral examination, facing five professors — giving a talk about my thesis and answering their questions, not only about my thesis but also whatever they thought that I should know. Geoff's pleasant and relaxed presence was very helpful. I gave my talk and answered all the questions — a pleasant surprise to myself. Then I stepped outside and waited. Soon they opened door. All are smiling, Geoff's was the biggest, as usual. Each professor congratulated me with a hand shake. Geoff said loudly, "Ling-Lie, I wish all my other students had done as well as you did!"[b]

---

[b]Only years later, unexpectedly Geoff revealed one of whom he had in mind. In 2007, David Gross was invited to give a public speech at UC Berkeley and the title was "The Coming Revolutions in Fundamental Physics." David was one of the thirteen Ph.D. students of Geoff's in the early 1960's as I was. He shared the 2004 Nobel Prize in Physics with H. David Politzer and Frank Wilczek, https://www.nobelprize.org/prizes/physics/2004/summary/. It was a highly anticipated event, to be held at the auditorium of the International House on the Berkeley campus. The then Physics Chair, Prof. Frances Hellman, Geoff and I were chatting at the pre-talk reception. She said quietly to Geoff, "I am going to introduce David before his talk. It will be nice to say something fun about David in my introduction." Without hesitation Geoff said with a big smile, "Yes, I have one for

After my Ph.D. in 1966, I stayed on in Berkeley as a researcher, and published three more papers,[c] till 1967 when I was offered a two-year postdoctoral position at Institute for Advanced Study without my having to apply. (Also without applying, I was offered an assistant professorship at a good university but I declined). I am sure that Geoff's reputation had helped. That started my streak of good luck of having never had to apply for a job.

Besides individual meetings with his research students, Geoff also held the weekly group meeting. At first the weekly meeting was held in the theory seminar room in the Lab. Soon many postdoctoral physicists also come to attend. Geoff then changed to hold the group meeting with his research students at his home. Those were most fun and inspiring meetings, in the beautiful home of Geoff and his first wife Ruth (sadly she died later too young because of an incurable health condition). The topics of discussion were not limited to physics, especially considering those were the days of the Free Speech Movement era at Berkeley. We all enjoyed and learned so much. Surely, it was major work for Geoff and Ruth to hold those meetings at their home. I will forever fondly remember those meetings.

In 1986 I was recruited to UC Davis, after two years at the Institute for Advanced Study 1967–69 and seventeen years at Brookhaven National Laboratory 1969–86. Geoff and his second wife Denyse gave me a wonderful welcome dinner party with the attendance of other physics guests, at their beautiful home overlooking the campus with the San Francisco Bay and the Golden Gate Bridge in view. They made me feel that I had returned home.

Davis is about one hour driving to Berkeley. Whenever I had a chance I would meet with Geoff, often at the Lab for lunch. His lunch was almost always just a big cup of soup with some crackers. Usually he was so occupied with talking about the physics he was working on that he forgot to eat, only hurried to finish at the end. This was the pattern how he ate his meals whenever I had a meal with him, no matter who else were present. He would happily entertain other topics that were injected into the conversation, but soon he would return to talking about what he had in mind about his physics. Once in a big banquet at the posh Ritz-Carlton Hotel in San Francisco, Geoff sat between Stanley (Mandelstam) and me among others at a table. The main course was a nice big steak. When most people at the table had finished, Geoff had eaten only a little bit and kept on with his talking. The waiter came, without asking, took his plate. I thought that was rude and

---

you. At David's oral examination he gave a fantastic talk about his thesis. However, during the questions, he could not answer any of them." There was a pause from Frances. I then quickly said, "I think I have a better one, more appropriate and more fun. I am sure David would be proud of it." I continued, "David was one of the 800 students got arrested during a sit-in demonstration in Sproul Hall during the Free Speech Movement!" Frances made a creative use of the one-sentence story in her introduction to David's talk and got loud applauds from the audience, as documented in the video https://www.youtube.com/watch?v=AM7SnUlw-DU, 0:02:30–0:03:30/1:39:09. David gave a brilliant talk, as he usually does, and also documented there 0:03:30–1:39:09/1:39:09.
[c]One of them was with Stanley Mandelstam. The interesting way the paper came about was told in Footnote 52 of my paper remembering Stanley published in the Memorial Volume for Stanley Mandelstam (World Scientific Publishing, 2017).

said quietly, "That's not right!" He just laughed and murmured, "It happens, it happens." Stanley kept his usual quiet with an understanding smile. I ran after the waiter with a smile and told him quietly that the professor had not yet finished. The waiter was embarrassed and brought Geoff a new plate of hot full steak!

I organized outings with Geoff and family and Stanley, e.g. the visit to the Oakland Museum when the treasures from the Forbidden City of China were shown there (a truly outstanding exhibit, many treasures were shown there that I had not seen in my several prior visits to the Forbidden City) and then the dinner at the Jack London Square; the visit to the Asian Museum in San Francisco; etc. Often Geoff's family members said, "Ling-Lie, thank you for inviting us to these outings. Usually Geoff would not do this sort of things!" One day I found out that there was a Jewish museum in Berkeley, The Magnes Collection of Jewish Art and Life. I organized a visit to it. While walking around there, Geoff learned that Stanley did not know the existence of the museum after having lived in Berkeley for so many years! Geoff laughed and said jokingly, not so quietly, "Stanley, Shame on you — being a Jew, you did not know about this museum?!" Stanley just smiled. Apparently Stanley also "usually would not do this sort of things". I am proud that not only did Geoff and Stanley agree to go along with the outings I organized, during one visit to SFMOMA in San Francisco, they even let me making them part of the arts![d]

In 1999, to mark the new millennium I began to host New Year's Eve dinner with Geoff, his family members, and Stanley. Geoff and Stanley had always attended. We had fun talking about everything, and Geoff would always go back to talk about his physics ideas and progresses. Sadly, Stanley passed away in 2016. With Geoff and family members we continued the New Year's Eve dinner get-together. Sadly Geoff passed away on April 11, 2019. Geoff always sent me a gracious thank-you email after the New Year's Eve dinner. His last one was:

2019-01-01, 09:17 Geoffrey Chew wrote:
Dear Ling-Lie,
Frank and I enormously enjoyed the treat you gave us last night. It will be long-remembered. As I review the paper that Pauline and Frank have helped me to write, I shall keep in mind your remarks about its current form.
Happy New Year!
Geoff and Frank
(Note: Pauline is Geoff's younger daughter of two and Frank his youngest son of three. The email was copied to both.)

I will forever remember the joyous times I had with Geoff, Stanley, and Geoff's family members.

For years, as far as I know from all my get-togethers with Geoff since 1986 when I moved to Davis, sometimes alone with him at the Lab and mostly together with

---

[d]See the photo, Fig. 7, of my paper in the Memorial Volume for Stanley Mandelstam (World Scientific, 2017).

Stanley, Geoff was looking for an all-encompassing theory for physics. However, I could not understand nor could Stanley what he meant to get done — but we always had fun getting together. Though he seldom talked about his S-matrix theory approach of his earlier years, he was proud of it. I remember clearly he said that among all his physics work he was most proud of his original idea of associating particles to the poles in scattering amplitudes. Following that he developed the bootstrap idea and the analytic S-matrix approach. Surely, his historical encounter with Stanley when Stanley was an unknown postdoctoral, his instantaneous recognition of the importance of Stanley's work of generalizing the single-dispersion expressions to double-dispersion expressions for scattering amplitudes and bringing Stanley to Berkeley[e] was a major event in both of their lives as well as in the development of theoretical physics. What he had advocated and worked so hard for the S-Matrix approach had laid the fertile ground for the development of the string theory, a theory that many have been advocated for decades to have the potential to become the theory of everything and are still working hard toward that goal. Stanley, a leading field and string theorist,[f] insightfully said in his talk at the conference celebrating Geoff's 60th birthday in 1984, published in the Proceedings "A Passion for Physics" published in 1985 by World Scientific, p. 97, *"Before I go further, let me emphasize one thing that will be well known to the older members of the audience; the string model originated as a model for the S matrix, and it may well not have been discovered if S-matrix had not been vigorously pursued at the time. It is therefore intimately related to the work of Geoffrey Chew. Of course this does not imply that he necessarily supports later developments after the model."* Indeed, Geoff had other ideas. He tried hard to develop them for decades, till the end of his life as indicated in his 2019-01-01 email to me copied above. Though he did not bring his ideas to a clear fruition, he had joy in developing them.

We miss Geoff dearly — his heartwarming big smiles and laughs and his passion for physics. His impact on physics and physicists will live on — in the continuing search for an all-encompassing fundamental theory for the physical universe.

---

[e]See Chew's paper in the Memorial Volume for Stanley Mandelstam (World Scientific, 2017).

[f]See the Memorial Volume for Stanley Mandelstam (World Scientific, 2017), my Physics Today Obituary for Stanley, May 2017 issue, http://chau.physics.ucdavis.edu/Chau-PhysicsToday-obi-Mandelstam-2017May.pdf, and its extended version as my paper in the Memorial Volume for Stanley, http://chau.physics.ucdavis.edu/Chau-biogSlctPapr-Mandelstam-Memo-WS-2017-dstPost.pdf.

Attachment: Photos

(Photo 1) The cover of the Proceedings of celebrating Geoff's 60th birthday in 1984 at UC Berkeley, published in 1985 by World Scientific.

(Photo 2) Five Chicago University Classmates in the mid-late 1940s: From left to
right, Lincoln Wolfenstein, Geoffrey Chew, Chen-Ning Yang, Marshall Rosenbluth,
and Jack Steinberger, taken at the Yang's Retirement Symposium, "Symmetry &
Modern Physics", at SUNY Stony Brook, May 21–22, 1999, and printed in the Proceedings published by World Scientific. I am proud and happy that I orchestrated
the taking of the photo. I got the idea of orchestrating such a photo when chatting
with Steinberg at the party and mentioned the idea to him. He rejected my idea.
So I kept quiet about it while walked around to chat with people. After a while,
Steinberg came back to me, saying "I've changed my mind. I now think it is a
good idea!" So I went about bringing them together. Notice that Chew, Yang, and
Rosenbluth were looking at me while I was orchestrating and taking a photo using
my camera. The official photographer of the conference took this one. I thank
World Scientific for permission to reprint the photo here.

(Photo 3) New Year's Eve dinner get-together (Dec 31, 2009); from left to right: Frank Chew (Geoff's youngest son of three), Stanley Mandelstam, Ling-Lie Chau, and Geoffrey Chew.

The Reggeon Field Theory, Describing Financial Markets in Crises, and Predicting Crises

Jan W. Dash

President, J. Dash Consultants LLC

The RFT was used in physics to describe high-energy diffraction scattering. Later an empirical relation between the critical RFT and behavior *inside* financial crises for diverse markets was discovered, with no arbitrary parameters. A method using scaling exponents was used to obtain probabilities of *future* equity crises, with success much better than chance. Here, I summarize this work.

1. Remarks about Geoff Chew

Along with many others, I was fortunate to be Geoff's PhD student at Berkeley (PhD, 1968). Then, phenomenology, combining theory with data analysis to determine parameters, was a modus operandi. After some years as a particle theorist, I changed fields to quantitative finance and risk management. Though the environment was different, Geoff's influence continued with me in a profound sense. Quantitative finance is phenomenology, using ideas from physics, with parameters mostly taken from data. This essay[a,1] outlines what I believed for many years[b] would be a fruitful area of research for finance, using the Reggeon Field Theory (RFT),[2] influential during the 1970's, and in which I worked. Geoff contributed to the origins of the RFT through his leadership in strong interaction physics, and his influence can be felt here. Thank you, Geoff.

---

[a] *Source*: Most of the material for this essay is adapted from Ref. 1, Chapter 46.

[b] *History of applying the RFT to Finance*: I worked on the RFT, calculating critical exponents and Green functions, and assessing relevance to experimental physics data. Later, I ran an Options Seminar at Merrill. At the first meeting in 1987, I said that the free diffusion behind the Black-Scholes model was too simple, more general scaling laws exist, and I mentioned the RFT as potentially relevant. This essay describes RFT-finance work finally done in 2011 and written up in 2013–016, when I was head risk quant at Bloomberg LP. I now work on climate change, our current crisis of risk and opportunity.

2. Introduction — Reggeon Field Theory (RFT) and Finance

In this essay I recount some of my work in the RFT in physics, and then applying the RFT and related methods to characterizing crises in finance.[c] The latter was an unusual research project. The project was documented in three working papers, in the public domain on SSRN.[3]

Standard financial models using ordinary diffusion are deficient because they do not describe the high-risk "fat tails" from large market moves. Standard diffusion implies a Gaussian distribution of financial returns,[d] with width an empirical "volatility", the standard deviation of returns. It is possible to use a "fat-tail volatility" to fit the empirical distribution roughly including the fat tails.[1]

The distribution in movements of underlying variables is critical for risk management, including financial crises. A potential use for the RFT lies in the description of fat tails in a more fundamental fashion than the description using fat-tail volatilities and ordinary diffusion. The nonlinear RFT interactions generate fat-tail events that modify standard finance theory. The idea of using the RFT to describe financial crises is greatly enhanced by the fact that the RFT reduces to familiar standard diffusion when the RFT nonlinearity is turned off, and so (with suitable translation) the RFT is a natural generalization of the standard diffusion or Brownian motion ubiquitously used in finance. Other models trying to describe fat tails do not generally have this desirable property.

It is important to note that the nonlinearities of the RFT have nothing to do with other "nonlinear" ideas, e.g. a nonlinear dependence of volatility on the underlying variable or a nonlinear transformation of the underlying variable. Rather, nonlinearity in RFT refers to a nonlinear dependence on a Green function itself in relevant equations. Different Green functions exist, coupled nonlinearly.

The critical exponents and scaling Green functions can be calculated in the RFT given the nonlinear interactions. The numerical result for an exponent was used, with some assumptions, to describe behavior of various markets inside crises, with no fitting or parameters, and surprisingly successfully.

Following the work describing markets in crises, related work was undertaken to go further. Empirical anomalous scaling exponents from non-crisis periods were used with a complicated noise filter to predict crises in equity markets, with success in backtesting much better than chance.

---

[c] *Acknowledgements*: I thank Xipei Yang for implementing the numerical comparison of crises to the RFT calculation, and for collaborating in formulating the method for probabilities of future crises. I also thank Adam Litke for much insightful help.
[d] *Returns*: A return of a financial variable (e.g. logarithm of stock price) is its difference over a given time interval.

3. Summary of the RFT in Physics

We begin with a summary of the RFT to remind people and establish notation. The RFT aims to describe high-energy diffractive scattering; the RFT is a theory for the Pomeron. The RFT without any interactions is free-field theory equivalent to free diffusion. The interactions generate non-linear diffusion.

The RFT was motivated by perturbation theory in the nonlinear interactions. The most common interaction assumed is the triple-Pomeron PPP coupling constant multiplied by $i = \sqrt{-1}$, denoted as ir.

The spatial variables in the RFT in physics correspond to the two transverse dimensions perpendicular to the incoming beam of particles in accelerator experiments. The time variable in the RFT is Fourier-conjugate to an "energy" variable that is really $E = 1 - j$ where j is a "cross-channel" angular momentum. The RFT is of interest in the infrared $E \approx 0$ or $j \approx 1$ region, which governs high-energy, low-momentum-transfer scattering.

3.1. *RFT formalism*

The RFT non-linear diffusion with the PPP coupling has the Lagrangian L:

$$L = \frac{1}{2}i\left[\psi^+\frac{\partial}{\partial t}\psi - \text{h.c.}\right] - \alpha_0'\boldsymbol{\nabla}_x\psi^+ \cdot \boldsymbol{\nabla}_x\psi - \frac{1}{2}ir_0\left(\psi^+\psi^2 + \text{h.c.}\right). \tag{1}$$

Bold type is used for vector notation. The field $\psi(\boldsymbol{x}, t)$ is a theoretical construct that depends on the spatial variable $\boldsymbol{x}$ and the "time" t, while ψ^+ is its hermitian conjugate (h.c.). The "bare" PPP coupling is r_0. The interaction part is the product of three fields. The "bare slope" α_0' is relevant to the first-order description of the high-energy scattering away from the forward direction.

3.2. *Non-interacting Free Diffusion (Brownian Motion) Limit*

If the interaction is set to zero, the Lagrangian has a free "non-relativistic" form. The corresponding dynamical equation is the same as the free diffusion equation, with imaginary time. The Green function is Gaussian with $\sqrt{(\Delta\boldsymbol{x})^2}/\sqrt{\Delta t}$ scaling for $\Delta\boldsymbol{x}$,

$$G_{\text{free}}(\Delta\boldsymbol{x}, \Delta t) = N\exp\left[i\frac{(\Delta\boldsymbol{x})^2}{4\alpha_0'\Delta t}\right]. \tag{2}$$

The interactions change the form of the free Green function and also the scaling behavior. Feynman rules summarize perturbation theory. However the applications of interest here are non-perturbative.

3.3. *Critical RFT calculations*

The non-perturbative RFT calculations use renormalization group equations.[2] In the critical dimension D_{crit}, the coupling constant is dimensionless. Here, $D_{\text{crit}} = 4$.

The physical dimension is $D_{phys} = 2$. We calculate for arbitrary D, at $\varepsilon = D_{crit} - D$. Then, $\varepsilon_{phys} = D_{crit} - D_{phys}$ is used, $\varepsilon_{phys} = 2$. The Gell-Mann-Low β-function must have a zero at a non-zero value of the coupling constant. Results are obtained for the critical exponents as power series in ε.

3.4. *Results of the RFT in physics*

The imaginary part of the scattering amplitude, $\text{Im}\, T_{ab}$ for elastic scattering $ab \to ab$ at Mandelstam variables s, t assuming the PPP interaction is[e]

$$\text{Im}\, T_{ab}(s,t) = \beta_a(t)\beta_b(t)s(\ln s)^{-\gamma} T^{(1,1)} \left[\frac{-\alpha_0' t}{k_2} (\ln s)^{1-\zeta/\alpha'} \right]. \tag{3}$$

Here γ and ζ/α' are the critical RFT exponents. They were calculated by Abarbanel and Bronzan in $O(\varepsilon)$, extended independently by Baker and by Bronzan and Dash in $O(\varepsilon^2)$. The critical exponent ζ/α' relevant here is

$$-\zeta/\alpha' = \frac{\varepsilon}{24} + \left(\frac{\varepsilon}{12}\right)^2 \left[\frac{59}{24} \ln \frac{4}{3} + \frac{79}{48}\right]. \tag{4}$$

A similar result for γ exists, and theoretically describes total cross sections at high energies. The form of the RFT result for $T^{(1,1)}$ in Eq. (3) is not simple to describe. For the $O(\varepsilon^2)$ result, see Baig, Bartels and Dash (Ref. 2).

4. Aspects of Applications of the RFT to Finance

The basic procedure of potential application of any physics to finance is to see which assumptions make sense, translate the variables, and check results.

Without interactions, and brute-force replacing the imaginary i by -1, the RFT reduces to standard model assumptions in finance. This is highly desirable.

RFT critical exponents describe deviations from square-root time volatility scaling of the standard Brownian assumption used in finance. That is, we get functions of $\sqrt{(\Delta x)^2}/(\Delta t)^{0.5+v}$, where v gives the deviation from $\sqrt{\text{time}}$ volatility scaling for Δx, due to nonlinear interactions.

4.1. *Mapping of the RFT to finance*

The assumed translation of the RFT in Eq. (3) to finance is $\ln s \to (\Delta t)$, $t \to 1/(\Delta x)^2$ where Δx in finance is the return[d] measured over time interval Δt. Hence, $t(\ln s)^{1-\zeta/\alpha'} \to (\Delta t)^{1-\zeta/\alpha'}/(\Delta x)^2$, so the finance volatility scaling variable is $\sqrt{(\Delta x)^2}/(\Delta t)^{0.5(1-\zeta/\alpha')}$. The anomalous critical volatility exponent, i.e. the difference in the power from square-root-time volatility scaling, is $-0.5\,\zeta/\alpha'$, which is a positive number. The equivalent statement for the variance, or volatility squared, is $(\Delta x)^2 \propto (\Delta t)^{1-\zeta/\alpha'}$. For finance, the postulated physical dimension corresponds to a single variable. So for finance $D_{phys} = 1$ (not 2). We get RFT variance scaling $(\Delta x)^2 \propto (\Delta t)^{1.27}$ replacing the standard $(\Delta x)^2 \propto (\Delta t)^1$.

---

[e] *Variables* (s,t): For diffraction scattering we have large s and small t. Unfortunately, the same letter t is also used for the RFT "time", which is really $\ln(s)$.

4.2. *General remarks on using critical exponents in finance*

Use of calculated quantities with no fitting or arbitrary parameters is attractive. Numerical values of theoretical critical exponents could be useful, in a limited sense. First, critical exponents from data vary by underlying, and also vary according to the data time window. Second, critical nonlinear theories are probably restricted to only a subset of behavior over time, so the notion of a critical exponent for an entire time series is only a rough approximation. Third, since the RFT dimension is only valid for $D \leq 4$ ($\varepsilon \geq 0$), use of an index (e.g. S&P500) is only possible as a single variable.

5. The RFT Applied to Financial Crises

There are two types of results[3]
1.   Approximate RFT scaling for various markets **already in crisis.**
2.   A related scaling-based method for **predicting crises** in equity markets.

5.1. *RFT-like scaling for various markets already in crisis*

Evidence of RFT-like scaling for various markets when **already in crisis** was discovered.[f] The empirical anomalous exponents are, on average, in surprisingly reasonable numerical agreement with the RFT two-loop result for the critical exponent, which for variance is twice that for volatility, $2v_{\mathrm{RFT}} = -\zeta/\alpha' \sim 0.27$. The summary comparative results are in Fig. 1, by market.[g,3] The markets examined were equities (major, other), FX/USD (up, down), commodities, emerging markets debt crises, and bonds, rates.[h]

5.2. *Rich-cheap analysis, the RFT, and markets already in crises*

The RFT critical behavior could be used as a benchmark for rich/cheap analysis of various markets in crisis. A variable whose empirical anomalous in-crisis exponent is above the RFT anomalous exponent 0.27 can be called "rich" within crisis. If

---

[f]*Definition of "Crisis" and "Scaling"*: A "crisis" for an underlying variable is defined using a recipe: down for equities, commodities and bond prices, either down or up for FX, up for rates and spreads — by 25% and 3 standard deviations. Scaling was evaluated between one week and one month. Only "big" moves were used within crises (confidence levels 70% to 95%). Trends were not extracted. See Ref. 3 for details.

[g]*Data Source*: Bloomberg LP; the derived quantities in Fig. 1 are public domain, Ref. 3.

[h]*Numerical RFT-Like Scaling Results for Various Markets already in Crisis*: The results were 0.35 using linear returns (0.35 includes the kinematic factor of 2, going from linear returns to quadratic variance returns). The result 0.17 was obtained using squared returns. The empirical interval (0.17, 0.35) brackets the RFT calculated value 0.27. The data were chosen up-front to be representative and diverse. Seven markets with 150 time series were included, going back as far as data exists, and up to 2011 when this work was done. There were around 200 cases, each with one or more crises. For example, the S&P 500 had 12 crisis periods between 1929 and 2011. Figure 1 has all cases, averaged for each market. For each case, the empirical anomalous exponents were averaged over crises.

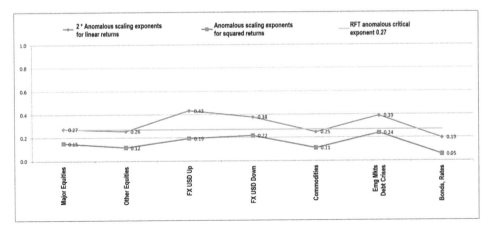

Fig. 1. *Comparison of the calculated critical exponent of the RFT (0.27) with empirical anomalous variance exponents during crises for various markets.*[3] *The lower (red) line is the anomalous empirical in-crisis exponent for squared returns, and the upper (blue) line is twice the anomalous empirical in-crisis exponent for linear returns (either is relevant). The straight (green) line is the RFT result. No fitting was done. The RFT calculation is in surprisingly reasonable numerical agreement on average.*

the empirical anomalous in-crisis exponent is below the RFT anomalous exponent 0.27, the variable can be called "cheap" within crisis.

5.3. *Predicting crises in equity markets and an earthquake analogy*

A method for **predicting** crises for individual equity markets before they occur was developed using scaling language similar to that of the RFT. This method worked, much better than chance and also a Weibull benchmark, for predicting crises in equity markets, with prediction dates up to 2011[3] (it was not possible to do the prediction analysis for other markets, or past 2011). A crisis "prediction" was defined as the probability for a crisis within a year in advance being above a certain level, using input information before the prediction time.[i]

It is significant that the *only* data used was the given price time series. That is, the patterns of prices themselves contain information about future crises.

The method involves a complex noise filter for the time series using "indicators". One especially useful indicator was the cumulative buildup of the empirical anomalous variance exponent during **non-crisis** periods. The *non-crisis* empirical anomalous exponent is generally smaller than the *crisis-period* empirical anomalous exponent (which makes sense due to the fat tails in crisis).

This predictive model has substantial memory effects. This is physically reasonable given "bubble-like buildups" that occur during rather long times before many crises. The analogy is that "stress" builds up and then is released during a

---

[i] *Results — Backtesting Crises:* The model successfully indicated around 70% of the tested crises within a year in advance, for 52 equity indices, worldwide.[3]

crisis, qualitatively rather like some types of earthquakes.[j] This is very different from standard zero-memory Brownian motion finance models.

5.4. Details of the crisis prediction method

The idea is to postulate various indicators $\vec{x}^{(\text{Indicator})}(t)$ as functions of real time depending on the time-series for the specific equity.[k] Then the probability of crisis $p(t)$ is calculated, occurring within time ΔT after the time t.

The crisis probability calculation uses a logistic fit during in-sample periods, namely with known crises. "S-shaped curves" are fit to step functions to determine logistic "beta" parameters $\vec{\beta}$ (these are independent of time). The step functions are crisis flags; $y(t) = 0$ or $y(t) = 1$ if a crisis doesn't or does occur within time ΔT after t. Here, $\Delta T = 1$ yr was used.

The beta parameters are varied to get the probabilities of crisis to match approximately the crisis flags during the in-sample test period (where crises are known). Then out-of-sample predictions for the probability of the next future crisis in time relative to the input are made with the same beta parameters.

Here is the method to determine the beta parameters. We discretize time as $\{t_i\}$. Define $\vec{x}_i \equiv \vec{x}^{(\text{Indicator})}(t_i)$ as the vector of indicators at time t_i. The probability p_i of a crisis within ΔT (1 year) after t_i is taken to be the logistic form with beta parameters $\vec{\beta}$, viz:

$$p_i(\vec{\beta}) = \exp\left(\vec{x}_i \cdot \vec{\beta}\right) / \left[1 + \exp\left(\vec{x}_i \cdot \vec{\beta}\right)\right]. \tag{5}$$

The crisis flag at t_i is $y_i = y(t_i)$. For in-sample testing or training, we know where the crises are, so we know the crisis flag values.

The likelihood function is defined as $L = \prod_i L_i$ with $p_i = p_i(\vec{\beta})$ in Eq. (5),

$$L_i(\vec{\beta}) = p_i^{y_i}\left(1 - p_i\right)^{1-y_i}. \tag{6}$$

The likelihood is maximized to get betas and crisis probabilities. See Fig. 2.[3]

6. Conclusions

Finance theory (no matter what the mathematics or the physics behind it) is phenomenology. This phenomenology is useful (and interesting), providing a common language for real-world quantitative finance and risk management.[1]

We showed that the critical RFT, plus related methods, with some assumptions, have empirical relevance for describing and even predicting financial crises. In chapter 1 of ref. 1, the first unsolved problem in finance and risk management listed is *"the inherent basic instability of financial systems, including crises"*. Maybe

---

[j] *Earthquakes:* There is no geophysics here, just a useful analogy - mainly to emphasize the cumulative indicator buildups before crises. Indicators were reset to zero after crises.

[k] *Scoring System, Smoothing:* A scoring system was constructed, including type I and type II errors. Some smoothing schemes were employed.

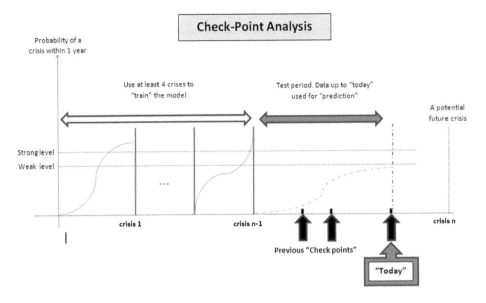

Fig. 2.   This shows the crisis prediction model as used in real time. Future crisis probabilities are calculated at "check-points"; the current one is the date "today". Indicators from data up to "today" are used, with the betas fixed from at least 4 previous crises. Two levels specify "strong" and "weak" predictions.

someone would like to update/expand/improve this RFT/crises work to get a better handle on crises, advancing risk management.

References

1. *Original Source:*
   J. W. Dash, *Quantitative Finance and Risk Management: A Physicist's Approach,* 2nd edn. (World Scientific Publishing, 2016), ISBN 9789814571234, https://doi.org/10.1142/9789814571241_0046

2. *Reggeon Field Theory in Physics:*
   H. D. I. Abarbanel and J. B. Bronzan, Structure of the Pomeranchuk singularity in Reggeon field theory, *Phys. Rev. D* **9**, 2397–2410 (1974); H. D. I. Abarbanel, J. D. Bronzan, R. L. Sugar and A. R. White, Reggeon field theory: Formulation and use, *Phys. Rep.* **21**, 119–182 (1975).
   J. B. Bronzan and J. W. Dash, Higher-order ε terms in Reggeon field theory, *Phys. Lett. B* **51**, 496–498 (1974); *Phys. Rev. D* **10**, 4208–4217 (1974); *Phys. Rev. D* **12**, 1850 (1975).
   M. Baker, The ε Expansion of Pomeron amplitudes, *Phys. Lett. B* **51**, 158 (1974); *Nucl. Phys. B* **80**, 62–76 (1974).
   M. Baig, J. Bartels and J. W. Dash, The complete $O(\varepsilon^2)$ Reggeon field theory scaling law $d\sigma_{el}/dt$ and its applications at collider energies, *Nucl. Phys. B* **237**, 502–524 (1984).

3. *Financial Crises and the Reggeon Field Theory:*
   J. W. Dash and X. Yang, *Market Crises, Earthquakes, and the Reggeon Field Theory*, Bloomberg LP working paper (2013):
   http://papers.ssrn.com/abstract=2274413.
   J. W. Dash and X. Yang, *Predicting Equity Crises, Critical Exponents, and Earthquakes - II*, Bloomberg LP working paper (2016):
   https://ssrn.com/abstract=2811719.
   J. W. Dash and X. Yang, *Describing Crises with a Critical Exponent of the Reggeon Field Theory*, Bloomberg LP working paper (2016):
   https://ssrn.com/abstract=2808149.

https://doi.org/10.1142/9789811219832_0016

The Pomeron — A Bootstrap Story

Chung-I Tan

Brown University, Providence, RI 02912, USA

In a contribution to the volume *A Passion for Physics*, a collection of essays in honor of Geoffrey Chew's sixtieth birthday, I wrote, together with A. Capella, Uday Sukhatme, and Tran Thanh Van *The Pomeron Story.* This is a follow-up to that contribution. This sequel also serves as an opportunity to acknowledge my gratitude to Geoff as a PhD student under his tutelage.

1. My Years at Berkeley

In our 1984 essay, *The Pomeron Story*, we wrote "In their seminal paper[1] Chew and Frautschi asserted that all hadrons are composite; they lie on Regge trajectories. ......" To account for the near constancy of hadronic total cross sections, it was necessary for them to postulate the existence of a Regge trajectory with vacuum quantum numbers and zero-energy intercept at $j = 1$, even though no physical particles were known at the time to lie on the trajectory. This Regge singularity, which was also proposed independently by Gribov, is known as the Pomeron.

Our understanding of the Pomeron has evolved since *The Pomeron Story* was written. The fundamental theory for the strong interactions, Quantum Chromodynamics (QCD), has been established. However, much of its tests involved perturbative aspects of QCD. Great progress has also been made via lattice studies. Yet, near-forward high energy scattering remains a challenge. What is the Pomeron, really?

As a student at Berkeley in the late sixties, one faced constant turmoil on social issues. We were all caught up with the Free Speech Movement, subscribing to *Ramparts*, etc. As a student of Geoff, we also found relief in moving daily to the Rad Lab on the hill. Geoff had a large group of graduate students, sharing a common interest in various aspects of strong interactions, ranging from purely phenomenological to foundational aspects of S-matrix theory. This group included Ling-Lie Chau, Jerry Finkelstein, Jiunn-Ming Wang, Farzam Arbab, Jan Dash, Huan Lee, Richard Brower, Michael Misheloff, Carleton DeTar, Dennis Sivers, Don Tow, and others.[a] Together with other students, e.g., Dick Slansky and Joe Weis, and also

---

[a]Students just finished included David Gross, John Schwarz, Shu-Yuan Chu, John Stack, etc. Other contemporary students include Dick Haymaker, Lay-Nam Chang, and others.

a large group of post-docs, a common theme shared by most was the question of consistency and uniqueness of the S-matrix, as exemplified by the bootstrap program.[2,3] Can all Regge parameters[4] be fixed within an analytic S-matrix? One could not avoid discussing the Pomeron and Bootstrap nearly on a daily basis.

Geoff allowed us sufficient flexibility in focussing on areas of particular interest of our own. I began my work when Geoff suggested that I take a look at various questions of bounds, and, in particular, the question of the Cerulus–Martin lower bound,[5] which had just gained certain interest from an experimental perspective. Asymptotic behavior of scattering amplitudes has already been recognized as an integral part of S-matrix consistency. The Regge hypothesis, which had become popular due to its application to experiment, was introduced as an integral component of bootstrap consistency.[2]

I was fortunate that Prof. Richard Eden of Cambridge came to Berkeley for a year's sabbatical, and I worked with him closely, which formed the main part of my thesis. Although my thesis work seemed formal, it connected well with the common interest of all my fellow students, as well as a large group of post-docs. Although we each had our own research focus, we all shared the excitement of the bootstrap philosophy, with regular weekly informal seminars.

The group grew steadily in size and one often could not find seats in the Building-50 seminar/coffee room. When Geoff and Ruth moved back to the Berkeley hills, Geoff invited his group of students for an evening get-together at their beautiful new home. As a house-warming gift, I organized and brought Chinese take-outs. It was a wonderful evening, the discussion invariably turning to physics, while we enjoyed the view of the Campanile and Golden Gate in the distance. Whether by design or not, this informal gathering turned into a regular weekly meeting in the Berkeley hills for his students.[b] I understand this practice continued after those first timers had left Berkeley after receiving their PhDs.

In looking back one cannot avoid concluding that the Pomeron remained the central topic of interest to most of us at that time. In this remembrance of Geoff, I will discuss how the issue of the Pomeron evolved over the past 50 years.

2. Rencontres de Moriond and the Soft Pomeron

The "Rencontres de Moriond" is a series of annual high energy physics meetings starting in 1966. The initial impetus for this series is "to promote fruitful collaboration by bringing together a small number of scientists in inspiring surroundings, to discuss recent findings and new ideas in physics." This series was organized by Prof. Tran Thanh Van of the University of Paris, Orsay. It has continued to thrive over more than half a century. As I will explain below for the period of 70s and

---

[b]In addition to discussing physics, other stories were also exchanged, e.g., Geoff's recollections of his student days at Chicago with Fermi, Goldberger, Yang, Lee, etc. I also remember his story about meeting Landau, who reminded Geoff of the significance of his being the first to emphasize the importance of the particle–pole correspondence.

80s, Pomeron Physics suffered a period of "crisis." However, during this period, these annual meetings served as a playground where the physics of the Pomeron flourished, from the experimental perspective. In particular, it provided support for the notion that the Pomeron is non-perturbative and exhibits aspects of string structure with a cylindrical topology, i.e., the exchange of a closed string.

Crisis for the Pomeron: There are several developments in the late 60s and early 70s which dealt unkindly to the study of the Pomeron. (a) Since total cross sections can be interpreted geometrically, where does the scale come from for the Pomeron coupling? (b) Total cross sections continued to increase with energy, albeit slowly. This is in conflict with the initial impetus of having the Pomeron with an exact intercept at $j = 1$. (c) As a collection of closed string excitations,[6,7] flat-space string theories led to a Pomeron intercept at $j = 2$. What are the low-lying particles on this trajectory? (d) Lastly, what is the Pomeron in QCD?

Let me comment on these aspects briefly, in reverse. QCD is a theory of quarks and gluons. Although confinement in the infrared (IR) remains to be demonstrated explicitly, most theorists believe this can eventually be accomplished. This belief is further supported by lattice studies, where the low-lying particle spectrum has been calculated. Presumably, particles on the Pomeron trajectory should be identified with a tensor glueball and its recurrences. As a non-perturbative phenomenon, the Pomeron is not considered a subject of immediate concern.

Having an intercept $j = 2$ for closed string excitations has led to the possibility of a theory for quantum gravity. This, of course, is based on string theory in flat space–time, albeit at a higher dimension. More seriously, string theories also suffer difficulty in introducing local currents. As a consequence, it is difficult to see how the experimental observation of power-behaved amplitudes at large momentum transfer can be accommodated.

The Topological Pomeron and Bootstrap: It has become increasingly clear that the increase in the total cross sections is driven by an increasing rate of in-elastic particle production. Since particle production can be successfully explained by a Regge mechanism, it remains consistent to assume that the Pomeron, through unitarity, should emerge as a sum of ladder-like diagram, e.g., so-called multiperiph-eral mechanism.[8] Indeed, this picture is consistent with string theories, with the Pomeron entering at a higher genus, e.g., exchanging a Pomeron has the topology of a cylinder. This picture was advocated by Veneziano, leading to the program of "dual topological unitarization"[9] (DTU). (See also the contribution by Veneziano in this volume.)

From a QCD perspective, one optimistically awaits for a future QCD string theory, propagating over a curved background. In the meantime, it is important to continue to search for evidence for such a topological structure via particle production. Notably, in a topological expansion, the Froissart bound no longer applies at the cylinder-level, which removes the proposition that the Pomeron has a unit intercept. In fact once this restriction is removed, the value of the Pomeron intercept, being greater than unity, becomes a challenging dynamical question. Experimentally it seems to be greater by an amount of the order of $0.1 \sim 0.2$. Can one provide an upper bound? Finally, the Froissart bound should ultimately be restored, e.g., via eikonalization. At what level should a geometrical scale enter in the discussion?

DTU: Much of particle production data during this period came from experiments at CERN. It was natural that new data was presented and discussed at Moriond, which was then held in the French alps. Due to its close proximity to Geneva, many theorists at CERN would regularly participate in these meetings, e.g., G. Veneziano, A. White, D. Amati, Chan Hong-Mo, etc. Theoretical interpretation for particle production was a regular topic of intense discussion.

During my first sabbatical at Orsay, working with A. Capella, U. Sukhatme, and Tran Thanh Van, we began an exploration of the consequences of the Pomeron having the desired topological structure, as suggested by string theory in general and DTU in particular. These studies were vigorously discussed and confronted with experiments. In particular, these were carried out regularly at annual Moriond meetings. Based on the initial success, a systematic calculus was formulated, which has subsequently dubbed the "Dual Parton Model" (DPM).[10] This is partly based on the general approach to the S-matrix bootstrap advocated by Veneziano[9] and also adopted by Geoff.[11] The details of this model were discussed in the contribution *The Pomeron Story*, and will not be repeated here. (See Ref. 10 and also the contribution of G. Veneziano in this volume.) Here we illustrate the key features by two schematics in Fig. 1.

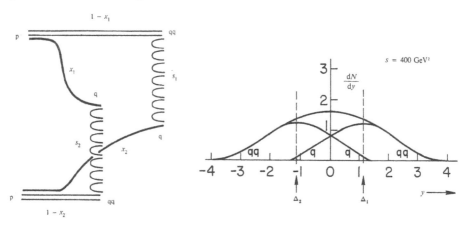

Fig. 1.   A cut-topological Pomeron for particle production and the resulting inclusive cross section.

Through these efforts, the notion of a topological Pomeron was firmly established phenomenologically. Since this identification is established via the production of particles, which tend to have limited transverse momenta, it has been dubbed a "soft Pomeron." This usage also indicates that further complications are lurking around the corner and would soon appear in the coming years.

3. A Perturbative vs a Non-Perturbative Pomeron

As hinted above, it soon became clear that jet production would become an important component of particle production. The question can then be posed, how to account for the total cross section resulting from such processes. However, it is unclear if this separation can be made meaningfully before a fully comprehensive confinement picture emerges. In any event, both during Moriond and also at other meetings, this issue became increasingly the center of discussion. One began to talk about "hard-Pomeron" versus "soft-Pomeron." Loosely speaking, hard-Pomeron should be calculable based on a perturbative approach, while soft-Pomeron remains non-perturbative. Should the bootstrap be restricted to the soft component only?

BFKL: Beginning in the mid-1970s, Lev Lipatov[12] and his collaborators (the Leningrad school), carried out a series of studies of the high-energy behavior of scattering amplitudes in QCD. These were serious and difficult analyses, which ultimately led to the emergence of a Pomeron-like structure. However, this study was carried out mostly by assuming conformal invariance, leading to a Pomeron being a branch-point, instead of a pole. Nevertheless, the location of this singularity was calculated. To first order in the 't Hooft coupling, λ, it is located above unity,

$$\alpha_P = 1 + (\ln 2/\pi^2)\lambda.$$

It should be emphasized that this result follows from summing gluon ladder graphs, which clearly ignores confinement: thus it is perturbative. Nevertheless, it has resulted in a robust research program, including an application to particle production. Furthermore, the BFKL program was able to lead to interesting predictions for deep-inelastic scattering (DIS) experiments at HERA. However, this picture clearly runs counter to that for DTU, responsible for a *soft Pomeron*. The corresponding BFKL Pomeron, is now identified with the *hard Pomeron*.

Donnachie–Landshoff: While the work of BFKL received increasing attention from both theoretical and experimental communities, a surprisingly simple parametrization was proposed by Donnachie and Landshoff for a soft-Pomeron.[13] They were able to explain experimental data for total cross sections for $\bar{p}p$, pp, π^-p, π^+p, K^-p, K^-p, and γp over a large energy range. It was a truly impressive fit to the data, with an effective Pomeron intercept of

$$\alpha_P \simeq 1.08.$$

This value is significantly lower than that expected for the BFKL-Pomeron. This analysis is purely phenomenological, (For more discussion of this development, see the contribution by Donnachie and Landshoff in this volume.) No suggestion on how this trajectory can be generated dynamically was provided, e.g., what states lie on this trajectory?

Russians are Coming — Moriond 1992: While the late 1960s was the period of social unrest and change in the US, the same can be said for the Soviet Union in the 1980s. Moriond had always been welcoming to physicists from eastern block countries. Beginning in late 1980s, increasingly more participants were able to come from the former Soviet Union. In particular, many from the Leningrad school began to come regularly. In 1992, while again on sabbatical at Orsay, I was fortunate to meet Genya Levin at Moriond. Genya is a contemporary of Lev Lipatov; both were students together working with Gribov. While I tried to convince him of the success of DTU, leading to a string-like topological-Pomeron, Genya would argue forcefully that the correct approach should be BFKL-Pomeron.[14]

After several days of argument, we both began to see the values of the opposite view. We clarified for each other how each respective Pomeron can be understood in terms of the picture of random walks. The difference between the two lies in the respective space in which this takes place. In the case of the Soft-Pomeron, the random walk occurs in the transverse impact space. In contrast, the BFKL Pomeron is generated primarily by random walks in "virtuality."

We kept up our communication after Moriond, and soon became good friends. We also came to the conclusion that a proper treatment of high energy scattering should incorporate both soft- and hard-features. We soon came up with a toy-model which would do precisely that. We coined the phrase "Heterotic Pomeron," and I presented a talk on this proposal at the July 1992 meeting for International Symposium on Multiparticle Dynamics (ISMD) at Santiago, Spain, with the intention of writing up a formal paper shortly afterwards. For various reasons, the project was delayed. (We finally were able to finish this project nearly ten years later.[15])

4. The Pomeron and Gauge–String Duality

A major development in our understanding of non-pertubative gauge theory occurred in 1990, the AdS/CFT conjecture of Maldacena. In particular, motivated by a suggestion of Witten, Jevicki *et al.* and independently Csaki *et al.* showed how scalar glueball masses can be calculated via AdS/CFT, once confinement is implemented. Working with Richard Brower, together with a then young colleague at MIT, Samir Mathur, we were able to carry out an analogous analysis generalized to tensor glueballs.[16] This is by itself an interesting exercise, which I learned greatly from Rich and Samir. Equally important is a paper by Polchinski and Strassler where they explained how point-like power behavior for scattering at large angle can emerge when AdS/CFT is implemented. Secretly, I was extremely excited; at long last, a promising theoretical framework lay ahead as the basis for the phenomenologically motivated heterotic Pomeron.

In 2007, Brower and I, together with J. Polchinski and M. Strassler,[17] provided a coherent first principles treatment of the Pomeron. In the large-N QCD-like theories, we used curved-space string-theory to describe simultaneously both the BFKL regime and the classical Regge regime. On the basis of gauge/string duality, this allows us to deal with high-energy small-angle scattering in QCD. The problem reduces to finding the spectrum of a single j-plane Schrödinger operator. Our results agree with expectations for the BFKL Pomeron at negative t, and with the expected glueball spectrum at positive t, but provide a framework in which they are unified.

In short, a dual-topological Pomeron is nothing but a reggeized graviton over a curved background, with the Pomeron intercept depending on the curvature. To first order in $1/\sqrt{\lambda}$, we find

$$\alpha_P \simeq 2 - 2/\sqrt{\lambda}.$$

This gauge-string motivated Pomeron has been referred to by others as the "BPST Pomeron". It is heterotic, as previously proposed. It is based on firm principles, and should be contrasted with weak-coupling BFKL Pomeron.

Further applications of this dual-topological Pomeron has been carried out in recent years.[18] Effects beyond the single Pomeron exchange have also been discussed.[19] Future developments and challenges undoubtedly await, e.g., the recent reported observation of the Odderon[20] requires further clarification from a string-dual perspective.[21] Similarly, the influence of the pion mass on the total and elastic cross sections has also been raised in the context of the topological Pomeron.[22] More intriguingly, can the value of the Pomeron intercept be fixed by the bootstrap?

5. Moriond 1984

After Berkeley, I moved to Princeton in 1968 as a post-doc, working primarily with Murph Goldberger, in particular on understanding the Pomeron from the perspective of the multiperipheral picture. After completing my post-doc, I moved to Brown as an Assistant Professor in 1970 and stayed since. Geoff and I kept up with our interactions during this period — in particular in 1970 when Geoff spent his sabbatical at Princeton. Equally rewarding is the fact that I was able to work with Carleton DeTar and Joe Weis,[23] my fellow graduate student colleagues at Berkeley, who had just moved to MIT. We had a fruitful collaboration, making use of Veneziano models for inclusive production via the multi-Regge picture. This is an extension of the work of Mueller, which strengthened the usefulness of a string-like picture for particle production.

I spent a fruitful sabbatical at Orsay in 1976–77, working with Capella, Sukhatme and Tran on DTU. After my return to Brown, my interests gradually shifted to the study of QCD non-perturbatively via the large-N expansion. I was greatly helped by my colleagues at Brown, in particular A. Jevicki, who introduced me to the collective-field approach. I was also fortunate in the following year to team up with my long-time graduate student friend, Rich Brower, who was spending a year at Harvard, in the study of matrix models. Together with a young post-doc at MIT, P. Rossi, we worked out and extended several nontrivial examples of coupled matrix models.[24] Rich also introduced me to the detailed working of lattice gauge theory, for which I am truly grateful, and he helped direct my graduate student, Kostas Orginos, who is by now an accomplished lattice gauge theorist at William and Mary. During this period, I did not have much direct contact with Geoff, although I have tried to keep up with his work when preprints arrived.

I had the fortune to overlap with Geoff again in 1984 when I spent my second sabbatical at Orsay. However, Geoff and I were located at different research institutes, and we ate at different canteens, preventing us from having regular get-togethers. At the same time my research interests had also diverged from his. I began to engage increasingly more with model studies in QCD at large N, while maintaining my continuing interests in phenomenological applications to multiparticle production based on the topological Pomeron. At the same time, Geoff focused on developing a topological approach to the S-matrix bootstrap.[25]

While he would occasionally listen to our explanations about how the Pomeron would fit in a topological setting, clearly he was less and less interested in phenomenological applications. Although we did not meet on a daily basis at Orsay, we tended to often catch the same commuter train back to Paris. During the short 30-minute ride, he would, in his usual enthusiastic fashion, expound on his latest thinking on physics. While I don't have a specific recollection of these conversations, I do remember one incident of interest: During one of the return trips, while we continued to chat at the transfer point at Denfert-Rochereau, Geoff realized that his commuter ticket was missing. After I had passed through the gate, Geoff, while looking briefly left and right for security guards, simply jumped over the barrier with a quick athletic move. We each moved quickly to the next quay to catch our respective transfers, Geoff with a smug smile on his face.

My last extended interaction with Geoff took place at the 1984 Moriond conference at les Arcs 1600, France. I recall Geoff expounding on his optimistic view of how unification of the standard model would emerge from a topological bootstrap approach.[25] He was at his usual best, trying to share his excitement on how new physics can be examined from a bootstrap perspective. After the 1984 Moriond meeting, my interactions with Geoff came less often.

6. Last Visits

I visited Geoff in 2010 while attending a conference in Berkeley. I had lunch with Geoff at the usual Rad Lab cafeteria, together with Geoff's longtime collaborators, Henry Stapp and Jerry Finkelstein. After lunch, we returned to their office, and Geoff tried to explain to me their latest research efforts. Although the visit was short it brought back memories of the wonderful graduate student days.

I had another chance to visit Geoff in September 2013. We had lunch together at his home in the Berkeley hills. How it brought back memory of fifty years ago when we went there for the first time after the house was built! Instead of Chinese take-out Geoff had ordered pizzas for lunch. He was pleased when I shared with him the resurgence of the Bootstrap program in current CFT studies as well as the heavy reliance on S-matrix principles for amplitude studies. I also updated him on the status of our topological Pomeron program. I learned from Geoff about his current goal in understanding cosmology from a bootstrap perspective.[26] He also talked more about his graduate student years with Fermi. I was also fascinated later by the story of how he had interacted with Von Neumann as an undergraduate at George Washington University and how G. Gamov likely helped start his physics career by sending him to join the Manhattan Project. When it was time to leave I had no idea that would be the last time I would see him.

Fig. 2.   Photo taken during my 2013 visit.

In March 2018, we had a chance to visit DAMTP where my former student, Kostas Orginos, was on sabbatical there. His wife, Lily, a classicist, was also on sabbatical at Cambridge. Through her connection they were staying at Clare Hall, where Richard Eden is a founding Fellow. The highlight of our visit was to have lunch with Richard. It was a pleasure to listen to him re-telling the story of his graduate days working with Dirac.

I started my research career more than 50 years ago by trying to understand the Pomeron and the Bootstrap. It remains a personal challenge and *The Pomeron Story* continues. It has been a wonderful and continuing journey. Thanks to Geoff, my fellow graduate students and Eden, for shaping my research interest and trajectory.

References

1. G. F. Chew and S. C. Frautschi, "Regge Trajectories and the Principle of Maximum Strength for Strong Interactions", *Phys. Rev. Lett.* **8**, 41 (1962).
2. G. F. Chew, "Hadron Bootstrap Hypothesis," *Phys. Rev. D* **4**, 2330 (1971); "Nuclear Democracy and Bootstrap Dynamics", UCRL-11163,12 (1963).
3. R. Dolen, D. Horn and C. Schmid, *Phys. Rev.* **166**, 1768 (1968).
4. S. Mandelstam, *Phys. Rev.* **166**, 1539 (1968).
5. F. A. Cerulus and A. Martin, *Phys. Lett.* **8**, 80 (1964); C. Chiu and C.-I. Tan, "Power Behavior of the Scattering Amplitude at Fixed Momentum Transfer and a Reconsideration of the Cerulus-Martin Lower Bound," *Phys. Rev.* **162**, 1701 (1967). This was my first serious paper. I was thrilled when it was cited in the seminal paper by Veneziano, (Footnote # 9, in Ref. 6).
6. G. Veneziano, "Construction of a crossing-symmetric, Regge-behaved amplitude for linearly rising trajectories," *Nuovo Cim. A* **57**, 190 (1968).
7. J. Scherk and J. H. Schwarz, *Nucl. Phys. B* **81**, 118 (1974).
8. N. F. Bali, G. F. Chew and A. Pignotti, *Phys. Rev. Lett.* **19**, 614 (1967); G. F. Chew, M. L. Goldberger and F. E. Low, *Phys. Rev. Lett.* **22**, 208 (1969).
9. G. Veneziano, "Duality and the Bootstrap," *Phys. Lett. B* **34**, 59–62 (1971).
10. A. Capella, U. Sukhatme, C.-I. Tan and J. Tran Thanh Van, "Dual Parton Model," *Phys. Rept.* **236**, 225–329 (1994); *Phys. Lett. B* **81**, 68 (1979); A. B. Kaidalov and K. A. Ter-Martirosian, *Phys. Lett. B* **117**, 247–251 (1982).
11. G. F. Chew and C. Rosenzweig, *Phys. Rept.* **41**, 263–327 (1978).
12. E. A. Kuraev, L. N. Lipatov and V. S. Fadin, *Sov. Phys. JETP* **45**, 199 (1977); I. I. Balitsky and L. N. Lipatov, *Sov. J. Nucl. Phys.* **28**, 822 (1978).
13. A. Donnachie anf P. V. Landshoff, *Phys. Lett. B* **296**, 227 (1992).
14. E. M. Levin and M. G. Ryskin, LENINGRAD-76-280.
15. E. Levin and C.-I Tan, "Heterotic pomeron: A Unified treatment of high-energy hadronic collisions in QCD," arXiv:hep-ph/9302308; S. Bondarenko, E. Levin and C.-I. Tan, *Nucl. Phys. A* **732**, 73 (2004).
16. R. C. Brower, S. D. Mathur and C.-I. Tan, "Glueball spectrum for QCD from AdS supergravity duality," *Nucl. Phys. B* **587**, 249 (2000).
17. R. Brower, J. Polchinski, M. Strassler and C.-I. Tan, "The Pomeron and gauge/string duality," *JHEP* **12**, 005 (2007).
18. R. C. Brower, M. Djuric, I. Sarcevic and C.-I. Tan, *JHEP* **1011**, 051 (2010).
19. R. Brower, M. Strassler and C.-I. Tan, *JHEP* **03**, 050 (2009); *JHEP* **03**, 092 (2009); L. Cornalba, M. S. Costa and J. Penedones, *JHEP* **0709**, 037 (2007).
20. TOTEM Collab. (G. Antchev *et al.*), *Eur. Phys. J. C* **76**, 661 (2016), arXiv:1610.00603; E. Martynov and B. Nicolescu, *EPJ Web Conf.* **206**, 06001 (2019).
21. R. Brower, M. Costa, M. Djuric, T. Raben and C.-I. Tan, "Strong Coupling Expansion for the Conformal Pomeron/Odderon Trajectories," *JHEP* **1502**, 104 (2015).
22. A. A. Anselm, V. N. Gribov, *Phys. Lett. B* **40**, 487 (1972); L. Jenkovszky, I. Szanyi and C.-I. Tan, "Shape of proton and the Pion Cloud," *Eur. Phys. J. A* **54**, 116 (2018).
23. C. DeTar, K. Kang, C.-I. Tan and J. Weis, *Phys. Rev. D* **4**, 257 (1971).
24. R. Brower, P. Rossi and C.-I. Tan, *Phys. Rev. D* **23**, 953 (1981).
25. G. F. Chew and V. Poenaru, "Topological Bootstrap Theory of Hadrons," *Z. Phys. C* **11**, 59 (1981); *Phys. Rev. D* **32**, 2683 (1985).
26. G. F. Chew, "Extended Lorentz Quantum Cosmology Symmetry Group," arXiv:1308.4366 [gen-ph]; G. F. Chew, "Unification of Gravity with Electromagnetism," arXiv:11209.6290 [gen-ph].

Back to the Future:
Reflection on Geoffrey Chew's Legacy

Richard C. Brower

Department of Physics, Boston University,
Boston, MA 02215, USA

This is a personal narrative on the enduring legacy of the bootstrap program of Geoffrey Chew and Stanley Mandelstam that sought to replace the short distance (UV) Feynman graph definition of local field theory with a direct assault on the long distance S-matrix (IR). This continues to guide theorists toward a deeper probe of fundamental non-perturbative physics beyond the standard model; even perhaps to the ultimate challenge of quantum gravity.

1. Introduction

The 60's in Berkeley were exciting times for a young graduate student, not just because of the anti-war and civil-rights protests, but also the challenge for elementary particle physics. Higher energy collisions of strongly interacting nucleons yielded a constant stream of new discoveries beyond any clear theoretical understanding. In the Fall of 1964, escaping my home in Cambridge, Massachusetts, with a BA in Physics and MA in Applied Mathematics from Harvard, I arrived on Berkeley Campus with protestors crowding into central Sproul Hall Plaza. My detour into applied mathematics was driven by my fascination with the non-perturbative problem of fluid dynamics but I decided to turn this interest to quantum physics and go west. The Berkeley group was an ideal destination, focussing on the strong nuclear force. I did not fully realize at the time what great luck I had in becoming a Ph.D. student of Geoffrey Chew. He provided an exceptionally open, bold and welcoming environment with a large number of students and postdoctoral fellows. Geoff's students would meet in his home spreading preprints on the floor to find projects to work on. My first modest paper with postdoctoral fellow, John Harte, on Kinematic Constraints for Infinitely Rising Regge Trajectories[1] was inspired by a 1966 preprint by Stanley Mandelstam on Dynamics Based on Rising Regge Trajectories.[2,3] My Harvard English professor father thought our title was poetic!

Here more than I could have imagined at the time, my subsequent graduate trajectory ran parallel to the dual-string revolution started by Gabriele Veneziano.[4]

For more details on early developments, the reader is referred to Gabriele's presentation and to the text on *The Birth of String Theory*.[5] These personal recollections are supplemented by the presentations of Carleton DeTar and Chung-I Tan with whom I continue to have lively collaborations. Beginning in graduate studies, I embarked with John Harte, in searching the Bateman manuscripts to find exact solutions to what was called Dolen-Horn-Schmid duality.[6] Our idea was simple. On the LBL blackboards one often saw the Regge equation for $\pi\pi$ scattering in Mandelstam variables,

$$A(s,t) \simeq \Gamma[1 - \alpha(t)]s^{\alpha(t)}, \quad \text{with} \quad \alpha(t) = \alpha_0 + \alpha't, \tag{1}$$

for $\pi\pi$ scattering dominated by the ρ trajectory, assuming Mandelstam's zero with linear trajectory approximation. The problem was to find an amplitude that satisfied this but was symmetric in s and t. I recall Geoff teasing me with the question: *Why are you looking for an exact solution to an approximation?* Apparently John Harte and I were looking at the wrong examples in the Bateman manuscript. A better-motivated example was found by the brilliant young contemporary Gabriele Veneziano[4] in Israel. I recall Mandelstam's immediate excitement. The rapid generalizations to higher-point functions, led Mandelstam to present at our so-called *secret seminars* (really just in-house informal discussions) to present an inspired guess on how zero width dual tree amplitudes could be iterated to provide loop amplitudes in a perturbative expansion.

To follow the vicissitudes of Chew's bootstrap ideas, it is useful to punctuate its trajectory with three watershed events: **(i)** The Veneziano bootstrap solution and the evolution from Dual Model to String Theory. **(ii)** The discovery of QCD gauge theory with both asymptotic freedom and confinement. **(iii)** The AdS/CFT conjecture or String/Gauge duality. To pick specific dates to start each epoch, I choose: **1963** with Chew's `Nuclear Democracy and Bootstrap Dynamics`;[7] **1973** with Gross and Wilczek's `Ultraviolet Behavior of Nonabelian Gauge Theories`[8] and **1997** with Maldacena's `The Large N Limit of Superconformal Field Theories and Supergravity`,[9] respectively.

2. The S-matrix Bootstrap Era

In the 60's the San Francisco Bay Area had two strong experimental and theoretical laboratories — at the Bevatron in Berkeley and at the Linear Accelerator in Stanford. Admittedly overstating the contrast, the Berkeley effort was to identify the resonance states in the S-matrix in the infrared (IR). Whereas at Stanford the deep inelastic microscope was revealing point-like partons fields in the ultraviolet (UV). In modern quark language, the Berkeley quest was to organize quark bound states as a rotational sequence on the Regge trajectories, including the Pomeron which was later identified with gluon bound states. Stanford was looking for the quark-gluon interactions in Feynman diagram of local fields. Needless to say at the time, there was little idea how these two pictures could co-exist. As a small effort to bridge the gap, Farzam Arbab and I hit on a nice proof showing in S-matrix language that the zero mass limit for spin 1 implied a photon coupled to the charge and for spin 2, a graviton coupled to the mass.[10]

In a more sustained effort in collaboration with a brilliant Mandelstam student, Joe Weis, I set out to incorporate local vector and axial currents into the on-shell Veneziano dual model.[11] This ambitious attempt was a dominant theme of my PhD thesis. Although we reproduced many current algebra results, with hindsight we were fighting the extreme softness of flat-space strings whose form factor (or momentum distribution) has a divergent rms radius, as emphasized later by Susskind. A real insight into how to proceed had to wait for 25 years with Maldacena's AdS/CFT revolution in which a 5d on-shell string S matrix maps to an off-shell local field at the 4d AdS boundary. Still the Dual Model provided a valuable tool with a concrete example of an analytic crossing symmetric multi-particle S-matrix that enabled a clear picture of the consistencies of the various multi-Regge limits, summarized in a review with Carleton DeTar and Joe Weis.[12]

Subsequently as a postdoctoral fellow at MIT, I was introduced to the full dual-amplitude spectrum by colleagues including postdoctoral fellow Gabriele Veneziano and Professor Sergio Fubini. A beautiful result by Del Giudice, Di Vecchia and Fubini[13] constructed $d-2$ transverse DDF vertex operators that commuted with the Virasoro algebra and generated only physical positive-norm states. In this era as postdoc at MIT, a Mandelstam student Charles Thorn and I embarked on trying to see if we could prove that the entire Hilbert space of physical states had a positive norm; e.g. the No Ghost Theorem or what in the **CFT back to the bootstrap** literature is referred to as unitarity. One surprise was the discovery that when you varied the number of space-time dimensions d above 26, negative norm (ghost) states in fact did appear. Fearing other surprises, while at CERN, with the help of the computer division, Peter Goddard and I ran a recursive algebraic computer program (written in Fortran) to enumerate all physical states to 30-th level finding no ghosts for $d \leq 26$. Finally returning to MIT, I was able to give a covariant extension of $d-2$ transverse DDF vertex operators by normal ordering a slightly off-shell conformal photon vertex operator,

$$A_n^\mu = \frac{1}{2\pi i} \oint \frac{dz}{z} : (\partial_t X^\mu)^{1+\alpha' k^2} e^{ik_\mu X^\mu} :, \tag{2}$$

followed by the zero-mass limit in light-cone momenta, $k_\mu = \{\vec{k}, k_+, k_-\}$ with $\vec{k} = 0$, $k_- = n, k_+ \to 0$. The transverse modes recover the DDF operator, $\vec{A}_n$, integrating by parts sets $A^{(-)}$ to zero, but surprisingly there are new longitudinal operators when the limit is taken carefully,

$$A_n^{(+)} = \frac{1}{2\pi i} \oint \frac{dz}{z} : \left[\partial_t X^+ + \frac{n^2}{2} \partial_t X^- \log(\partial_t X^-) \right] e^{ik_\mu X^\mu} : . \tag{3}$$

Defining $A_{-n} = A_n^\dagger$, one finds that the commutator algebra,

$$[A_n^i, A_m^j] = n\delta_{ij}\delta_{n+m,0}, \qquad [A_n^i, A_m^{(+)}] = nA_{n+m}^i,$$
$$[A_n^{(+)}, A_m^{(+)}] = (n-m)A_{n+m}^{(+)} + 2n^3\delta_{n+m,0} \tag{4}$$

is closed. The new $A_n^{(+)}$ operators complete the spectrum-generator algebra for all physical state with positive norm. Thus the *No-ghost Theorem*[14,15] followed for any dimensions **less than or equal** to the critical dimension ($d \leq 26$) for the Bosonic string and ($d \leq 10$) for the super string. This was consistent with the beautiful Lorentz invariant quantization of the Bosonic string in light-cone gauge for $d = 26$ given by Goddard, Goldstone, Rebbi and Thorn.[16] Saturating the critical dimension is still, to me, an unresolved peculiarity of string perturbation theory, probably not required by a full unitary realization. The $A_n^{(+)}$ operators, which produce null states that decoupled at $d = 26$, provide additional longitudinal modes to restore Lorentz invariance for $d < 26$ or $d < 10$ for Bosonic and super string respectively. A kind of stringy Higgs phenomena in the planar (large N) limit.

At this time I hoped that the Dual Model would rapidly lead to the still-illusive QCD string. Indeed in the planar limit the Neveu-Schwarz super string, originally suggested for pion scattering amplitudes, does give a remarkably good description when moving the intercept to fit the ρ mass, exactly reproducing the low energy chiral Lagrangian at tree level. Perhaps this is a QCD string S-matrix in the IR regime, deep in the interior for an unknown deformed AdS space with correction as it moves to the UV boundary consistent with asymptotic freedom. A tantalizing hint of the still-illusive QCD string expansion.

3. QCD Gauge Theory vs Super Strings

In **1973** with the discovery of asymptotic freedom and the candidate QCD, a dramatic split took place between super-string theory and strong-coupling hadronic physics. Clearly Chew's hadronic bootstrap did not provide a complete framework. Additional details of parton interactions in the UV must be specified. The break with conventional field theory axioms prior to QCD is no longer seen as dramatic as it was at the time. Fields in QCD are local but the physical (gauge invariant) Hilbert space of the S-matrix is not given by the Feynman propagators. Moreover when combined with the surprising new phenomena of color confinement, QCD does explain some of Chew's basic bootstrap axioms. Through the magic of confinement, still only established by numerical simulation, all asymptotic states are not elementary fields but composites lying on Regge trajectories, and 't Hoofts observation that QCD perturbative diagrams order by order in the $1/N_c$ color expansion map onto the Riemann surface anticipated in Chew's topological bootstrap as well as string perturbation theory.

Phenomenologically, the incompatibility between QCD and flat-space string theory hinged on the complete lack of hard scattering at short distances. The simplest example of very soft string scattering properties was noticed soon after the Veneziano proposal in terms of the exponential fall off

$$\lim_{s \to \infty} \frac{\Gamma[1 - \alpha(t)]\Gamma[1 - \alpha(s)]}{\Gamma[1 - \alpha(t) - \alpha(s)]} \simeq e^{-2\alpha' s f(\theta) \log(f(\theta))}, \tag{5}$$

at fixed angle scattering. This fact alone discouraged almost everyone from seeking a QCD string and led to the idea that the original motivation for the discovery of string theory was an 'accidental', if not illegitimate, child of hadronic phenomenology. The problem is that the super string in flat space does not exhibit the UV short distance physics of QCD with its asymptotic freedom and parton properties. Incidentally, this soft ultraviolet behavior of string theory is central to its success as an example of a finite perturbative expansion for quantum gravity.

Like many others, to pursue non-perturbative QCD, I found myself drawn to the Wilson-lattice formulation, which has developed into a vast computational program, rigorously converging to the Euclidian path integral with amazingly accurate prediction of hadronic states, scattering and matrix elements, crucial to precision tests of the standard model.[17]

4. AdS/CFT Reunification with String/Gauge Duality

Remarkably, in **1997** again the picture changed abruptly with the extensive study of new non-perturbative string effects in curved space culminating in the beautiful example of AdS/CFT or string/gauge duality by Juan Maldacena for $\mathcal{N} = 4$ super symmetry.[9] Finally a synthesis of local field theory and the string S-matrix appeared possible. To me this immediately brought me back to my graduate days with Geoff Chew. The string S-matrix in AdS^5 mapped into a local field theory at the 4d AdS boundary. Very roughly one may say that the on-shell string states $(p_5^2 \sim p^\mu p_\mu + m^2)$ in 5d are conjugate to 4d space-time (x_μ) for off-shell local currents — a possibility way beyond my imagination in my PhD thesis. Also it was soon realized that deformations (or branes) in the IR give a phenomenological model of confinement in gauge theory and hard scattering for string fluctuations close to the UV boundary. Subsequently in nice paper with Polchinski, Strassler and Tan [18] we introduced the similar limiting method in Eq. (3) for the closed-string graviton in AdS space. This gave a strong-coupling Pomeron with a unified hard (UV) gluonic and soft (IR) Regge features. The hard BFKL string at the boundary moves into the bulk as the Regge trajectory for glueballs — again removing another barrier to the possible existence of a bootstrapped topological expansion for a QCD string. While a dual QCD string is no longer paradoxical, actually constructing a mathematically precise string expansion remains a challenging problem in the Chew bootstrap agenda.

Returning to Maldacena's AdS/CFT conjecture, another feature is the special role of conformal field theory (CFT). Remarkably, for $\mathcal{N} = 4$ super symmetry, the string that is characterized by a single mass scale for the Regge slope is mapped to a scale-invariant theory on the boundary. This, in turn, increased interest in exactly-conformal field theories and the development of the *back to the bootstrap* program[19] for CFT's with vastly more powerful constraints than Chew's original hadronic bootstrap. Why? Apparently this is due to the lack of UV structure in a CFT. The correlator, $\langle \phi(x_1)\phi(x_2)\rangle = c(x_1 - x_2)^{-2\Delta}$, for primaries (or eigenstate of dilatons) is an exact power all the way from the UV to IR. In a sense there is only IR physics — a great place for the revival of Chew's bootstrap, as it turns out. The operator product expansion (OPE) in coordinate space (not momentum space of the hadronic bootstrap) implies a pure tree-diagram expansion with tri-linear vertices exactly obeying a dual-bootstrap crossing relation. The full CFT data are the primary operator dimensions and trilinear coupling constants. This is too technical to explain in detail, but the exact CFT algebraic constraints bear a striking relation to the old dual-bootstrap relations that gave birth to string theory.

Beyond that Mack[20] has taken this analogy further to conjecture that the planar limit of string theory in Mellin space (the analogue of momentum space for CFTs) may even have an exact map to conformal field theory in any space-time d. This leads me back to the No-Ghost Theorem [14,15] that in the planar limit was proven for any dimension less than or equal to the critical dimension. Could this give an exact CFT bootstrap construction? Finally, proposals have made to use CFT bootstrap data, supplemented by breaking terms, as a new basis to compute the S matrix for unitary evolution in Minkowski space[21] — the return of bootstrap to S-matrix theory.

It is far too early to understand where all this will lead but what is clear is that this "back to the future" of the bootstrap is a testament to Chew's insis-tence to pose the fundamental, ancient questions of the ingredients of the physical world. Should one seek a quantum extension of the the Greek's atomic hypothesis of elementary constituents at short distance or look for global consistency, such as information preservation (unitarity) and basic geometric symmetries to define the physical world? The great reward to Chew's cohort of graduate students was to be challenged by this basic tension — one that appears to be more relevant today with the realization that a theory of quantum gravity (if it exists) must give rise to space time itself in the IR, starting from more primitive axiomatic principles beyond local field theory. The hadronic bootstrap not only gave birth to string theory,[5] but its legacy continues to pose fundamental questions to guide future research.

References

1. Richard C. Brower and John Harte. Kinematic constraints for infinitely rising regge trajectories. *Phys. Rev.*, 164:1841–1844, Dec 1967.
2. S. Mandelstam. Dynamics based on indefinitely rising Regge trajectories. *Conf. Proc.*, C670828:605–620, 1967.
3. Stanley Mandelstam. Dynamics Based on Rising Regge Trajectories. *Phys. Rev.*, 166:1539–1552, 1968.
4. G. Veneziano. Construction of a crossing-symmetric, Regge behaved amplitude for linearly rising trajectories. *Nuovo Cim. A*, 57:190–197, 1968.
5. Andrea Cappelli, Elena Castellani, Filippo Colomo, and Paolo Di Vecchia, editors. *The Birth of String Theory*. Cambridge University Press, Cambridge, UK, 2012.
6. R. Dolen, D. Horn, and C. Schmid. Finite energy sum rules and their application to pi N charge exchange. *Phys. Rev.*, 166:1768–1781, 1968.

7. Geoffrey F. Chew. Nuclear Democracy and Bootstrap Dynamics, 1963. LBNL Report #: UCRL-11163.

8. David J. Gross and Frank Wilczek. Ultraviolet Behavior of Nonabelian Gauge Theories. *Phys. Rev. Lett.*, 30:1343–1346, 1973.

9. Juan Martin Maldacena. The Large N limit of superconformal field theories and supergravity. *Int. J. Theor. Phys.*, 38:1113–1133, 1999. [Adv. Theor. Math. Phys.2,231(1998)].

10. Farzam Arbab and Richard C. Brower. Zero-mass bosons in s-matrix theory. *Phys. Rev.*, 178:2470–2477, Feb 1969.

11. R. C. Brower and J. H. Weis. Vector currents and current algebra. III. Dual-resonance model with universally coupled vector mesons. *Phys. Rev. D*, 3:451–461, 1971.

12. R.C. Brower, Carleton E. DeTar, and J.H. Weis. Regge Theory for Multiparticle Amplitudes. *Phys. Rept.*, 14:257, 1974.

13. E. Del Giudice, P. Di Vecchia, and S. Fubini. General properties of the dual resonance model. *Annals Phys.*, 70:378–398, 1972.

14. Richard C. Brower. Spectrum generating algebra and no ghost theorem for the dual model. *Phys. Rev. D*, 6:1655–1662, 1972.

15. R. C. Brower and K. A. Friedman. Spectrum Generating Algebra and No Ghost Theorem for the Neveu-Schwarz Model. *Phys. Rev. D*, 7:535–539, 1973.

16. P. Goddard, J. Goldstone, C. Rebbi, and Charles B. Thorn. Quantum dynamics of a massless relativistic string. *Nucl. Phys. B*, 56:109–135, 1973.

17. S. Aoki et al. FLAG Review 2019: Flavour Lattice Averaging Group (FLAG). *Eur. Phys. J. C*, 80(2):113, 2020.

18. Richard C. Brower, Joseph Polchinski, Matthew J. Strassler, and Chung-I Tan. The Pomeron and gauge/string duality. *JHEP*, 12:005, 2007.

19. David Poland, Slava Rychkov, and Alessandro Vichi. The Conformal Bootstrap: Theory, Numerical Techniques, and Applications. *Rev. Mod. Phys.*, 91:015002, 2019.

20. Gerhard Mack. D-dimensional Conformal Field Theories with Anomalous Dimensions as Dual Resonance Models. *Bulg. J. Phys.*, 36:214–226, 2009.

21. Nikhil Anand, A. Liam Fitzpatrick, Emanuel Katz, Zuhair U. Khandker, Matthew T. Walters, and Yuan Xin. Introduction to Lightcone Conformal Truncation: QFT Dynamics from CFT Data. 5 2020.

Recollections of Life as a Berkeley Chew PhD Student in the Late 1960s

Carleton DeTar

Department of Physics, University of Utah,
Salt Lake City, UT 84112, USA

I had the good fortune to have been a PhD student of Geoffrey Chew's in the late 1960s. Here is a brief personal account of my experience and of Geoff's influence on my career.

I received my undergraduate degree from Harvard College in 1966. At that time a "National Defense and Education Act" provided for graduate fellowships in physics. I managed to obtain one and decided to take it to Berkeley. Steven Weinberg was moving to Harvard from Berkeley at the time. I ran into him in the physics library. When I mentioned that I was Berkeley bound, he suggested that if I studied with Chew, I would undoubtedly be tasked with "Reggeizing" some scattering amplitude. At the time, I had no idea what that meant.

At Berkeley, I took the usual set of physics graduate courses, including an introduction to quantum field theory. That year Eyvind Wichmann taught it. His approach was formal and meticulously precise. It covered the Poincaré group and its representations, but left the traditional elementary calculations of Møller and Bhabha scattering for us to work out on our own without first having seen Wick's theorem. Not being Freeman Dyson, we all found it quite challenging. Classmates Joe Weis, Michael Misheloff, and I often worked together. We managed to solve the scattering problem together. Later on, I audited Stanley Mandelstam's version of the course and found it refreshingly orthogonal with an emphasis on the nuts and bolts of Feynman diagram calculations, with less concern about the more formal issues. But the course that I found most seductive was Geoff Chew's introduction to elementary particle physics. He based it on his small book on the S matrix. It offered an appealing path to understanding strong hadronic interactions without the machinery of perturbation theory. Eventually, I approached Geoff, asking if he would take me on as a PhD student. I had been warned that he might turn me down because he already had several students. Instead, he was very welcoming, as he was to all of his students. I joined his group the year after David Gross and John Schwarz had earned their PhDs with him. Contemporary Chew students included Richard Brower, Jan Dash, Michael Misheloff, Chung-I Tan, Farzam Arbab,

Charles Chiu, Jiunn-Ming Wang, and Ling-Lie Chau Wang. Theory postdocs at the time included Jerome Finkelstein, Dale Snider, Joel Yellin, and Marcello Ciafaloni.

Geoff's appeal was his intense passion for high energy physics, and a desire to understand it in terms of his world view that was grounded in the bootstrap idea. Its basic tenet was that everything would emerge in a self-consistent way from the axioms of analyticity, unitarity, and Poincaré invariance. It expressly rejected the fundamentalist view, accepted by many others, that hadrons are composed of elementary particles, such as the quarks of Murray Gell-Mann and George Zweig or, later on, Richard Feynman's more agnostic "parton" model. Instead, for Geoff, through the bootstrap process, hadrons should emerge in a self-consistent manner. The Chew–Low model developed by Geoff and Francis Low for pion–nucleon scattering was a paradigm for the bootstrap.[1] In that model nucleon exchange provides a force between the pion and nucleon, and that force gives rise to the nucleon bound state and the delta resonance. Thus hadrons give rise to themselves in a self-consistent manner. The N/D formalism, based on unitarity and analyticity, that Geoff developed later with Mandelstam, provided a formal realization of the idea.[2] A shortcoming of the Chew–Low model was that it was necessary to introduce an adjustable cutoff parameter to obtain the desired results. However, the hope was that enlarging the scope of the model would make the cutoff unnecessary. Indeed, the ambitious hope was that the bootstrap would eventually be found to account for everything, including, perhaps, even quantum electrodynamics and the dimensionality of spacetime in order to achieve complete self-consistency.

The N/D method could be thought of as enlarging on the potential model, and both produced a Regge trajectory that linked the increasing angular momenta of a series of excited states as a function of their masses. Regge trajectories controlled the high energy behavior of scattering amplitudes and, thus, differential cross sections. At the time I became a student of Geoff's, he had moved beyond "Reggeizing" scattering amplitudes and was interested in multiparticle production. So I was spared the fate that Weinberg had foretold.

In the late 1960s the strong interations were a challenge and poorly understood. Before the rise of the Standard Model, including quantum chromodynamics (QCD), and before the discovery of asymptotic freedom in QCD, there were no convincing fundamentalist alternatives to the bootstrap. Despite extensive searches, quarks had never been isolated in an experiment. For all we knew at the time, quarks might have proven to be only a mathematical device for understanding flavor symmetries. At least that role appealed to Geoff. So most theorists at the time were content to work on pieces of the strong interaction puzzle, focusing, for example, on flavor symmetries and current algebra.

The bootstrap idea was, of course, not widely accepted. The complete bootstrap without approximation was criticized as being unscientific in that it offered no practical way to make precise, experimentally testable predictions. In its ultimate expression it states that the world is as it is because it is the only way for it to be consistent within the axioms of unitarity, Poincaré invariance, and analyticity. Thus one would have to solve the whole consistency problem in order to test uniqueness and correctness. Nonetheless, with Geoff, it was practically a religion. I am sure he rankled his unconverted contemporaries with published statements such as this: "Only after fundamentalists repeatedly fail to explain more than part of the hadron spectrum will they gradually accept the self-consistent particle model of the bootstrappers.[3]"

At the time, experiments at the Berkeley Bevatron and the Brookhaven AGS were producing results for multihadron production that offered theorists opportunities for an explanation. For this purpose, Geoff was particularly fond of the ABFST model (named after its authors, Amati, Bertocchi, Fubini, Stanghelini, and Tonin). Geoff must have liked it because it was based entirely on hadrons with no elementary particles. From a diagrammatic point of view, the calculation of total cross sections, summed over multiparticle final states in that model resembled cross-channel ladder diagrams in potential theory, so it led to a series of excited states with increasing angular momentum, i.e., a Regge trajectory — in this case the "Pomeron", named after Isaak Yakovlevich Pomeranchuk. My PhD thesis treated this summation in a manner resembling the Bethe–Salpeter equation. The ABSFT multiperipheral model led to a justification of "limiting fragmentation" for inclusive particle production, a precursor to Feynman's parton-model treatment.[4]

Geoff's students were assigned desks in shared offices in Building 50A of the Lawrence Radiation Lab (now Lawrence Berkeley National Lab), on the same floor as his office. Geoff was very accessible and always eager to meet with his students, which was especially remarkable, given how many there were. Even when he suggested a problem of his own devising for a student to work on, once the student picked it up, he thereafter referred to it as the student's problem — thus conferring a sense of ownership and commitment.

The Rad Lab had been the site of some nuclear weapons research under the aegis of the Atomic Energy Commission (now absorbed into the Department of Energy). Building 50 had a vault for storing classified documents under the stairway leading to the conference room. At the time this vault was guarded by a uniformed officer, which impressed us students. Later all classified work at the Rad Lab ceased.

The Rad Lab was a perfect place to immerse oneself in high-energy physics. Student, postdoc, and faculty offices were on the same floor, promoting ease of interaction. Experiments were underway at the Bevatron, so it was easy to interact with fellow graduate students engaged in experiments, such as Henry Frisch and Dennis Smith. Theory students of David Jackson, including Chris Quigg and Robert Cahn, students of Stanley Mandelstam, including Joe Weis, and students of Geoff Chew often ate lunch together at the tables outside the Rad-Lab cafeteria, braced to the often somewhat chilly Berkeley summer eucalyptus-scented breezes. When it was particularly chilly, Chris Quigg was fond of asserting brightly that it was just like a "fine English summer day".

Geoff organized weekly informal "secret" seminars for all the theorists at the Rad Lab. They were secret only because the title and speaker were revealed only at the beginning of the seminar. This strategy allowed last-minute changes in speakers without embarrassment, and may have improved attendance, because no one could decide in advance that the topic was uninteresting. For beginning students, the seminars were especially useful to us for picking up the concepts and language of contemporary high energy theory. William Rarita of Rarita–Schwinger fame was a regular attendee, always assuming the role of cheerleader, praising students uniformly for their accomplishments.

Geoff and his then wife Ruth regularly invited all of his students in the evening to their home high in the Berkeley hills. We would sit in comfortable chairs and sofas in his living room, dazzled by the view of the city lights, and discuss ideas and current preprints. This was an exciting time when new experimental discoveries were coming at a rapid rate, so we were never short of topics. The discussion was almost always qualitative. Geoff did purchase a blackboard once for his home, perhaps thinking we would actually write equations or diagrams to aid the discussion. But we almost never used it. This shows the strong emphasis with Geoff on ideas as opposed to formalism.

The 1960s were a time of student unrest at Berkeley. In the mid-1960s the Free Speech Movement insisted that students have the right to engage in political organizing on campus. The war in Vietnam was intensifying and with it, a strong antiwar sentiment, and the Black Power movement was beginning. The University administration was concerned that political activity would anger state lawmakers, so they tried to suppress it and brought in police to put down the ensuing demonstrations. Nonviolent protest activities included nearly daily rallies in Sproul Plaza, occupying buildings (sit-ins), teach-ins, strikes, etc. Although he was not directly involved, Geoff was sympathetic with the goals of the protesters.

President Ronald Reagan, then governor of California, used Berkeley as a political foil to boost his conservative, law-and-order credentials, so he insisted on firm measures. There were confrontations with Berkeley city police, Alameda County sherriff's deputies and, after the People's Park battles, even the National Guard. There were frequent surveillance helicopters and tear gas. I shared an apartment with Joe Weis close to campus on College Avenue. At the time this was going on, we had to pass through National-Guard lines to walk up the hill to the Rad Lab. Still, we were much more comfortable with the National Guard than with sherriff's deputies, because the deputies seemed to relish the opportunity to rough up immature intellectuals, whereas the Guard was composed of youth our age trying to avoid being sent to Vietnam. My occasional antiwar involvement was more traditional — manning tables at shopping centers and visiting churches to talk to parishioners, but even when concentrating on graduate studies, it was difficult to escape completely. I recall once sitting at a window during one of Geoff's secret seminars at the Rad Lab and catching an irritating whiff of tear gas wafting up from campus.

Geoff's bootstrap philosophy and its grand scope compelled him continually to improve his comprehensive view of the strong-interactions. He admitted that his present limited understanding often required working on small parts of the bootstrap edifice, but, even so, he never lost sight of the big picture. To this end he occasionally had to supplement the bootstrap with additional requirements, such as "second degree analyticity", i.e., particle poles should always lie on Regge trajectories. But the entire edifice also evolved.[5] He was remarkable in his ability to sieze upon and incorporate new developments.

In the late 1960s, I believe the best example of Geoff's ability to assimilate came from Gabriele Veneziano's dual resonance model.[6] When first introduced, it was a simple Euler beta function, but when written in terms of the Mandelstam channel invariants s and t and interpreted as an elementary model of a two-particle elastic scattering amplitude, it had some remarkable mathematical properties. It has Regge trajectories. That is, in the limit of large s and fixed t, it behaves as $s^{\alpha(t)}$, where $\alpha(t)$ interpolates the angular momenta of the leading states in the t channel, as Regge trajectories are supposed to do. The trajectories were even linearly rising. And it did this without the need for a potential. The beta function is analytic, except for an infinite series of poles in either channel. It could be reconstructed by summing all of its poles in only one of those channels. Thus, beside its natural Regge behavior, it had another desirable bootstrap feature, namely, that the series of poles in one channel gave rise to all the poles in the other and vice versa — hence the name "dual resonance model". So here were hadrons giving birth to each other without the apparent need for a force. Geoff was thrilled and seemed to prefer this bootstrap approach to the more potential-model-inspired N/D approach.

Shortly afterwards, Miguel Virasoro and Korkut Bardakçi and Henri Ruegg generalized Veneziano's model to a five-point amplitude with all the desirable Regge-pole features.[7,8] It was subsequently generalized to any number of particles. The dual resonance model was found to have remarkable mathematical properties that later led to the development of string theory, thanks to work by former Chew student John Schwarz among others.

In the late 1960s, an MIT-SLAC collaboration at the Stanford Linear Accelerator Center collided high-energy electrons with protons and measured the deep-inelastic electron scattering cross section. In effect, in this experiment, the SLAC accelerator was operating as a very powerful electron microscope that could look inside the proton. It revealed point-like constituents, namely, quarks. This result and subsequent deep-inelastic neutrino scattering experiments obviously gave strong support for the fundamentalist approach. Nonetheless, I recall at least one evening discussion at Geoff's house that explored an alternative explanation in terms of a beta-function-like series of poles in the invariant momentum transfer to the proton, but it was obvious that J.D. Bjorken's and Emmanuel Paschos's explanation in terms of partons was much simpler.[9]

The MIT-SLAC experiment, the 1973 discovery of asymptotic freedom by former Chew student David Gross with Frank Wilcek and simultaneously by David Politzer, and the discovery of charmonium, a bound state of a charm quark and antiquark, all established quantum chromodynamics, the gauge theory of quarks and gluons, as the fundamental theory of the strong interactions. Thus the fundamentalists had prevailed, but in a remarkable way. The fundamental theory is described by a Lagrangian function of quark and gluon fields resembling quantum electrodynamics with its fundamental electron and photon fields. But strong confinement prevents the quarks and gluons from ever emerging as asymptotic states. Confinement could be understood only through nonperturbative methods. Kenneth Wilson's lattice regularization of QCD with its nonperturbative treatment provided a simple way to see how this comes about, starting from the QCD Lagrangian.[10]

Since they could not emerge as asymptotic states, quarks and gluons could not give rise to poles in the S matrix. Only the color-singlet combinations that formed hadrons could do that. And because they were not S-matrix poles, Geoff wouldn't accept their particle nature. By the mid 1970s Geoff's thinking had evolved and it became much more diagrammatic. He seized upon a topological approach he called "dual topological unitarity". It bore a resemblance to Gerard 't Hooft's classification of Feynman diagrams involving elementary quarks and gluons as an expansion in powers of $1/N_c$, the inverse of the number of colors. That expansion provided a topological classification of QCD Feynman diagrams. But Geoff had an entirely different use for them. For him the topological expansion provided a pictorial representation of contributions to the S matrix. Ribbons in the diagrams corresponded to physical hadrons. The edges of the ribbons carried quark quantum numbers, but they didn't actually represent particles — just convenient labels for the color and flavor routing. The topological classification started with planar diagrams, but through unitarity, one obtained diagrams of higher order in $1/N_c$ and of higher topological genus. One could associate the planar diagrams with the dual resonance model and the low-lying mesons. The next-order iteration of unitarity, produced the Pomeron. At a pictorial level, it all fit neatly together and seemed a nice starting point for visualizing the bootstrap.

After completing a PhD at Berkeley I took up a postdoctoral position at MIT as did Rich Brower and Joe Weis. Francis Low was on the MIT Physics faculty at the time. His strong association with Geoff undoubtedly helped us get those jobs. The three of us shared an office in the MIT Center for Theoretical Physics, which was organized with faculty offices on the main floor and student and postdoc offices on the floor above. This "upstairs–downstairs" arrangement created a hierarchical divide that was in stark contrast with the more democratic layout at the Rad Lab. (A more recent remodel with an open, interior staircase has helped reduce this separation.) We continued to work on the dual resonance model. Joe Weis and I with Chung-I Tan and Kyungsik Kang at Brown University used it to describe inclusive particle production with particular emphasis on limiting fragmentation and the relationship between leading Regge trajectories and the leading dependence on momentum fraction.[11] Later Joe, Rich, and I wrote a Physics Reports review article on multiparticle production based on the dual-resonance model.[12]

I continued as a junior faculty member at MIT and later moved to the University of Utah. My research interests drifted away from the bootstrap. Alan Chodos, Robert Jaffe, Kenneth Johnson, Charles Thorn, and Victor Weisskopf were developing the MIT bag model of confined, which definitely had fundamental quarks and gluons.[13] I used it to work out its implications for the nuclear force that binds the deuteron. I was slow to come around to that way of thinking — Chew's influence had been so strong.

Apart from exchanging annual Christmas greetings, I had little contact with Geoff in later years. His work had progressed to much more esoteric considerations of the bootstrap, and I had switched to studying high-temperature quantum chromodynamics and deconfinement using lattice QCD. Here was a case where one expected that the elementary nature of quarks and gluons would be clearly manifest. However, thanks to Geoff, my bootstrap instincts aided my understanding. At extremely high temperature, the running QCD coupling is small, and one could hope to study the quark–gluon plasma using perturbation theory. However, one could show that perturbation theory breaks down at higher orders because there are still vestiges of confinement, no matter what the temperature. These confining effects are most apparent in the determination of the spatial screening lengths in the plasma, which are controlled by analogs of confined, low-lying mesons. Many of us calculated their temperature-dependent inverse screening lengths (similar to masses) using lattice QCD. A further argument for confinement at all temperatures came from the phase diagram. One could show, at least numerically, that when quarks (of nonzero mass) were included in the thermal ensemble, there was no phase transition separating a low-temperature "confined" phase and a high-temperature "deconfined" phase. Without a phase boundary, plasma correlation functions had to be continuous functions of temperature. Thus continuity implied that excitations of the plasma had to be associated with zero-temperature color singlets, despite the fact that a description in terms of a free Stefan–Boltzmann gas was a convenient approximation.[14] When I gave a seminar about this work at the Lawrence Berkeley Laboratory some time around 1990, Geoff was in attendance. I spoke with him afterwards, and he was clearly happy with the argument that even in a "deconfined" thermal plasma where the fundamental nature of quarks and gluons seemed to emerge, poles and branch cuts in physical plasma correlation functions still had to be associated with color-singlet hadronic states — no quark poles and no quark–antiquark branch points.

The story of Geoff Chew's passionate attachment to the bootstrap idea as the guiding principle for a comprehensive theory of the strong interactions forms a fascinating chapter in the pre-quantum-chromodynamics history of high-energy physics. It was exciting to have experienced it first-hand.

References

1. G.F. Chew and F.E. Low. Effective range approach to the low-energy p wave pion - nucleon interaction. *Phys. Rev.*, 101:1570–1579, 1956.
2. Geoffrey F. Chew and Stanley Mandelstam. Theory of low-energy pion pion interactions. *Phys. Rev.*, 119:467–477, 1960.
3. G.F. Chew. Hadron bootstrap - triumph or frustration? *Phys. Today*, 23N10:23–28, 1970.
4. Carleton E. DeTar. Momentum spectrum of hadronic secondaries in the multiperipheral model. *Phys. Rev. D*, 3:128–144, 1971.
5. G.F. Chew. *bootstrap* - a scientific idea? *Science*, 161:762–765, 1968.
6. G. Veneziano. Construction of a crossing - symmetric, Regge behaved amplitude for linearly rising trajectories. *Nuovo Cim. A*, 57:190–197, 1968.
7. M.A. Virasoro. Generalization of veneziano's formula for the five-point function. *Phys. Rev. Lett.*, 22:37–39, 1969.
8. K. Bardakçi and H. Ruegg. Meson resonance couplings in a five-point veneziano model. *Phys. Lett. B*, 28:671–675, 1969.
9. J.D. Bjorken and Emmanuel A. Paschos. Inelastic Electron Proton and gamma Proton Scattering, and the Structure of the Nucleon. *Phys. Rev.*, 185:1975–1982, 1969.
10. Kenneth G. Wilson. Confinement of Quarks. *Phys. Rev. D*, 10:45–59, 2 1974.
11. Carleton E. DeTar, J.H. Weis, K. Kang, and Chung-I Tan. Duality and single-particle production. *Phys. Rev. D*, 4:425–439, 1971.
12. R.C. Brower, Carleton E. DeTar, and J.H. Weis. Regge Theory for Multiparticle Amplitudes. *Phys. Rept.*, 14:257, 1974.
13. A. Chodos, R.L. Jaffe, K. Johnson, Charles B. Thorn, and V.F. Weisskopf. A New Extended Model of Hadrons. *Phys. Rev. D*, 9:3471–3495, 1974.
14. Carleton E. DeTar. Pseudodeconfinement and Dynamical Confinement in the Quark Plasma. *Phys. Rev. D*, 37:2328, 1988.

© 2022 World Scientific Publishing Company
https://doi.org/10.1142/9789811219832_0019

Recollections of Geoff

R. Shankar

Sloane Physics Lab, Yale University,
217 Prospect St, New Haven, CT 06520, USA

I share some informal recollections of our dear friend and role model, Goeff Chew.

It was my good fortune to have, for nearly 50 years, Goeff Chew as my mentor, friend and role model.

I was a student at Berkeley between 1969 and 1974. I had very little interaction with Geoff in my first two years when I was focused on getting ready for the qualifier exams, taking courses from Jackson, Wichmann, Suzuki, Steiner and others. This was essential for me as I had no real degree in physics at that point, having majored in electrical engineering. All I knew was that I wanted to do physics and at that point had not decided which area to jump into. I knew that if I was going to leave the practical world of engineering it had to be for something as removed from it as possible. I naturally ended up in particle physics, which in addition was in a very exciting stage at that point.

The next step was to find an adviser. I had no knowledge of field theory and I was told this would be a great plus if I wanted to work with Geoff. So I popped into Geoff's office to broach the subject and received a very friendly greeting, accompanied by the usual twinkle in his eyes and kind smile. Geoff then went off the Princeton for his sabbatical and I took my qualifiers. After passing them I wrote to him asking if he could take me on. He wrote after some weeks explaining the delay: his wife had died rather suddenly and his life was upside now. But then he went on to say that he had no objection to giving it a try, to see how it suited both parties.

The next step must be very familiar to all his students: attend large gatherings of his students at 10 Maybeck Twin Drive in the evenings. I could not follow anything being said. It was one thing to flaunt my ignorance of QFT but another to try the same with Regge theory. So I kept quiet most of the time except for one occasion when the topic of the representations of the Lorentz group came up, and I launched into a long discourse that I subsequently learned had impressed Geoff. I attended seminars at the Rad Lab, facing a similar initial shock but slowly things began to make sense. I remember two things from this period. Geoff, who would doze off during talks on weak interactions, once woke up with a start when the speaker said that at some point weak interactions could become strong. Geoff asked him "when?" and the speaker said "after 300 GeV", he relaxed into a smile and went back to his nap. The other was a trick I learned from Mandelstam to trip up any field theory speaker: say "All this is true in perturbation theory of course" and the speaker invariably had to sheepishly confess that this was so.

Geoff was ever ready to help if I dropped in to his office. Being a zero in QFT did not mean S-matrix theory came easily to me. So Geoff would patiently explain the basic strategy. When I said I was still foggy, he would smile his beatific smile and tell me it would all become clear in due course. (This continued till the week before I graduated.) Anyway I soon learned enough to write my maiden paper and Geoff went through it line by line and I finally graduated in 1974.

At the time I graduated, the war between S-matrix and QFT approaches was rapidly ending, following the discovery of quarks as the uncontested elementary particles and of QCD as their unique theory of interactions. Geoff obtained for me a position as junior fellow at Harvard. (I compare this to superman's father putting little superman on a rocket ship to earth as Krypton was in danger of extinction.) He must have known I might switch teams (as David Gross had done a few years ago). But that never stopped Geoff from helping me with my career. I remember Steve Weinberg telling me that he argued for his (Steve's) tenure over an S-matricist, even though Steve was not then, or later, a fan of the S-matrix approach. So off I went to Harvard, now brandishing my total ignorance of S-matrix theory to win friends among the field theorists there. Three years later I landed in Yale and never left.

I kept in touch afterwards. When I got married I took my wife Uma to meet him and later my son AJ spent hours interviewing Geoff on his life and times. My family l knew adored him, and could easily see why.

I once tried to get him excited about the work I had done with Witten on the exact S-matrices where general principles led to a unique answer. He listened eagerly till I blurted out that it was in two dimensions and could not engage him after that slip.

In the pragmatic scene of American physics Geoff was a rare philosopher. He went about his own way. For him the photon was "the photon concept" while *the* pomeron was as as real as anything in Rosenfeld's Particle Data book. For many years his interests were aligned with those of the broader community and for many years not. This never mattered to him and did not affect in the least his vigorous pursuit of his vision. He would write to me sometimes asking for clarification, say, on condensed matter physics, to which I had switched after coming to Yale. He kept working and communicating even after suffering serious physics handicaps near the end.

Despite my wisecracks above, I learned a lot of physics from him, and most of all, lessons on how to pursue one's goals without waiting for public validation and how to be a cheerful, principled and helpful human being.

A Passion for Physics*

Marvin L. Goldberger

Since I stopped doing physics completely in 1978 I felt it was necessary for me to take up another, at least part-time, profession. These are not so easy to come by, especially if you insist on a certain amount of ego gratification. However, I have found one that will satisfy me for a few years until I can find something else. Namely, I do gigs at my physics friends' 60th-birthday parties.

Let me explain how the ego thing comes about. You are asked to give the after-dinner speech for an old friend's 60th-birthday worship. There is often a scientific program associated with the celebration, but since you have nothing of any scientific interest whatsoever to say, you are relieved not to have to fake something or talk about "my life in physics", etc., and still grateful that people remember you and invite you to the party. That's the first reason to feel good. Next, you have an opportunity to wallow in the nostalgia of your youth. Nostalgia, you know, is edited amnesia. This, too, is very satisfying. The honoree is beginning, like you, to lose his memory and will believe anything about the "good old days that never really were" so long as you flatter him (there are very few hers in our generation) enough. The young attendees don't know anything, and by the time the after-dinner speech comes, they are either drunk or asleep; and anyhow, they don't matter very much. They can wait for their own time to come.

The last part of the process is the most gratifying. It is where you talk about your long, personal association with the birthday boy, your history of scientific collaboration and its profound bearing on whatever distinction your great old friend and co-worker has attained. You do this all in such a way that the perceptive listener wonders why they even bothered to come to a party for such an unworthy person. This must be done with taste and tact so that the 60 year-old boob doesn't know it's happening.

I should probably not have revealed all this in advance, but having done so makes it more of a challenge to see if I can pull it off one more time.

I've entitled this talk for Geoff, *A Passion for Physics*. I don't think that anyone

---

*Reprinted from *A Passion for Physics: Essays in Honour of Geoffrey Chew, including an Interview with Chew*, eds. C. DeTar, J. Finkelstein and C.-I. Tan (World Scientific, 1985).

who knows him could deny that he has a real thing about physics. In spite of the fact that we've known each other for nearly forty years, I don't know how this spark was kindled. But it was clearly apparent when we first met in the spring of 1946 at Chicago. Geoff came from Los Alamos where he had been on the Manhattan Project, a protégé of Edward Teller and Emil Konopinski. I had been on the project in Chicago and I was interviewed by Edward for suitability to be a graduate student. Admission to graduate school was a little more cavalier in those days than it is now.

Geoff and I met, as I recall, as students in Edward's course on quantum mechanics. The course was a disaster. Edward never prepared a lecture and had as a backup teacher Maria Mayer, one of the all-time rotten lecturers. Occasionally, Bob Christy, who lived in Edward's garage, gave a reasonable lecture. Frank Yang, who even then was terribly precocious, was the grader for the course. But the fact that that particular course was bad, or the one we took in solid state physics from Clarence Zener was in its own way even more grotesque, was not particularly important. The really important thing was that we were a part of a small, but unusually talented group of students, most of whom had been working for two or three years as real physicists and were thus far more knowledgeable and sophisticated than normal first year graduate students.

It was truly a remarkable time. We had been working side by side with the very best scientists in the world: Fermi, Wigner, Bohr, Teller, Weisskopf, Bethe, Alvarez, Peierls, Serber, Oppenheimer, Segre, Urey, Seaborg, Christy, Kistiakowsky, Rabi — even spies like Hans Fuchs and Alan Nun May. These people were all shockingly young — middle thirties, early forties — in retrospect it's hard to imagine anyone taking such whippersnappers seriously. But we worked with and were on a first-name basis with all of these people. And, we were wild about physics and bursting to get going.

We were unusually close as students at Chicago. We shared offices on the fourth floor of Eckhart Hall and we ate together, studied together and taught each other physics. I remember many things about the learning process — like field theory — Wentzel's book had just appeared in German and we struggled to understand it.

As we gradually established ourselves in the pecking order as students are wont to do, it quickly became clear that Geoff was very close to the top. He was not always the quickest or the most articulate or glib, but somehow or other he always got the homework done and exuded an air of quiet confidence.

For the most part the graduate students were married and our families became quite close. I was the only one who had a car and I recall many expeditions with the Chews, the Chamberlains and the Wattenbergs in that miserable 1936 Plymouth. We were also all extremely poor, but it didn't matter. We were young and full of beans.

Geoff and I began scurrying around for some kind of thesis problem in summer and fall of 1947 under the general aegis of Edward Teller. Teller, however, was so preoccupied with weapons work and concerned over the future of Los Alamos that he suggested that Geoff and I approach Fermi and ask him if he would take us on. Fermi had not ever had any theoretical students before, but much to our pleasure he agreed to accept us. After a number of false starts we finally got into serious thesis work. Fermi had a very refreshing view about Ph.D. theses: He said "You don't have to show your thesis to your grandchildren." His own thesis had consisted of something like putting an iron cross in an electron beam and observing the shadow on a fluorescent screen. At any rate, we did get through, I doing a Monte Carlo calculation of neutrons passing through a nucle.us, and Geoff working on neutron-deuteron scattering.

We got finished and began the process of thinking about jobs. We both ended up going to the Radiation Laboratory at Berkeley. I know how I made the decision to go there — an extraordinarily interesting story — but quite irrelevant to this occasion, and I no longer remember why Geoff made his decision. However made, it was an exciting environment for both of us to be in Bob Serber's theoretical group. Bob, overshadowed for a time by the glamorous Oppenheimer, was a brilliant physicist who could talk equally well with theorists and experimentalists. Berkeley was where it was at in high energy physics, and Geoff and I quickly became involved in a series of interesting projects related to nucleon-nucleon scattering — a preoccupation of the times — and had our introduction to applied field theory trying to understand photo-meson production.

We were both offered assistant professorships at Berkeley; and Geoff, in fact, accepted. He was unfortunately swept into the maelstrom of the loyalty oath issue at UC and, along with many others, left after a rather turbulent year. After an interlude at Brookhaven he went to Illinois and I went to Chicago. I mention this because we began collaborating again by telephone and by frequent visits, a practice which continued until I left for Princeton in 1957.

It was at Illinois that Geoff started a mode of doing physics that I always envied. He had, at any given time, a program. He became interested in the static model of pion-nucleon interactions and he just studied the hell out of it. There never seemed to be any indecision about what he would be working on next week, next month, or next year. I don't mean to imply that Geoff was immune to new ideas, but that there never seemed to be any ambiguity for him about what he would be doing at $t + \Delta t$.

The Illinois years were very good for Geoff. He had a tremendously fruitful collaboration with Francis Low who is, of course, much older than Geoff and me, and at whose 60th-birthday worship I had the honor of working, too. I can save some time and agony for the audience by referring to my brilliant speech recorded in Francis' festschrift instead of telling once again about the profoundly important work the three of us did in plasma physics and in collaboration with Nambu, on pion-nucleon and photo-pion production.

We all fled the mid-west at about the same time — Francis to MIT, I to Princeton, ang Geoff to Berkeley. I think Geoff's return to Berkeley was a very important thing for that University. He had left in protest, had been very active as head of the Federation of American Scientists during the McCarthy era, and had obviously not abandoned his principles. He totally revitalized the high energy theory group that had been decimated by the loss of Serber and Wick. He captured Stanley Mandelstam, embarked on a big program of exploitation of Stanley's double dispersion relations, and began a real production line of outstanding graduate students, many of whom are now grown-ups and are here tonight to honor their inspirational teacher.

Let me interrupt the beautiful continuity of this narrative to remark that Geoff had an uncanny ability to extract the best from his collaborators and his students. His positive mental attitude had a great deal to do with this, in my opinion. His natural inclination was to say yes, to be enthusiastic and to look for redeeming virtue in things people said. He was never a nay-sayer and his general attitude was very supportive.

It was around this time in the very early sixties that Geoff became enamored with the work of Regge and with Steve Frautschi, and made important observations about the relation of poles in the angular momentum plane to elementary particles. This work led to the concept, long championed by Geoff, of nuclear democracy, the idea that none of the observed long-lived particles should be regarded as more fundamental than any others; that everything is part of an incestuous mess. These notions led him to a rejection of quantum field theory to the extent that such theories appeared to contain arbitrary parameters and to the belief that an S-Matrix theory could be constructed based on concepts of unitarity and analyticity of various. kinds that would be unique and have no free parameters. I don't know what Geoff's present position on these issues is.

At any rate, Geoff and I continued to interact scientifically, albeit less frequently. One summer, 1968, Geoff, Francis and I got together in Aspen and worked on a Regge-exchange model of the old multiperipheral model of Amati, Fubini and Stanghellini. Geoff insisted at one time in describing this work in a terribly erudite and complicated way he had been taught by his student Carleton DeTar which I recall Francis and I had a terrible time understanding. Perhaps I'm being unfair to Francis — I can certainly say it for myself. Geoff and I came together again in Princeton in 1969-70 where, with Henry Abarbanel and Leslie Saunders, we wrote a number of papers on multiperipheral things, pomerons, etc. This was a year of terrible tragedy for Geoff when his wife Ruth was struck down by a massive stroke. His inner strength, and I believe in part his dedication to physics, carried him through this unbelievably difficult period. He had the enormous good fortune to meet Denyse at a critical time in his life.

In the ensuing years I have drifted further away from physics and have had fewer opportunities to interact with Geoff and I have not kept up with his work. I will therefore conclude by returning briefly to the title of this talk, *A Passion for Physics*.

Aside from the fact that Geoff really wanted to be a professional baseball player and was frustrated in this desire by a bad back, in all of my experience with him, physics seemed to be the dominant force in his life. Of course, he was terribly involved in the world around him in a variety of ways. I'm sure he even read books other than Goldberger and Watson — though why would one need anything more? But it was physics with a capital F, as our friend Bernardini used to say, that was and still seems to be Geoff's obsession. There are worse kinds of infatuations, I've noticed.

Perhaps if my own passion had been greater I wouldn't now have to be dealing with 275 faculty maniacs, all of whom think they know how to do my job better than I.

I have to say that I'm glad that Geoff had that back problem. I can't really see him (he was a catcher) as a WASP Yogi Berra chewing tobacco and screaming at umpires — perhaps a Gary Cooperesque Lou Gehrig character. But baseball's loss was physics' gain. I became a much better physicist, indeed a better human being because of Geoff, as I know did all of his collaborators. The world is much richer for his deep contributions to our field. Eugene Wigner used to occasionally ask the question, "What is so-and-so's second fault?" I can answer this about Geoff. He's too good to be true. It's also his first fault, his third fault, etc.

Happy 60th Birthday, Geoff.

Part II

Architect of the Bootstrap

Recollections of Geoffrey Chew

David Kaiser

Department of Physics and Program in Science, Technology, and Society
Massachusetts Institute of Technology
77 Massachusetts Avenue, Cambridge, MA 02139, USA

I first met Geoffrey Chew during the winter of 1992, soon after I began an undergraduate internship at the Lawrence Berkeley National Laboratory. I joked with friends that moving to Berkeley was my "semester abroad": as a kid from New Jersey who was in college on the East coast, Berkeley seemed just as exotic and unfamiliar to me as the tales my friends shared of their semesters in Edinburgh, Florence, and Tokyo. My first day on Berkeley's main campus, down the hill from the lab, student protesters had hoisted themselves up to the top of the iconic Campanile — a three-hundred-foot-tall clock tower — and refused to come down until all experiments on campus involving laboratory animals had stopped. And *that* was before I had discovered Telegraph Avenue. I might as well have moved to Mars.[1]

My internship at the lab was with the Solenoidal Detector Collaboration, a sprawling, international collaboration of experimental particle physicists. The group was building a huge detector to be used at the Superconducting Supercollider, then still under construction in Texas. I had a desk in a tiny corner of the big Building 50 complex, tucked away among the experimentalists. Before long I began wandering the halls one or two floors above my office, spending more and more time in the Theory division.

I had learned a bit about Chew's contributions to high-energy theory before arriving in Berkeley. Alongside my studies in physics I had also become fascinated by the history of science. Not long before my internship, I had read James Cushing's detailed account, *Theory Construction and Selection in Modern Physics: The S Matrix*.[2] Cushing had begun his career in theoretical high-energy physics, working on ideas that grew directly from Chew's S-matrix program for the strong interactions. From Cushing's historical account I had begun to learn a bit about crossing symmetry, Mandelstam variables, duality, Regge poles, and bootstrap techniques. It was heady stuff — conceptually ambitious while also steeped in empirical results — and yet it seemed worlds away from the little bits of quantum field theory I had begun to study in my other classes.

Cushing's book had piqued my interest, and now I found myself right at the spot where so much of Chew's work had taken shape. I screwed up the courage to knock on Professor Chew's office door, and quickly learned why he was so admired by his many students and advisees. He welcomed me into his office and sat patiently as I fumbled my way through ill-formed questions about particle physics and its recent history. He gave generously of his time, answered each of my questions with care, and — most important — encouraged me to keep asking similar questions. My job that semester was to tinker on a tiny corner of an endcap calorimeter, but my head quickly filled with contortions of scattering amplitudes throughout the complex plane.

My Berkeley internship solidified my interest in high-energy physics as well as the history of science. It also convinced me that experimentalists were better off — and probably safer — without me. I threw myself into studies of early-universe cosmology while also working hard to understand how Chew and his group had forged their impressive, rival approach to quantum field theory by creatively refashioning tools like Feynman diagrams. I pursued both of these projects in graduate school, becoming ever more fascinated by the shifting methods that physicists have forged to make sense of our cosmos.

The allure of Berkeley proved difficult to resist. I returned in 1998, a few weeks after defending my physics dissertation. I spent a marvelous year immersed in archival research for my dissertation in the history of science. Interspersed between long sessions in the Bancroft Library's beautiful reading room, I met some more with Professor Chew and several of his colleagues and former students. During these later interactions I learned additional details about Chew's S-matrix program as well as his courageous efforts beyond the narrow confines of particle physics: his principled stand during the California loyalty oath controversy in the late 1940s; his leadership within the Federation of American Scientists throughout the 1950s; and his active efforts to build a supportive community among his large group of students. Chew's contributions in each of these domains became central to my historical study.[3] He remained as generous and engaged with my questions as when I had first met him, and offered helpful comments on some of my early drafts.

How exciting to see renewed attention from high-energy theorists in recent years to the analytic properties of scattering amplitudes. These days the physics preprint server arXiv.org teems with new papers about crossing symmetry, dualities, and self-consistent bootstrap techniques. Everything old is new again — a fitting testament to how creative, ambitious, and generative Geoffrey Chew's framework continues to be, guiding new generations as they work hard to understand the universe around us.

References

1. D. Kaiser, *Quantum Legacies: Dispatches from an Uncertain World* (University of Chicago Press, Chicago, 2020), pp. 156–157.
2. J. T. Cushing, *Theory Construction and Selection in Modern Physics: The S Matrix* (Cambridge University Press, New York, 1990).
3. D. Kaiser, *Drawing Theories Apart: The Dispersion of Feynman Diagrams in Postwar Physics* (University of Chicago Press, Chicago, 2005), Chaps. 8–9.

Nuclear Democracy

The following article (pages 138–177) is reproduced by permission from "Nuclear Democracy: Political Engagement, Pedagogical Reform, and Particle Physics in Postwar America" by David Kaiser. Originally published in Isis 93, no. 2 (June 2002): 229–268. (c) 2002 by The History of Science Society. Published by The University of Chicago Press. All rights reserved.

Nuclear Democracy

Political Engagement, Pedagogical Reform, and Particle Physics in Postwar America

By David Kaiser*

ABSTRACT

The influential Berkeley theoretical physicist Geoffrey Chew renounced the reigning approach to the study of subatomic particles in the early 1960s. The standard approach relied on a rigid division between elementary and composite particles. Partly on the basis of his new interpretation of Feynman diagrams, Chew called instead for a "nuclear democracy" that would erase this division, treating all nuclear particles on an equal footing. In developing his rival approach, which came to dominate studies of the strong nuclear force throughout the 1960s, Chew drew on intellectual resources culled from his own political activities and his attempts to reform how graduate students in physics would be trained.

INTRODUCTION: MCCARTHYISM AND THE WORLD OF IDEAS

H istorians have studied several examples in which scientists framed details of their work in explicitly political language. German physiologists and physical scientists such as Rudolf Virchow, Ernst von Brücke, Emil Du Bois-Reymond, Hermann von Helmholtz, and others endeavored self-consciously to keep their dreams of political unity alive even after the crushing defeat of 1848 by pursuing methodological and epistemological unity within scientific knowledge. Just a few years later, the Swiss-French chemist Charles Frédéric Gerhardt suggested a "chemical democracy" in which all atoms within molecules

* Program in Science, Technology, and Society and Department of Physics, Building E51-185, Massachusetts Institute of Technology, Cambridge, Massachusetts 02139.

My thanks to Stephen Adler, Louis Balázs, E. E. Bergmann, John Bronzan, Geoffrey Chew, Jerome Finkelstein, William Frazer, Carl Helmholz, Francis Low, Stanley Mandelstam, Howard Shugart, Henry Stapp, Kip Thorne, and Eyvind Wichmann for their interviews and correspondence with me. Thanks also to the Office for History of Science and Technology at the University of California, Berkeley, for its warm hospitality while most of the research for this paper was conducted, and to David Farrell of the Bancroft Library for his assistance with several collections. This paper has benefited from comments and suggestions from Cathryn Carson, Patrick Catt, Geoffrey Chew, James Cushing, Peter Galison, Tracy Gleason, Michael Gordin, Stephen Gordon, Loren Graham, Kristen Haring, Kenji Ito, Matt Jones, Alexei Kojevnikov, Mary Jo Nye, Elizabeth Paris, Sam Schweber, Jessica Wang, and five anonymous *Isis* referees.

Isis, 2002, 93:229–268

would be treated as equals, a notion the German organic chemist Hermann Kolbe countered with a hierarchical, "autocratic" model of molecular structure. More recently, several Soviet theoretical physicists, including Yakov Frenkel, Igor Tamm, and Lev Landau, drew explicitly on their own life experiences under Stalin's rule—experiences that often included extended prison terms—when describing solid-state physics, referring, for example, to the "freedom" of electrons in a metal and particles' other "collectivist" behavior. The Indian astrophysicist Meghnad Saha emphasized during the 1920s that various chemical elements within the stars would respond to the same "stimulus" in varying ways, an analysis of atoms' agency deeply resonant with his own social and political struggles against caste hierarchies. Meanwhile, during the postwar period, the Japanese particle theorist Shoichi Sakata, thinking along explicitly Marxist lines, favored a strict hierarchy among subatomic particles, finding in such an arrangement appropriate base-superstructure relations.[1] Social metaphors abound within the physical sciences.

Like others in these earlier periods of tumult and turmoil, American scientists working after World War II experienced dramatic and fast-moving political currents. McCarthyism in America meant sweeping violations of civil liberties for thousands of citizens; blacklists and unfair firings from all manner of jobs affected people in and out of academia. But it also meant more than this: historians must supplement the tallies of such injustices with attention to the intellectual legacy of McCarthyism.[2] This essay explores the interplay during the early postwar decades between changing assumptions about political engagement, effective pedagogical approaches, and ideas about the behavior of subatomic particles. In particular, I will focus on the work of the prominent Berkeley particle theorist Geoffrey Chew and his concept of "nuclear democracy."

Beginning in the early 1960s, Chew railed against physicists' reigning approach to particle physics, quantum field theory, arguing that this framework offered no help for understanding the strong nuclear forces that kept atomic nuclei bound together. As he emphasized with great gusto at a June 1961 conference in La Jolla, California, quantum field theory was as "sterile" as "an old soldier" when it came to treating the strong interaction and hence was "destined not to die but just to fade away"—a memorable pronouncement that many of his peers repeated over the next several months.[3] In its place,

---

[1] On the German physiologists and physical scientists see Keith Anderton, "The Limits of Science: A Social, Political, and Moral Agenda for Epistemology in Nineteenth Century Germany" (Ph.D. diss., Harvard Univ., 1993), Ch. 2; cf. Timothy Lenoir, "Social Interests and the Organic Physics of 1847," in *Science in Reflection,* ed. Edna Ullmann-Margalit (Dordrecht: Kluwer, 1988), pp. 169–191. On Gerhardt's and Kolbe's positions see Alan Rocke, *The Quiet Revolution: Hermann Kolbe and the Science of Organic Chemistry* (Berkeley: Univ. California Press, 1993), pp. 208, 325 (my thanks to Michael Gordin for bringing this reference to my attention). On the Soviet physicists' outlook see Alexei Kojevnikov, "Freedom, Collectivism, and Quasiparticles: Social Metaphors in Quantum Physics," *Historical Studies in the Physical and Biological Sciences,* 1999, *29*:295–331; cf. Karl Hall, "Purely Practical Revolutionaries: A History of Stalinist Theoretical Physics" (Ph.D. diss., Harvard Univ., 1999). On Saha see Abha Sur, "Egalitarianism in a World of Difference: Identity and Ideology in the Science of Meghnad Saha," unpublished MS. On Sakata see "Philosophical and Methodological Problems in Physics," *Progress of Theoretical Physics,* 1971, *50*(Suppl.):1–248; Shunkichi Hirokawa and Shûzô Ogawa, "Shôichi Sakata—His Physics and Methodology," *Historia Scientiarum,* 1989, *36:*67–81; and Ziro Maki, "The Development of Elementary Particle Theory in Japan—Methodological Aspects of the Formation of the Sakata and Nagoya Models," *ibid.,* pp. 83–95 (my thanks to Masakatsu Yamazaki for discussions of Sakata's work).

[2] Cf. Loren Graham's suggestive work on the sometimes fruitful, generative appropriations of dialectical materialism by Soviet scientists: Loren Graham, *Science, Philosophy, and Human Behavior in the Soviet Union* (New York: Columbia Univ. Press, 1987); Graham, *What Have We Learned about Science and Technology from the Russian Experience?* (Stanford, Calif.: Stanford Univ. Press, 1998); and Graham, "Do Mathematical Equations Display Social Attributes?" *Mathematical Intelligencer,* 2000, *22:*31–36.

[3] A preprint of Chew's talk at the 1961 La Jolla conference is quoted in James Cushing, *Theory Construction and Selection in Modern Physics: The S Matrix* (New York: Cambridge Univ. Press, 1990) (hereafter cited as

Chew aimed to erect a new program based directly on the so-called scattering matrix, or *S* matrix, which encoded mathematical relations between incoming and outgoing particles while eschewing many of the specific assumptions and techniques of quantum field theory.

The single most novel conjecture of Chew's developing *S*-matrix program, and its most radical break from the field-theoretic approach, was that all nuclear particles should be treated "democratically"—Chew's word. The traditional field-theory approach, against which Chew now spoke out, posited a core set of "fundamental" or "elementary" particles that acted like building blocks, out of which more complex, composite particles could be made. As we will see in Section I, Chew and his young collaborators argued against this division into "elementary" and "composite" camps, instead picturing each particle as a kind of bound-state composite of all others; none was inherently any more "fundamental" or special than any other. Deuterons, for example, treated by field theorists as bound states of more "elementary" protons and neutrons, were to be analyzed within Chew's program in exactly the same way as protons and neutrons themselves; the "democracy" extended, in principle, all the way up to uranium nuclei. Chew described this notion by the colorful phrase "nuclear democracy."[4]

The larger physics community reacted swiftly to the string of early calculational successes that Chew and his Berkeley group produced under the "democratic" banner, one example of which I will examine in the next section. Ten months after his initial "call to arms" in La Jolla, Chew was elected a member of the National Academy of Sciences, an honor he attained before his thirty-eighth birthday. There followed a string of coveted invited papers at National Academy and American Physical Society meetings. "Such lectures invariably drew capacity crowds," Chew's department chair gloated to a dean in 1964, "since Chew is generally recognized as the outstanding exponent of a particular approach to the theory of elementary particles known as the S-matrix."[5] Earlier that year Murray Gell-Mann had introduced his "quark" hypothesis, whose proposed core, fundamental building blocks look to our eyes today like the very antithesis of Chew's nuclear democracy. Yet in his first papers on the quark hypothesis, Gell-Mann took pains to re-

Cushing, *Theory Construction*), p. 143. See also Murray Gell-Mann, "Particle Theory from *S*-Matrix to Quarks," in *Symmetries in Physics (1600–1980)*, ed. M. G. Doncel, A. Hermann, L. Michel, and A. Pais (Barcelona: Bellaterra, 1987), pp. 479–497. This portion of Chew's unpublished talk was incorporated verbatim in the introduction to his 1961 textbook, *S-Matrix Theory of Strong Interactions* (New York: Benjamin, 1961), pp. 1–2. Several physicists cited Chew's unpublished La Jolla talk during the October 1961 Solvay conference in Brussels, as recorded in the Solvay conference proceedings: R. Stoops, ed., *The Quantum Theory of Fields* (New York: Interscience, 1961), pp. 88, 132, 142, 179–180, 192–195, 214–215, 222–224. In a recent interview, Chew likened his strongly worded talk at the 1961 La Jolla meeting to a "coming out of the closet" speech: Geoffrey Chew, interview with Stephen Gordon, Dec. 1997, quoted in Stephen Gordon, "Strong Interactions: Particles, Passion, and the Rise and Fall of Nuclear Democracy" (A.B. thesis, Harvard Univ., 1998) (hereafter cited as **Gordon, "Strong Interactions"**), p. 32.

[4] Geoffrey Chew, "Nuclear Democracy and Bootstrap Dynamics," in Maurice Jacob and Chew, *Strong-Interaction Physics: A Lecture Note Volume* (New York: Benjamin, 1964), pp. 103–152, on p. 105. For more on Chew's S-matrix program see esp. Cushing, *Theory Construction;* Tian Yu Cao, "The Reggeization Program, 1962–1982: Attempts at Reconciling Quantum Field Theory with *S*-Matrix Theory," *Archive for History of Exact Sciences,* 1991, *41*:239–282; and David Kaiser, "Do Feynman Diagrams Endorse a Particle Ontology? The Roles of Feynman Diagrams in *S*-Matrix Theory," in *Conceptual Foundations of Quantum Field Theory,* ed. Cao (New York: Cambridge Univ. Press, 1999), pp. 343–356.

[5] Burt Moyer to Dean W. B. Fretter, 30 Dec. 1964, quoted in Raymond Birge, "History of the Physics Department, University of California, Berkeley," 5 vols., ca. 1966–1970 (unpublished; copies are in the Bancroft and Physics Department Libraries) (hereafter cited as **Birge, "History"**), Vol. 5, Ch. 14, p. 50. On Chew's election to the NAS and his other invited lectures see *ibid.,* pp. 49–52; "Physics Professor Wins Prize," *Daily Californian* [Berkeley student newspaper], 3 Jan. 1963, p. 5; and Stanley Schmidt, "The 'Basic' Particle— It's Out of Date," *ibid.,* 23 Oct. 1963, pp. 1, 12.

assure his readers that quarks were fully compatible with Chew's "democratic" framework.[6] By the mid 1960s, Chew and his fast-growing group at Berkeley had changed the way most theoretical physicists approached the strong interaction.

Though the idea of nuclear democracy was put forth repeatedly throughout Chew's 1961 lectures, it took several more years before Chew could, by his own lights, escape "the conservative influence of Lagrangian field theory." His updated lectures, published in 1964, were distinguished from the older ones, Chew explained, by their "unequivocal adoption of nuclear democracy as a guiding principle." He began by contrasting at some length "the aristocratic structure of atomic physics as governed by quantum electrodynamics" with the "revolutionary character of nuclear particle democracy." Chew left his students and readers with little doubt: "My standpoint here ... is that every nuclear particle should receive equal treatment under the law."[7]

All of this bluster about "conservative" field theory versus "revolutionary" nuclear democracy might inspire a knee-jerk *Zeitgeist* interpretation: as Chew sat in Berkeley, with the 1964 "Free Speech Movement" taking flight all around him, a particular vision of "democracy" floated freely from Telegraph Avenue into the Radiation Laboratory. It is no coincidence, one might conclude, that "democracy" was "in the air" among increasingly radical Berkeley activists and students—and *therefore* also permeated the seemingly rarefied project of Chew and his students. Like most knee-jerk reactions, however, this one is both too hasty and misplaced: whatever Chew was up to, his physics sprang from more than a vague "spirit of the times." Many roots of his new program extended back to theoretical developments from the mid 1950s. Chew's work on nuclear democracy cannot be read, in other words, simply as another instantiation of the strong Forman thesis, with Chew "capitulating" or "accommodating" his physics to a particular "hostile external environment."[8]

Leaving aside claims of strong external determination, we are nonetheless left with a puzzle: Why did Geoffrey Chew combine these various theoretical developments, stretching as they did over the better part of a decade, into his particular "democratic" interpretation? Why, moreover, did physicists working further and further from Chew's immediate group in Berkeley attribute different meanings to Chew's "democratic" physics? If we take a longer view of Chew's activities, intriguing questions and associations arise. In both his developing political activism (as discussed in Section II) and his unusual attempts to reform the training of graduate students (Section III), certain specific meanings of "democracy"

---

[6] Murray Gell-Mann, "A Schematic Model of Baryons and Mesons," *Physics Letters,* 1964, 8:214–215; and Gell-Mann, "The Symmetry Group of Vector and Axial Vector Currents," *Physics,* 1964, 1:63–75. See also George Johnson, *Strange Beauty: Murray Gell-Mann and the Revolution in Twentieth-Century Physics* (New York: Knopf, 1999), pp. 209–214, 225–227, 234. After the tide had shifted away from the *S*-matrix program and back to (gauge) quantum field theories, it was Chew who suggested that quarks and the nuclear-democratic formulation could be made compatible: Geoffrey Chew, "Impasse for the Elementary-Particle Concept," in *The Great Ideas Today,* ed. Robert Hutchins and Mortimer Adler (Chicago: Encyclopaedia Britannica, 1974), pp. 92–125, on pp. 124–125.

[7] Chew, "Nuclear Democracy and Bootstrap Dynamics" (cit. n. 4), pp. 104–106.

[8] Paul Forman's famous argument regarding the acceptance of acausal quantum mechanics in Weimar Germany may be found in Paul Forman, "Weimar Culture, Causality, and Quantum Theory, 1918–1927: Adaptation by German Physicists and Mathematicians to a Hostile Intellectual Environment," *Historical Studies in the Physical Sciences,* 1971, 3:1–115. James Cushing correctly dismisses an overly simplistic reading of Chew's "nuclear democracy" vis-à-vis the Free Speech Movement in *Theory Construction,* p. 217; cf. Gordon, "Strong Interactions," pp. 35–37. On the Free Speech Movement in Berkeley see esp. W. J. Rorabaugh, *Berkeley at War: The 1960s* (New York: Oxford Univ. Press, 1989), Ch. 1; and Todd Gitlin, *The Sixties: Years of Hope, Days of Rage,* rev. ed. (1987; New York: Bantam, 1993), pp. 162–166.

recurred: no one should be singled out as special, either for privileges or for penalties; all should be entitled to participate equally. In short, as he described his beloved nuclear particles in 1964, all should receive "equal treatment under the law." Weaving in and out of a series of specific contexts, these particular elements of "democracy" found explicit expression again and again. As evidenced by Chew's continuity of vocabulary when discussing his work in these various domains—a vocabulary that he took great care to hone—it appears that while developing his new program for particle physics, Chew drew on intellectual resources culled from his own efforts at political and pedagogical reform. Laboring to ensure a democracy among particles was only one way in which Chew's work took shape after the war.

Chew's work on particle theory during the 1950s and 1960s thus highlights how certain ideas about "democracy," brought to the surface by changing political and cultural conditions, could infuse a new vision both of how graduate students should be trained and of how nuclear particles interact. The story of "nuclear democracy" can thereby reveal certain connections—labyrinthine and indirect, to be sure, but connections nonetheless—between the Cold War national security state and changing ideas within theoretical physics. In the process, we learn why certain approaches and ideas within theoretical physics might have had special appeal or salience for Chew, as well as why these ideas found varying interpretations or associations outside of his immediate group.

It is important to realize, at the same time, that this is not a story about a well-formulated "ideology" or political philosophy steering the course of physical research. For one thing, Chew left no direct statements announcing that his interesting and influential ideas about particle physics were caused by his political convictions—nor would it make much sense to expect such pronouncements. American physicists after the war came to fashion themselves as eminently practical people, pragmatic tinkerers rather than philosopher-kings. The handful of theorists who did champion strong political or philosophical positions rarely acted on them with any consistency, as Sam Schweber's recent work on J. Robert Oppenheimer has shown. Moreover, those even rarer theorists who claimed to mingle such political commitments with their physical theorizing—such as David Bohm, dismissed from Princeton after pleading the Fifth Amendment before the House Un-American Activities Committee in 1949, or Sakata's Marxist group of particle theorists in Kyoto—were easily marginalized by mainstream American physicists, written off as "doctrinaire."[9]

Yet high-sounding principles and elaborate political philosophies are hardly prerequi-

---

[9] On Oppenheimer see Silvan Schweber, *In the Shadow of the Bomb: Oppenheimer, Bethe, and the Moral Responsibility of the Scientist* (Princeton, N.J.: Princeton Univ. Press, 2000). On Bohm see Russell Olwell, "Physical Isolation and Marginalization in Physics: David Bohm's Cold War Exile," *Isis,* 1999, *90:*738–756; Shawn Mullet, "Political Science: The Red Scare as the Hidden Variable in the Bohmian Interpretation of Quantum Theory" (B.A. thesis, Univ. Texas, Austin, 1999); and Alexei Kojevnikov, "David Bohm and Collective Movement," unpublished MS. On Sakata's school and their dismissal by American theorists see the sources in note 1, above; Robert Crease and Charles Mann, *The Second Creation: Makers of the Revolution in Twentieth-Century Physics* (New York: Macmillan, 1986), pp. 261–262, 295–296; and Johnson, *Strange Beauty* (cit. n. 6), pp. 202, 231–232. On postwar American physicists' views of themselves see Paul Forman, "Social Niche and Self-Image of the American Physicist," in *The Restructuring of Physical Sciences in Europe and the United States, 1945–1960,* ed. Michelangelo de Maria *et al.* (Singapore: World Scientific, 1989), pp. 96–104; David Kaiser, "The Postwar Suburbanization of American Physics," unpublished MS; and Schweber, "The Empiricist Temper Regnant: Theoretical Physics in the United States, 1920–1950," *Hist. Stud. Phys. Biol. Sci.,* 1986, *17:*55–98. American physicists' lack of explicit political philosophizing fit within broader trends during the American 1950s. See Daniel Bell, *The End of Ideology: On the Exhaustion of Political Ideas in the Fifties* (New York: Free Press, 1960); cf. Alan Brinkley, *Liberalism and Its Discontents* (Cambridge, Mass.: Harvard Univ. Press, 1998).

sites for political action. Rather than formulating an explicit political philosophy of democracy, Chew undertook a series of actions throughout the late 1940s and 1950s to fight for a (perhaps vague) notion of fair play and equal treatment. The fact that at the time these actions were sometimes interpreted quite differently by his peers, or that we might evaluate them today as something other than obviously or inherently "democratic," in no way diminishes the importance Chew himself invested in such experiences or the significance they carried for him. The absence of explicit pronouncements tying his political interests and physical research together in a neat and tidy package—the kind of "smoking gun" that is almost never to be found in historical investigations—hardly relieves us of the job of interrogating such episodes to tease apart the intellectual residue of McCarthyism in the world of ideas.

I. DIAGRAMMATIC BOOTSTRAPPING AND A DEMOCRACY OF PARTICLES

Much of Geoffrey Chew's work during the late 1950s and throughout the 1960s centered on new ways of interpreting and calculating with Feynman diagrams. Richard Feynman had first introduced the stick-figure line drawings that bear his name in the late 1940s. The diagrams were designed for service as a handy bookkeeping device when making lengthy calculations within quantum electrodynamics, the physicists' theory for how electrons interact with light. The full calculation of electron-electron scattering, for example, could be broken up into an infinite series of more and more complicated terms, each associated with more and more complicated ways in which the two incoming electrons could scatter. In the simplest contribution, one electron could emit a photon or quantum of light, which would then be absorbed by the other electron, with each electron thereby changing from its original momentum.[10] Feynman illustrated this process with the diagram in Figure 1. Uniquely associated with this diagram, Feynman continued, was a mathematical expression that yielded the probability for two electrons to scatter in this way.

The process illustrated in Figure 1, however, was only the start of the calculation; the two electrons could scatter in all kinds of more complicated ways, and these correction terms had to be included systematically as well. Feynman thus used his line drawings as a shorthand to keep track of these correction terms; examples of the diagrams that entered at the next round of approximation are shown in Figure 2. The key to the diagrams' use, as Feynman and his young collaborator Freeman Dyson emphasized, was the unique one-to-one relation between each element of the diagram and each mathematical term in the accompanying equation. Dyson, in particular, demonstrated that these calculational relations between diagram elements and mathematical expressions could all be derived rigorously from the foundations of quantum field theory. Soon after Feynman introduced the diagrams, however, theorists like Chew began to adopt them—and subtly adapt them—for applications well beyond the domain of electromagnetic interactions.[11]

---

[10] Richard Feynman, "Space-Time Approach to Quantum Electrodynamics," *Physical Review,* 1949, *76:*769–789; F. J. Dyson, "The Radiation Theories of Tomonaga, Schwinger, and Feynman," *ibid.,* 1949, *75:*486–502; and Dyson, "The *S* Matrix in Quantum Electrodynamics," *ibid.,* 1949, *75:*1736–1755. "Virtual" particles, such as the photon exchanged in Figure 1, "borrow" excess energy and momentum for very short periods of time (as compared with the energy they would carry as free, noninteracting particles), as allowed by Heisenberg's uncertainty relation. According to quantum field theory, all interactions arise from the exchange of virtual particles.

[11] On the introduction and use of Feynman diagrams within quantum electrodynamics see Silvan Schweber, *QED and the Men Who Made It: Dyson, Feynman, Schwinger, and Tomonaga* (Princeton, N.J.: Princeton Univ. Press, 1994). On the dispersion of Feynman diagrams during the 1950s and 1960s for new and different types of calculations see David Kaiser, "Making Theory: Producing Physics and Physicists in Postwar America" (Ph.D. diss., Harvard Univ., 2000), pp. 271–470; and Kaiser, "Stick-Figure Realism: Conventions, Reification, and the Persistence of Feynman Diagrams, 1948–1964," *Representations,* Spring 2000, *70:*49–86.

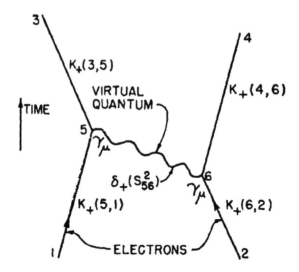

Figure 1. *The simplest Feynman diagram for electron-electron scattering. From R. P. Feynman, "Space-Time Approach to Quantum Electrodynamics," Physical Review, 1949, 76:769–789, on p. 772. Reprinted with kind permission of the American Physical Society.*

Chew pushed his new interpretation of the diagrams further than most of his peers by proclaiming that the diagrams' content, meaning, and calculational role could be severed entirely from the original field-theoretic framework in which they had been introduced. He emphasized in his published 1961 lectures, for example, that many of the new results of his developing program were "couched in the language of Feynman diagrams," even though, contrary to first appearances, they did not "rest heavily on field theory." "It appears to me," he further prophesied, "likely that the essence of the diagrammatic approach will eventually be divorced from field theory" altogether. Whereas field theorists had lectured to their students for over a decade that the lines within Feynman diagrams could only represent elementary particles—that is, the particles whose associated quantum fields appeared in the governing unit or basic interaction term—Chew countered that this distinction was not borne out by the diagrams themselves. As he expounded in his lecture notes, the diagrams themselves "contain not the slightest hint of a criterion for distinguishing elementary particles [from composite ones]. . . . If one can calculate the S matrix without distinguishing elementary particles, why introduce such a notion?"[12] He emphasized the next year, during his summer-school lectures at Cargèse, that the conjecture underlying

[12] Chew, *S-Matrix Theory of Strong Interactions* (cit. n. 3), pp. 3, 5. See also Chew's 1962 lectures at the Cargèse summer school: Geoffrey Chew, "Strong-Interaction *S*-Matrix Theory without Elementary Particles," in *1962 Cargèse Lectures in Theoretical Physics,* ed. Maurice Lévy (New York: Benjamin), Lecture 11, pp. 1–37, on p. 8. Chew was referring in particular to Lev Landau's 1959 modified rules for using the diagrams, which in turn had been based on Chew's 1958 work on the so-called particle-pole conjecture. The particle-pole conjecture stipulated that the only pole-like singularities within the scattering amplitude for a given process would occur uniquely at the values of mass and momentum corresponding to an exchanged particle. Landau used this speculation to derive general rules for isolating the singularities within generic scattering amplitudes. See Chew, "Proposal for Determining the Pion-Nucleon Coupling Constant from the Angular Distribution for Nucleon-Nucleon Scattering," *Phys. Rev.,* 1958, *112*:1380–1383; Lev Landau, "On Analytic Properties of Vertex Parts in Quantum Field Theory," *Nuclear Physics,* 1959, *13*:181–192; Cushing, *Theory Construction,* pp. 109–113, 129–131; and Kaiser, "Do Feynman Diagrams Endorse a Particle Ontology?" (cit. n. 4), pp. 345–349.

236   NUCLEAR DEMOCRACY

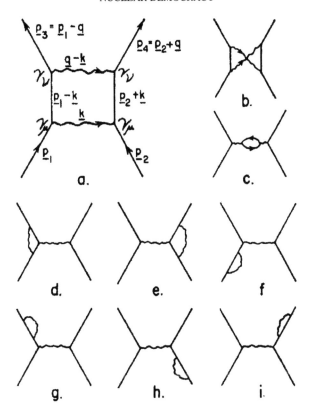

Figure 2. *Feynman diagrams for electron-electron scattering correction terms. From R. P. Feynman, "Space-Time Approach to Quantum Electrodynamics,"* Physical Review, *1949, 76:769–789, on p. 787. Reprinted with kind permission of the American Physical Society.*

this diagrammatic equality "grew out of field theory, particularly from Feynman graphs, but it is now believed that the principle can be formulated completely within the framework of the S matrix."[13]

Drawing liberally on a smattering of theoretical developments from the mid and late 1950s, Chew pointed to the diagrams when formulating his new "democratic" conclusion.[14]

[13] Chew, "Strong-Interaction *S*-Matrix Theory without Elementary Particles," pp. 6–7. Here Chew was again referring to his own 1958 particle-pole conjecture. He later emphasized the "decisive" role played by his new interpretation of Feynman diagrams, writing that by the end of the 1950s it had become clear to him "that graphs of the type invented by Richard Feynman for perturbative evaluation of a Lagrangian field theory are relevant to the analytic S matrix, independent of any approximation based on a small coupling constant." Geoffrey Chew, "Particles as *S*-Matrix Poles: Hadron Democracy," in *Pions to Quarks: Particle Physics in the 1950s,* ed. Laurie Brown, Max Dresden, and Lillian Hoddeson (New York: Cambridge Univ. Press, 1989), pp. 600–607, on p. 601.

[14] In addition to his particle-pole conjecture and Landau's rules for Feynman diagrams, Chew drew especially on Feynman diagrams' crossing symmetry (as established in 1954), single- and double-variable dispersion relations (which had been formulated during 1954–1958), and Tullio Regge's analysis of complex angular momenta (from 1959). On the dispersion relations work see Marvin Goldberger, "Introduction to the Theory and Application of Dispersion Relations," in *Relations de dispersion et particules élémentaires* [Proceedings of the 1960 Les Houches Summer School], ed. C. de Witt and R. Omnès (Paris: Hermann, 1960), pp. 15–157; and J. D. Jackson, "Introduction to Dispersion Relation Techniques," in *Dispersion Relations: Scottish Universities' Summer School, 1960,* ed. G. R. Screaton (New York: Interscience, 1961), pp. 1–63.

A particle that in one orientation of a Feynman diagram looked like the fundamental building-block constituent of a more complicated particle would appear, upon various rotations of the diagrams, as either the exchanged particle responsible for a force between other particles or as the end-state composite of other constituents. Two of Chew's young colleagues at Berkeley, Frederik Zachariasen and Charles Zemach, built directly on this work in their 1961–1962 treatment of the ρ meson, a heavy, unstable particle that decayed via the strong force relatively rapidly into pairs of pions. The ρ had been discovered by Berkeley experimentalists at the famous cyclotron in 1960, though its existence had actually been predicted earlier by a pair of Chew's graduate students.[15] Zachariasen and Zemach leaned on a series of simple Feynman diagrams, interpreted now along Chew's lines, to guide their calculations.[16] Figure 3*a* depicted a ρ meson being exchanged between two incoming pions, much as the two electrons in Feynman's Figure 1 exchanged a photon. The exchange of the ρ in this case would give rise to an attractive force between the pions. Figure 3*b*, on the other hand, showed two pions coming together to form a new bound-state composite particle, the ρ meson, which, being unstable, later decayed into a new pair of pions. If two pions could create a ρ meson, however, then interactions like that shown in Figure 3*c* had to be considered as well, in which a ρ meson acted just like an "elementary" particle, scattering with an incoming pion; in Figure 3*c*, meanwhile, a pion appeared as the exchanged force-carrying particle. The field theorists' labels of "elementary," "force-carrier," and "composite" swapped places with each rotation of a given Feynman diagram, Chew and his young colleagues charged. The only consistent theoretical framework, they therefore maintained, was a "democratic" one that made no distinctions between "elementary" and "composite" particles.

From this reinterpretation of Feynman diagrams, only a short step led Chew to the essential point of his autonomous *S*-matrix program: not only were composite particles to be treated as equivalent to elementary ones, but all (strongly interacting) particles could in fact be seen as composites of each other. Diagrammatic initiatives akin to those in Figure 3 were central to the new scheme, as Chew explained in his summer-school lectures during 1960:

> The forces producing a certain reaction are due to the intermediate states that occur in the two "crossed" reactions belonging *to the same diagram*. The range of a given part of the force is determined by the mass of the intermediate state producing it, and the strength of the force by the matrix elements connecting that state to the initial and final states of the crossed reaction. By considering all three channels [i.e., orientations of the Feynman diagram] on this basis we

---

Cushing treats the relation of these various ideas and approaches to Chew's *S*-matrix program in *Theory Construction*, Chs. 3–5.

[15] The experimental discovery of the ρ meson as a resonance in pion scattering was announced in A. R. Erwin, R. March, W. D. Walker, and E. West, "Evidence for a π-π Resonance in the $I = 1$, $J = 1$ State," *Physical Review Letters*, 1961, *6*:628–630. It had been predicted earlier in William Frazer and José Fulco, "Effect of a Pion-Nucleon Scattering Resonance on Nucleon Structure," *ibid.*, 1959, *2*:365–368; and Frazer and Fulco, "Partial-Wave Dispersion Relations for the Process $\pi + \pi \rightarrow N + \underline{N}$," *Phys. Rev.*, 1960, *117*:1603–1608. Frazer completed his dissertation under Chew's direction in 1959, and Chew helped to advise Fulco's dissertation, completed in 1962 in Buenos Aires. Years later, Frazer recalled that "Geoff advised us every step of the way" with this work "but generously decided not to put his name on the paper": Frazer, "The Analytic and Unitary S-Matrix," in *A Passion for Physics: Essays in Honor of Geoffrey Chew*, ed. Carleton DeTar, J. Finklestein, and Chung-I Tan (Singapore: World Scientific, 1985) (hereafter cited as **DeTar** *et al.*, eds., *Passion for Physics*), pp. 1–8, on p. 4.

[16] Frederik Zachariasen, "Self-Consistent Calculation of the Mass and Width of the $J = 1$, $T = 1$ $\pi\pi$ Resonance," *Phys. Rev. Lett.*, 1961, *7*:112–113; erratum, *ibid.*, p. 268; and Zachariasen and Charles Zemach, "Pion Resonances," *Phys. Rev.*, 1962, *128*:849–858.

238   NUCLEAR DEMOCRACY

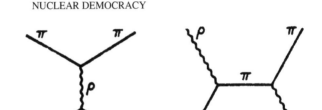

Figure 3. *From Frederik Zachariasen and Charles Zemach, "Pion Resonances," Physical Review, 1962, 128:849–858, on pp. 850–851, 857. Reprinted with kind permission of the American Physical Society.*

have a self-determining situation. One channel provides forces for the other two—which in turn generate the first.[17]

In this way, Chew explained his notion of a particle "bootstrap," saying simply that "each particle helps to generate other particles which in turn generate it."[18] Consider again the case of the ρ meson, one of the earliest successes for Chew's bootstrap model. In its force-carrying mode, as in Figure 3*a*, the exchange of the ρ would create an attractive force between the two incoming pions. Thus drawn together owing to this attractive force, the two pions could combine to produce a resonance—a new bound state or composite particle, as in Figure 3*b*. Next, the "self-determining" feature would come in: Chew and his young collaborators looked for self-consistent solutions such that the force-carrying process produced a resonance whose properties were precisely those of the force-carrying particle itself. If such a solution existed, then the ρ could, all by itself, bring the pions together so that they could produce a ρ. The ρ meson, in this series of diagrams, would

[17] Chew, *S-Matrix Theory of Strong Interactions* (cit. n. 3), p. 32 (emphasis added). The same paragraph appears in his summer-school lectures from 1960—Geoffrey Chew, "Double Dispersion Relations and Unitarity as the Basis of a Dynamical Theory of Strong Interactions," in *Dispersion Relations,* ed. Screaton (cit. n. 14), pp. 167–226, on p. 185—and in the proceedings of the 1960 Les Houches summer school—*Relations de dispersion et particules élémentaires,* ed. de Witt and Omnès (cit. n. 14), pp. 455–514. The duplication is not surprising since, as Chew explained, the 1961 lecture note volume "originated in lectures given at summer schools at Les Houches and Edinburgh in 1960": Chew, *S-Matrix Theory of Strong Interactions,* p. vi.

[18] Chew, "Nuclear Democracy and Bootstrap Dynamics" (cit. n. 4), p. 106. Chew's former postdoc and partner in the early bootstrap work, Steven Frautschi, explained simply in lectures from the 1961/1962 academic year that "bootstrap calculations lean heavily on 'crossing,'" that is, on the symmetries obeyed by scattering amplitudes as the associated Feynman diagrams underwent various rotations: Steven Frautschi, *Regge Poles and S-Matrix Theory* (New York: Benjamin, 1963), p. 176. The bootstrap notion was illustrated with the aid of crossed Feynman diagrams in Chew, "Nuclear Democracy and Bootstrap Dynamics," pp. 134, 136, and also in a text by one of Chew's former students: William Frazer, *Elementary Particles* (Englewood Cliffs, N.J.: Prentice Hall, 1966), p. 134. For more on Chew's bootstrap see James Cushing, "Is There Just One Possible World? Contingency vs. the Bootstrap," *Studies in History and Philosophy of Science,* 1985, *16*:31–48; Cushing, *Theory Construction,* Ch. 6; Cao, "Reggeization Program" (cit. n. 4); Kaiser, "Do Feynman Diagrams Endorse a Particle Ontology?" (cit. n. 4); and Yehudah Freundlich, "Theory Evaluation and the Bootstrap Hypothesis," *Stud. Hist. Phil. Sci.,* 1980, *11*:267–277. Chew expanded on his bootstrap idea in several popular pieces written after the idea had fallen from favor for most particle theorists. See Geoffrey Chew, "'Bootstrap': A Scientific Idea?" *Science,* 23 Aug. 1968, *161*:762–765; Chew, "Hadron Bootstrap: Triumph or Frustration?" *Physics Today,* Oct. 1970, *23*:23–28; and Chew, "Impasse for the Elementary-Particle Concept" (cit. n. 6). See also the interview of Chew by Fritjof Capra: "Bootstrap Physics: A Conversation with Geoffrey Chew," in *Passion for Physics,* ed. DeTar *et al.,* pp. 247–286.

have pulled itself up by its own bootstraps.[19] In Zachariasen and Zemach's numerical analysis, the self-consistent solutions for the mass and coupling constant of the ρ meson were surprisingly close to recent experimental results—this at a time when theorists working within more traditional field-theoretic traditions remained completely stymied in their attempts to analyze the reams of data pouring forth from the nation's accelerators.[20]

Chew and his postdoctoral student Steven Frautschi wondered whether every particle arose in this way: their bootstrap conjecture held that every strongly interacting particle was a composite particle, composed of just those other particles that were brought together by exchanging the first particle as a force. Chew elaborated on this point in his 1964 lecture notes:

> The bootstrap concept is tightly bound up with the notion of a democracy governed by dynamics. Each nuclear particle is conjectured to be a bound state of those S-matrix channels with which it communicates, arising from forces associated with the exchange of particles that communicate with "crossed" channels. . . . Each of these latter particles in turn owes *its* existence to a set of forces to which the original particle makes a contribution.

The bootstrap thus offered Chew the ultimate nuclear democracy: elementary particles deserved no special treatment separate from composite ones; in fact, there might not even exist any "aristocratic," elementary particles, standing above the composite fray. It was in this sense that Chew concluded that "every nuclear particle should receive equal treatment under the law."[21]

Feynman diagrams could be—and, indeed, were—deployed in a host of different ways throughout the 1950s and 1960s; the diagrams themselves did not dictate how physicists would use and interpret them. Chew built directly on his new and unprecedented interpretation of the diagrams to proclaim that all particles should be treated equally. Several of the ingredients for these new types of calculations had been forged over the previous decade (many of them by Chew himself); yet no one before Chew had put these particular elements of the calculation together in the same way. No one, moreover, had gleaned quite the same lesson about "democracy" from the structure of rotated Feynman diagrams. As we will see in Section IV, theorists working at Princeton at this time picked up a number of Chew's new calculational techniques. Yet they broke with him over the calculations' ultimate theoretical implications. Given the tremendous plasticity with which the diagrams had been appropriated ever since Feynman and Dyson originally introduced them, can we

---

[19] See G. F. Chew and S. C. Frautschi, "Unified Approach to High- and Low-Energy Strong Interactions on the Basis of the Mandelstam Representation," *Phys. Rev. Lett.,* 1960, *5:*580–583; Chew and Frautschi, "Principle of Equivalence for All Strongly-Interacting Particles within the S-Matrix Framework," *ibid.,* 1961, *7:*394–397; Chew and S. Mandelstam, "Theory of Low-Energy Pion-Pion Interaction," *Phys. Rev.,* 1960, *119:*467–477; Chew and Mandelstam, "Theory of Low-Energy Pion-Pion Interaction, II," *Nuovo Cimento,* 1961, *19:*752–776; and Chew, Frautschi, and Mandelstam, "Regge Poles in ππ Scattering," *Phys. Rev.,* 1962, *126:*1202–1208. See also the references cited in note 18, above, as well as Frederik Zachariasen, "Lectures on Bootstraps," in *Recent Developments in Particle Physics,* ed. Michael Moravcsik (New York: Gordon & Breach, 1966), pp. 86–151, and references therein.

[20] Zachariasen, "Self-Consistent Calculation" (cit. n. 16); and Zachariasen and Zemach, "Pion Resonances" (cit. n. 16). For further details on the steps within Zachariasen and Zemach's calculation see Kaiser, "Making Theory" (cit. n. 11), pp. 442–444. Zachariasen and Zemach improved on the closeness between their result and experimental data by including the exchange of three other particles in addition to the ρ in their calculation; Zachariasen's original calculation included only the pion-ρ interaction. Cushing treats some of the other early successes of Chew's S-matrix program in *Theory Construction,* pp. 145–151.

[21] Chew, "Nuclear Democracy and Bootstrap Dynamics" (cit. n. 4), p. 106.

understand what lay behind Chew's specific reading of them—let alone the fervor with which he turned the diagrams against their original field-theoretic birthplace?

Chew noted in his 1961 lectures that, more than anything else, "general philosophical convictions" (the details of which he left unstated) helped to guide him in his democratic reading of Feynman diagrams and in his conclusion that no particles were truly elementary. In these lecture notes, as elsewhere, Chew struggled to find a vocabulary that would support the conceptual breaks he aimed to make with quantum field theory. Finding the right terminology was no mean feat: "The language just didn't exist in physics," he later recalled. He remembers being "annoyed" by other physicists' sloppy terminology—it made him "gag"—"because the language wasn't there, there were no words, there was no way to say it."[22] The language he did produce to express his new physical concepts—his recurring incantations of "nuclear democracy," "equal treatment," and "equal participation"—therefore had to come from somewhere other than the stable repertoire of his fellow physicists. Guided in part by Chew's conscious choices of language, I will attempt in the next two sections to clarify what he called his "general philosophical convictions" and to study their evolution over time. In order to unpack Chew's nuclear democracy, we must begin not with his flipped Feynman diagrams of the early 1960s but, rather, with the battles fought in Berkeley over domestic anti-Communism, beginning in 1948.

II. GEOFFREY CHEW: A SCIENTIST'S POLITICS OF DEMOCRACY IN 1950s AMERICA

Geoffrey Chew, born in 1924, came of age in a generation of American theoretical physicists just after that of Richard Feynman and Julian Schwinger. Growing up in Washington, D.C., where his father worked in the U.S. Department of Agriculture, Chew graduated from high school at the age of sixteen. Four years later he completed his college education with a straight-A record from George Washington University. It being 1944, one of his undergraduate advisors, George Gamow, helped to arrange for Chew to head straight to Los Alamos, where he joined Edward Teller's special theoretical division, working on early ideas for a hydrogen "superbomb." Entering graduate school in February 1946 at the University of Chicago, Chew completed his doctorate in less than two and a half years, under the tutelage of Enrico Fermi. Ph.D. in hand, he and his fellow graduate student Marvin "Murph" Goldberger headed off to postdocs under Robert Serber at Berkeley's Radiation Laboratory. Before long, Berkeley's physics department took notice of Chew, who was already developing a reputation for both brilliance and clarity. The physics department appointed Chew as an assistant professor, to begin in the fall of 1949.[23] Chew's transit from the East Coast to the West Coast included some of the best stops along the way for young physicists at that time, and by the age of twenty-five he had arrived, this physicist's manifest destiny complete. (See Figure 4.)

---

[22] Chew, *S-Matrix Theory of Strong Interactions* (cit. n. 3), p. 4 ("general philosophical convictions"); and Chew, interview with Gordon, Dec. 1997, quoted in Gordon, "Strong Interactions," pp. 31–32.

[23] These biographical details are taken from Raymond Birge to Dean A. R. Davis, 27 Feb. 1949, in Raymond Thayer Birge Correspondence and Papers, call number 73/79c, Bancroft Library, University of California, Berkeley (hereafter cited as **Birge Papers**); and A. C. Helmholz to Dean Lincoln Constance, 25 Mar. 1957, Birge Papers, Box 40, Folder "Letters written by Birge, January–May 1957." Some clarification may be helpful here. Letters written by Birge are filed chronologically. The items cited in this essay are from Boxes 39 and 40; explicit folder titles will not be cited. Letters written to Birge (and other pertinent materials in this collection) will be cited with box number and folder titles. For further biographical information see also Birge, "History," Vol. 5, Ch. 19, pp. 43–51; and DeTar *et al.*, eds., *Passion for Physics*.

Figure 4. *Geoffrey Chew, circa 1960. Lawrence Berkeley National Laboratory, University of California, Berkeley. Courtesy of the Emilio Segrè Visual Archives, American Institute of Physics.*

This simple picture of professional progress grew complicated, however, nearly as soon as Chew was hired. Chew became increasingly engaged with political issues in the late 1940s, continuing with greater and greater intensity throughout the 1950s. His activities took him on an extended orbit that began in Berkeley and, seven years later, brought him back there again. His reasons for leaving Berkeley in 1950 can be understood only if we consider the political situation in which physicists found themselves soon after World War II and how Berkeley physicists, in particular, experienced the early years of the Cold War. The fast-moving descent into McCarthyism affected daily life in Berkeley's Radiation Laboratory and Department of Physics, shaping hallway discussions, straining old friendships, and altering many young physicists' career paths.

Politics and Physics at Berkeley, 1949–1954

Few American physics departments experienced the pains of transition to the postwar political scene more abruptly, or more publicly, than that at the Berkeley campus of the University of California. The House Un-American Activities Committee (HUAC) turned its sights on atomic espionage directly on the heels of its sensational probe of alleged Communists in the film industry.[24] Its first stop: Berkeley's Radiation Laboratory, built up to international prominence during the 1930s by Ernest O. Lawrence. During the war, Lawrence's famed laboratory on the hill had been staffed with teams endeavoring to separate the scarce, fissionable uranium-235 isotope from its more ubiquitous cousin, uranium-238. Some of this staff, HUAC began to insinuate nearly as soon as Chew arrived at the Rad Lab as a postdoc in 1948, had been "red."

Startling headlines greeted students and faculty returning to Berkeley in September 1948: a physicist who had worked at the Radiation Laboratory during the war, identified by HUAC only as "Scientist X," purportedly had leaked vital atomic secrets to the Soviets. Five physicists who had worked in the wartime Radiation Laboratory were singled out for intensive questioning. Though no evidence of espionage at the Rad Lab was ever uncovered, all were convicted of contempt of Congress and one of perjury, based on his testimony. Each lost his job immediately upon being indicted by Congress.[25]

By the time HUAC began its investigation, only one of these Rad Lab physicists remained in Berkeley. In September 1949 his case grabbed local attention when Berkeley's student newspaper, the *Daily Californian,* reported the front-page story, "T.A. Queried on Communist Ties." I. David Fox, the teaching assistant in question, had worked at the Radiation Laboratory during the war and was currently a graduate student in Berkeley's physics department. Fox refused to name names for his HUAC investigators, invoking the

---

[24] The literature on HUAC and McCarthyism is, of course, vast. On McCarthyism and American higher education in particular see esp. Ellen Schrecker, *No Ivory Tower: McCarthyism and the Universities* (New York: Oxford Univ. Press, 1986); Sigmund Diamond, *Compromised Campus: The Collaboration of Universities with the Intelligence Community, 1945–1955* (New York: Oxford Univ. Press, 1992); James Hershberg, *James B. Conant: Harvard to Hiroshima and the Making of the Nuclear Age* (Stanford, Calif.: Stanford Univ. Press, 1993), Chs. 19, 21–23, 31; Noam Chomsky *et al.,* eds., *The Cold War and the University: Toward an Intellectual History of the Postwar Years* (New York: New Press, 1997); Catharine M. Hornby, "Harvard Astronomy in the Age of McCarthyism" (A.B. thesis, Harvard Univ., 1997), esp. Ch. 2; Jessica Wang, *American Science in an Age of Anxiety: Scientists, Anticommunism, and the Cold War* (Chapel Hill: Univ. North Carolina Press, 1999); Lawrence Badash, "Science and McCarthyism," *Minerva,* 2000, *38:*53–80; Naomi Oreskes and Ronald Rainger, "Science and Security before the Atomic Bomb: The Loyalty Case of Harald U. Sverdrup," *Studies in History and Philosophy of Modern Physics,* 2000, *31:*309–369; and Schweber, *Shadow of the Bomb* (cit. n. 9). On the politicization of American scientists before World War II see Peter J. Kuznick, *Beyond the Laboratory: Scientists as Political Activists in 1930s America* (Chicago: Univ. Chicago Press, 1987).

[25] On the establishment of the Radiation Laboratory see John Heilbron and Robert Seidel, *Lawrence and His Laboratory: A History of the Lawrence Berkeley Laboratory,* Vol. 1 (Berkeley: Univ. California Press, 1990). On the HUAC investigation of the Rad Lab physicists see the San Francisco–area newspaper clippings in Department of Physics, University of California, Berkeley, Records, ca. 1920–1962, call number CU-68, Bancroft Library (hereafter cited as **Dept. Physics, Berkeley, Records, ca. 1920–1962**), Folder 4:12; "Thomas Committee Calls Ex-Instructor," *Daily Californian,* 22 Sept. 1948, p. 5; Louis Bell, "'No Great Surprise'; Identity of Scientist X Suspected Here," *ibid.,* 3 Oct. 1949, p. 1; "Ex-Physicist at U.C. Held for Perjury," *ibid.,* 26 May 1952, p. 1; and Schrecker, *No Ivory Tower,* pp. 126–148. The specifics of the legal case against the Rad Lab physicists are outlined in Carl Beck, *Contempt of Congress* (New Orleans, La.: Hauser, 1959), pp. 65–70. Two other cases involving Berkeley scientists also caught the attention of HUAC and the national media in 1948: those of Martin Kamen and E. U. Condon. See Martin Kamen, *Radiant Science, Dark Politics: A Memoir of the Nuclear Age* (Berkeley: Univ. California Press, 1985), Chs. 11, 12; and Jessica Wang, "Science, Security, and the Cold War: The Case of E. U. Condon," *Isis,* 1992, *83:*238–269. Condon was at the time the head of the National Bureau of Standards, having formerly been an undergraduate and graduate student at Berkeley and a consultant at the Berkeley Rad Lab during World War II. See also Schweber, *Shadow of the Bomb.*

Fifth Amendment twenty-five times. Three months later, the Board of Regents of the University of California dismissed Fox, without making any formal charges against him.[26] Even before Senator Joseph McCarthy had begun the anti-Communist activities with which his name forever will be associated, physicists in Berkeley were feeling the brunt of McCarthyism.

The politics of domestic anti-Communism invaded physics departments far beyond Berkeley in the ensuing months and years. Physicists across the country debated the new proposal, in 1949, to require full background security checks for all recipients of Atomic Energy Commission graduate student fellowships. Thirty-four Berkeley physics graduate students wrote to the *San Francisco Chronicle* to protest what they saw as the proposed exclusion from training of "those among us who hold unpopular viewpoints," arguing that "education in a democracy must be available to everyone." The next year, after the final, contentious establishment of the National Science Foundation (NSF), prominent scientists such as Berkeley's iconic physics department chair Raymond Birge objected to congressional discussion that considered prohibiting NSF grants to members, past or present, of any organization listed on the attorney general's "subversive" list.[27]

With the outbreak of fighting in Korea that June, department chairs such as Birge found more and more of their time devoted to draft deferments for their students and young faculty and to problems involving personnel security clearances.[28] Berkeley physicists, like their colleagues across the country, found themselves routinely denied passports for foreign travel; meanwhile, foreign physicists experienced insulting delays or even rejections of their applications for visitors' visas. The Rad Lab, since 1947 under the auspices of the Atomic Energy Commission, was no longer permitted to have foreign scientists conduct any work there—classified or unclassified, paid or not.[29]

---

[26] "T.A. Queried on Communist Ties," *Daily Californian*, 28 Sept. 1949 [misprinted as 1948], p. 1 (see also the *Daily Californian* stories on 4 Jan. 1950, p. 1; 27 Mar. 1950, p. 7; and 31 Mar. 1950, pp. 1, 4); Schrecker, *No Ivory Tower*, pp. 126–127; and David Gardner, *The California Oath Controversy* (Berkeley: Univ. California Press, 1967), pp. 91–94.

[27] For the graduate students' protest see Letter to the Editor, *San Francisco Chronicle*, 27 May 1949; I. David Fox was among the signers. See also Birge to Lyman Spitzer, Jr., 26 May 1949, Birge Papers; E.R. [Eugene Rabinowitch], "The 'Cleansing' of AEC Fellowships," *Bulletin of the Atomic Scientists*, June–July 1949, 5:161–162; "The Fellowship Program: Testimony before the Joint Committee," *ibid.*, pp. 166–178; "Loyalty Tests Cause Cut in AEC Fellowship Program," *ibid.*, Jan. 1950, 6:32; "The Curtailment of the AEC Fellowship Program," *ibid.*, pp. 34, 62–63; "Loyalty Tests for Science Students?" *ibid.*, Apr. 1950, 6:98; and Wang, *American Science in an Age of Anxiety* (cit. n. 24), Ch. 7. On the debates over the founding of the NSF see Daniel Kevles, "The National Science Foundation and the Debate over Postwar Research Policy, 1942–1945," *Isis*, 1977, 68:5–26; Kevles, *The Physicists: The History of a Scientific Community in Modern America*, 2nd ed. (1978; Cambridge, Mass.: Harvard Univ. Press, 1987), Chs. 11–12; Nathan Reingold, "Vannevar Bush's New Deal for Research; or, The Triumph of the Old Order," *Hist. Stud. Phys. Biol. Sci.*, 1987, 17:299–344; and Jessica Wang, "Liberals, the Progressive Left, and the Political Economy of Postwar American Science: The National Science Foundation Debate Revisited," *Hist. Stud. Phys. Sci.*, 1995, 26:139–166. On the issue of the attorney general's list of "subversive" organizations and NSF grants see Birge to Robert G. Sproul, 14 Mar. 1950, Birge Papers.

[28] On draft deferments see Birge to R. C. Gibbs, 10 Aug. 1950, and Birge to Local Board No. 62, Santa Clara County, 8 June 1953, Birge Papers; and David Kaiser, "Putting the 'Big' in 'Big Science': Cold War Requisitions, Scientific Manpower, and the Production of American Physicists after World War II," unpublished MS. The concerns with draft deferments for physics students persisted well after fighting had ceased in Korea; see the correspondence from 1958 in the American Institute of Physics, Education and Manpower Division, Records, 1951–1973, Box 4, Folder "Scientific Manpower Commission, Washington, D.C." These records are held in the American Institute of Physics, Niels Bohr Library, College Park, Maryland, call number AR15. On security clearance troubles see Birge to K. K. Darrow, 11 Jan. 1955, Birge Papers; Wang, *American Science in an Age of Anxiety;* Ellen Schrecker, *The Age of McCarthyism: A Brief History with Documents* (Boston: Bedford, 1994), pp. 37–40, 150–164; and Adam Yarmolinsky, ed., *Case Studies in Personnel Security* (Washington, D.C.: Bureau of National Affairs, 1955).

[29] Regarding passport and visa problems see Birge to Darrow, 26 May 1955; Birge to Congressman Francis

Not all of the reactions to these fast-moving events were glum. Five years after the HUAC investigation of purported espionage, Berkeley student reporters concluded their five-part series on the Radiation Laboratory in December 1953 with the light-hearted story "Espionage at the Rad Lab—Naw!" After describing dozens of "scattered instruments painted a bright red with the white letters 'USSR' printed on them," the reporters explained that the letters stood for "United States surplus reserve," an old joke up at the laboratory. The gag, and visitors' stunned reactions to it, could still evoke "guffaws" from "the six foot, amiable, loose-jointed physicist" who gave these student reporters their tour.[30]

Still, the levity could only go so far. Birge remained more somber, hesitating before accepting his nomination as vice-president of the American Physical Society (APS). Writing in May 1953 to the society's secretary, Karl Darrow, he recalled a former "time when the American Physical Society was concerned only with physics. At the present time, however, I am afraid it is concerned almost as much with politics as it is with physics and I must say I do not like politics." Almost exactly one year later, just as Birge was preparing to assume the presidency of the APS, he and his Berkeley department were stunned over the decision by the Atomic Energy Commission to deprive J. Robert Oppenheimer, Berkeley's former star theorist and the world-famous "father of the atomic bomb," of his security clearance.[31]

These many developments shaped physicists' experiences across the country in the late 1940s and 1950s; reactions ranged from graduate student protests to the hand-wringing of the president of the American Physical Society. At Berkeley, however, all of these events took shape in the shadow of a local situation that dominated hallway discussions and faculty meetings for the better part of a decade: the "loyalty oath" controversy at the University of California. As we will see, the loyalty oath helped to prompt Geoffrey Chew's own political engagement as he struggled to define a working definition of "democracy" for scientists in Cold War America.

Walter, 15 July 1955; and Birge to Senator Harley Kilgore, 17 Nov. 1955: Birge Papers. The Federation of American Scientists focused on passport and visa problems; many of their efforts were reported in the *Bulletin of the Atomic Scientists*, a special issue of which (Oct. 1952, 8) was dedicated especially to these problems. On the exclusion of foreign scientists from the Rad Lab see Birge to Walter Thirring, 8 Jan. 1952, Dept. Physics, Berkeley, Records, ca. 1920–1962, Folder 5:117. On the establishment of the AEC national laboratory system see Richard Hewlett and Francis Duncan, *A History of the United States Atomic Energy Commission*, Vol. 2: *Atomic Shield, 1947–1952* (University Park: Pennsylvania State Univ. Press, 1969), Ch. 8; Robert Seidel, "A Home for Big Science: The Atomic Energy Commission's Laboratory System," *Hist. Stud. Phys. Sci.,* 1986, *16*:135–175; and Peter Westwick, "The National Laboratory System in the U.S., 1947–1962" (Ph.D. diss., Univ. California, Berkeley, 1999).

[30] Sandra Littlewood and Skip Garretson, "Espionage at the Rad Lab—Naw!" *Daily Californian,* 11 Dec. 1953, p. 8. It is interesting to note that this was the only segment of the five-part series that Birge clipped and saved with his other newspaper clippings, which may be found in Dept. Physics, Berkeley, Records, ca. 1920–1962, Folder 4:12.

[31] Birge to Darrow, 22 May 1953, Birge Papers; Atomic Energy Commission, *In the Matter of J. Robert Oppenheimer: Transcript of Hearing before Personnel Security Board* (Washington, D.C.: Atomic Energy Commission, 1954); Philip Stern, *The Oppenheimer Case: Security on Trial* (New York: Harper & Row, 1969); John Major, *The Oppenheimer Hearing* (New York: Stein & Day, 1971); and Barton J. Bernstein, "'In the Matter of J. Robert Oppenheimer,'" *Hist. Stud. Phys. Sci.,* 1982, *12*:195–252. Of course, not all members of the Berkeley department were shocked by the news; some had even lobbied behind the scenes to ensure the outcome. On Berkeley involvement in and reactions to the hearing see "Oppenheimer Conflict: Former Professor Center of Dispute," *Daily Californian,* 15 Apr. 1954, p. 1; Birge to Edwin A. Uehling, 28 Mar. 1955, Birge Papers; Birge, "History," Vol. 5, Ch. 17; Nuel Pharr Davis, *Lawrence and Oppenheimer* (New York: Simon & Schuster, 1968), Chs. 8–10; Luis W. Alvarez, *Alvarez: Adventures of a Physicist* (New York: Basic, 1987), pp. 179–181; and A. Carl Helmholz with Graham Hale and Ann Lage, *Faculty Governance and Physics at the University of California, Berkeley, 1937–1990* (Berkeley: Regional Oral History Office, Bancroft Library, 1993), pp. 152–157, 276–279.

The California Loyalty Oath

Only a few months before David Fox was dismissed from his Berkeley teaching-assistant position, the Board of Regents of the University of California imposed a new "loyalty oath" on all university employees. This anti-Communist oath was drafted hastily during the lunch break of an otherwise routine monthly meeting of the regents on 25 March 1949, with only eleven of the twenty-four regents in attendance; these eleven adopted it unanimously. As historians such as David Gardner and Ellen Schrecker have emphasized, the new oath sprang more from questions regarding self-governance than from overriding fears on most of the regents' part about actual Communist infiltration of the university system. The California State Legislature was at that moment considering a proposal to wrest control over adjudicating the "loyalty" of the university faculty from the regents; enacting the new oath offered one way for the regents to show the legislature that it could manage such issues on its own. Soon the oath set off a long and bitter struggle between the regents and the faculty over this same question of self-governance: most members of the faculty claimed that the regents' act violated the faculty's traditional role of choosing its own members and, when necessary, keeping them in line.[32]

Adding to the furor, the regents failed to inform the faculty of the new requirement until mid-June 1949—over two months after the oath officially had been adopted and at just the time when many faculty were leaving Berkeley for the summer. The oath required all university employees—from janitorial staff to graduate student teaching assistants to tenured faculty—to swear that they were not members of the Communist Party; each signing had to be witnessed by a notary public. Letters to faculty members soon revealed that their "reappointment," and the payment of their salaries, was now conditional upon their signing the oath—even for those who believed that tenure had already removed all questions of reappointment.[33]

A vocal minority of the faculty, including several professors who had fled European dictatorships before and during World War II, immediately decried the oath as an infringement on academic freedom. Yet during the fall of 1949, in the midst of reports of the Soviets' detonation of their own atomic bomb and the "fall" of China to Communist leaders, the objections of most faculty members did not concern the ban on Communists per se. Instead, they opposed the fact that university employees had been singled out and held to an oath more stringent than that required of other state employees. Hundreds of faculty members at first refused to sign the oath in protest, a recalcitrance that, in turn, only strengthened the view of certain key regents that university faculties could not be trusted to govern themselves. Crowds of fifty to two hundred professors met each week at Berkeley's faculty club throughout the academic year 1949/1950 to discuss tactics and strategies.[34]

---

[32] Gardner, *California Oath Controversy* (cit. n. 26); and Schrecker, *No Ivory Tower* (cit. n. 24), pp. 116–125. See also George Stewart, *The Year of the Oath* (New York: Doubleday, 1950); Birge, "History," Vol. 5, Ch. 19; and Helmholz with Hale and Lage, *Faculty Governance*, pp. 96–97, 152–157. Gardner cautions against reading the regents' enactment of the oath too narrowly as a direct reaction to the state legislature's proposals, which, he concludes, provided only a clearly articulated measure of the "mood" of the times: Gardner, *California Oath Controversy*, p. 10.

[33] See the forms and notices in Dept. Physics, Berkeley, Records, ca. 1920–1962, Folder 3:41; and Gardner, *California Oath Controversy*, pp. 52–54.

[34] One of the early faculty leaders who spoke out against the oath because of what he perceived to be ties to fascist practices was Ernst Kantorowicz, a German-born medieval historian. See Grover Sales, Jr., "The Scholar and the Loyalty Oath," *San Francisco Chronicle*, 8 Dec. 1963, pp. 27–30, written soon after Kantorowicz's death. Robert Serber recalled that Gian Carlo Wick, a physicist at Berkeley who refused to sign the oath, said

Meanwhile, by February 1950 the regents' line had hardened: any university employee who had not signed the oath by the end of April would be dismissed. Alumni and the president of the university managed to pry a compromise from the regents, who agreed that the cases of all nonsigners would be reviewed on an individual basis by the faculty's Committee on Privilege and Tenure. Yet when North Korea invaded South Korea in late June 1950 and the United States entered the conflict, most of the remaining faculty holdouts simply signed the oath. Dismissing the faculty committee's recommendations, the regents then fired the remaining thirty-one "non-signers" on 25 August 1950. Even though none of those fired had ever been accused by the faculty or the regents of Communist Party membership or sympathy, their fates drew national attention to the question (and, as many proclaimed, the danger) of Communists in the classroom.[35]

The ensuing seven-year battle between the fired nonsigners and the regents, weaving in and out of the California State Supreme Court, left no department on the California campuses untouched. Yet few felt the full brunt of the controversy more, or in more ways, than Berkeley's Department of Physics. It is difficult to overestimate the effects of the oath controversy on daily life within the department. The department secretary had to rush off lists of graduate students who had signed the oath after an initial "oversight" to prevent their termination as course graders or teaching assistants. The examining committees for several students' dissertation defenses had to be rearranged at the last minute, since the regents ruled that faculty nonsigners could no longer serve on such committees. Birge circulated a memorandum to the department's faculty in early April 1950, cautioning them against putting their personal opinions regarding the oath in writing or even discussing them "at a meeting of a fairly large group." Writing to university president Robert Sproul in the midst of the controversy, Birge found himself wondering whether the "wave of hysteria now sweeping the country," as evidenced locally by the loyalty oath disaster, might even put "the entire democratic structure of this country . . . in some danger."[36] Whether interpreted ultimately as an anti-Communist witch-hunt, a principled fight over academic freedom, or a local power play pitting faculty against administration, the effects of the loyalty oath controversy on mundane daily life were palpable to students and faculty alike.

Several members of the physics department did more than just absorb the oath's after-

---

at the time that "he had been coerced into taking an oath once before in Italy, where he had to swear loyalty to Mussolini; he said he'd regretted it ever since and wasn't going to make the same mistake twice." Robert Serber with Robert Crease, *Peace and War: Reminiscences of a Life on the Frontiers of Science* (New York: Columbia Univ. Press, 1998), p. 171. Another Berkeley physicist, Emilio Segrè, on the other hand, later wrote that he had signed at least fifteen loyalty oaths while in Mussolini's Italy; thus he found them all meaningless, and therefore, as a practical matter, he saw little reason not to sign the California oath as well. In fact, as he put it, "I even remembered a pronouncement by Pope Pius XI, elicited by a Fascist oath, explicitly stating that under certain circumstances one could take such oaths with mental reservations that made them void. I dug the papal document out in the library and translated it, and some colleagues to whom I had sent it posted it in Los Alamos, which administratively depended on the Regents of the University of California. At Berkeley it circulated less openly." Emilio Segrè, *A Mind Always in Motion: The Autobiography of Emilio Segrè* (Berkeley: Univ. California Press, 1993), pp. 235–236. On the weekly meetings in the faculty club see Gardner, *California Oath Controversy*, p. 87.

[35] Gardner, *California Oath Controversy*, Chs. 5–6; and Schrecker, *No Ivory Tower* (cit. n. 24), pp. 120–122.

[36] O. Lundberg, University Controller, memo to "chairmen of departments, administrative officers, and others concerned," 27 Nov. 1950; RLY [Rebekah Young], Physics Department Secretary, to Lundberg, 30 Nov. 1950; M. A. Stewart, Associate Dean of the Graduate Division, memo to Physics Department Graduate Advisers, 14 Dec. 1950: Dept. Physics, Berkeley, Records, ca. 1920–1962, Folder 3:41. Rebekah Young to Robert Serber, 18 July 1951; and Serber to Young, 25 July, 1951: Dept. Physics, Berkeley, Records, ca. 1920–1962, Folder 3:4. Birge, "Memorandum to Members of the Physics Department Staff," 6 Apr. 1950; and Birge to Sproul, 14 Mar. 1950: Birge Papers.

effects. From the start, the department included active members on all sides of the controversy. The teaching assistant David Fox—who, ironically, had signed the oath—was nonetheless touted by certain regents as an all-too-visible reminder of the need to enforce loyalty among the faculty. Professor Francis Jenkins joined the "Operating Committee of Seven" to help coordinate faculty opposition to the oath and later served on the faculty's Committee on Privilege and Tenure. Robert Brode, also a senior professor in the department, served as the official custodian of the funds raised by the Operating Committee of Seven, which were intended in part to help pay the expenses of the nonsigners who had stopped receiving their salaries. And department chair Raymond Birge often played the part of back-room negotiator, pushing certain people (such as Wendell Stanley, later a Nobel laureate in biology) into visible positions on faculty committees, while presenting a series of prepared remarks against the regents' actions before Academic Senate and departmental meetings.[37]

On the other side, the physics department's Victor Lenzen, chair of the Committee on Privilege and Tenure for the northern section of the Academic Senate, helped to engineer the composition of this crucial committee by removing all nonsigners before agreeing to retire from the committee himself. In less explicit ways, Ernest Lawrence, Edwin McMillan, Luis Alvarez, and others, while remaining "aloof" from the campus-wide discussions of the controversy, created an "atmosphere" at the Radiation Laboratory that convinced several physicists that neither nonsigners nor their supporters would be welcome there. For young physicists at the Rad Lab, there was little room left for doubt: postdocs who had not signed the oath found notes on their desks on 30 June 1950, informing them that they had to turn in their badges and keys, clear off their desks, and leave by the end of that day.[38]

These many different types of participation in the controversy complicated daily life in the physics department. Even more concretely, Berkeley's physics department fell victim to the oath in losing six faculty members within one year. Two professors, Harold Lewis and Gian Carlo Wick, allowed themselves to be fired in August 1950, when the regents finally dismissed all remaining nonsigners. By June 1951 four other professors—Robert Serber, Wolfgang "Pief" Panofsky, Howard Wilcox, and Geoffrey Chew—had resigned in protest. Among those who left were all of the department's theoretical physicists. The very first to resign from the physics department over the issue—and perhaps the first in the entire university—was Geoffrey Chew.[39]

[37] Birge, "History," Vol. 5, Ch. 19, pp. 1–15; and Gardner, *California Oath Controversy* (cit. n. 26), pp. 123–124, 171–172. Some of Birge's speeches at Academic Senate and physics department faculty meetings are reprinted in Birge, "History," Vol. 5, Ch. 19, pp. 8–12.

[38] Gardner, *California Oath Controversy,* p. 248; reference to the "atmosphere" at the Rad Lab comes from Geoffrey Chew's letter of resignation, quoted below. See also Helmholz with Hale and Lage, *Faculty Governance* (cit. n. 31), pp. 96–97, 153; Serber with Crease, *Peace and War* (cit. n. 34), pp. 171–172; and Segrè, *Mind Always in Motion* (cit. n. 34), pp. 234–237. Jack Steinberger was one of the postdocs dismissed on 30 June 1950 for not signing the oath. See Jack Steinberger, "A Particular View of Particle Physics in the Fifties," in *Pions to Quarks,* ed. Brown *et al.* (cit. n. 13), pp. 307–330, esp. p. 311; and Steinberger, "Early Particles," *Annual Review of Nuclear and Particle Science,* 1997, *47:*xiii–xlii, esp. pp. xxxix–xl.

[39] Birge suggests that Chew was "apparently" the first professor to resign from all of the University of California over the oath controversy in Birge, "History," Vol. 5, Ch. 19, p. 45. Robert Serber recounts his own decision to leave in Serber with Crease, *Peace and War,* pp. 171–172. Serber had endured an extended, and at times hostile, personal security review in 1948, though perhaps because of his close affiliation with Ernest Lawrence his case did not stir the same media attention as the HUAC Rad Lab investigation did (*ibid.,* pp. 162–165). For more on Serber's continuing security woes see Barton Bernstein, "Interpreting the Elusive Robert Serber: What Serber Says and What Serber Does Not Explicitly Say," *Stud. Hist. Phil. Mod. Phys.,* 2001, *32:*443–486. Wolfgang Pauli kept abreast of the developments in Berkeley's physics department via his friend Erwin

Geoffrey Chew and the Politics of Democracy

Though no one on the faculty could have known it at the time, Chew's official letter of appointment to his assistant professorship arrived exactly one week after the regents secretly passed the new loyalty oath requirement.[40] Chew, who still maintained Q clearance to work on classified nuclear weapons projects, refused to sign what he called, in a letter to Oppenheimer, "the objectionable part of the new contract," which seemed to him to threaten "the right of privacy in political belief." He became further frustrated with what he saw as weak attempts by the rest of the faculty to fight the oath. At the end of his very first year of teaching Chew acted on his convictions, becoming the first person to resign from the physics department over the issue. As he explained to Birge in July 1950, one month before the regents finally dismissed the remaining nonsigners, Chew had decided "to get away from an intimidating and precarious situation."[41]

The firing of David Fox, Chew told Birge, had shown beyond doubt that the regents seemed bent on removing from the faculty the "right" to "maintain its own qualifications." The regents' actions with the oath, furthermore, aimed at nothing less than to "root out the last resistance" among the faculty. The few signs of "faculty solidarity" with the nonsigners had all but vanished when fighting broke out in Korea in June 1950. As Chew pressed Birge one month later, "In a war-time situation, what security can a non-conformist have?" With the outbreak of fighting, the few remaining nonsigners on campus "have now become lepers who must keep out of sight." On top of this, Chew reported that the Radiation Laboratory, which was "the chief stimulus" of his scientific work, had made it clear that it "does not welcome non-signers. Even if I were allowed to maintain my affiliation [with the laboratory], the unsympathetic atmosphere would not be pleasant. This would be a more subtle form of intimidation." Though Chew found it a difficult decision, he left Berkeley in July 1950 and accepted a position at the University of Illinois in Urbana. He was promoted from assistant to associate professor within a year and became a full professor at Urbana in 1955, at the age of thirty-one.[42]

Panofsky, a senior art historian at the Institute for Advanced Study, whose son Wolfgang was one of the experimentalists to leave Berkeley's department because of the loyalty oath. In October 1950 Pauli forwarded to the elder Panofsky news that he had heard from the young theorist J. M. Luttinger. Pauli quoted from Luttinger in his letter to Panofsky: "Apart from Physics, the atmosphere is very unpleasant in Berkeley. Both [Gian Carlo] Wick and [Harold] Lewis have been fired for refusing to sign a Loyalty Oath, and both (so far as I know) are fighting the case in court. They have only a very slim chance of winning—on the whole it is a degrading business. *In addition to that the lab is full of secret work, and is overrun by petty officials and bureaucracts of all kinds.*" Wolfgang Pauli to Erwin Panofsky, 23 Oct. 1950, in Wolfgang Pauli, *Wissenschaftlicher Briefwechsel*, ed. Karl von Meyenn (New York: Springer, 1996), Vol. 4, Pt. 1, p. 179.

[40] Birge notes that Dean A. R. Davis sent the official letter of appointment to Chew on 1 Apr. 1949 (Birge, "History," Vol. 5, Ch. 19, p. 45); the regents enacted the new loyalty oath in their meeting on 25 Mar. 1949. While still a postdoc at the Rad Lab, Chew had delivered several talks on his research to the physics department, both formal and informal, so that Birge could introduce Chew as already "well known" at the Sept. 1949 departmental meeting. See Birge's handwritten notes, "First Dept. meeting, Wed., Sept. 28, 1949," Dept. Physics, Berkeley, Records, ca. 1920–1962, Folder 2:4.

[41] Geoffrey Chew to J. Robert Oppenheimer, 11 May 1950, quoted in Birge, "History," Vol. 5, Ch. 19, p. 45 (Birge also notes discussions with Chew over his frustration with Academic Senate resolutions regarding the oath); and Chew to Birge, 24 July 1950, Birge Papers, Box 5, Folder "Chew, Geoffrey Foucar, 1924–."

[42] Chew to Birge, 24 July 1950; and Birge, "History," Vol. 5, Ch. 19, p. 47. Chew's emphasis on the importance of the Korean war is echoed in several historians' recent studies of postwar American science policy. See Daniel Kevles, "Cold War and Hot Physics: Science, Security, and the American State, 1945–56," *His. Stud. Phys. Biol. Sci.*, 1990, 20:239–264; Hershberg, *James B. Conant* (cit. n. 24), Chs. 27–28; Wang, *American Science in an Age of Anxiety* (cit. n. 24), Ch. 8; and Kaiser, "Putting the 'Big' in 'Big Science'" (cit. n. 28). The University of Illinois also had a standing loyalty oath requirement at the time Chew accepted his job there, but one that did not mention the Communist Party or any other group by name. Birge and University of California president

A few months after he left Berkeley, and after the regents dismissed the thirty-one remaining nonsigners, Chew reported on the struggle in the *Bulletin of the Atomic Scientists*. Like most of the faculty at the University of California, Chew objected that the faculty had been singled out and subjected to a more specific loyalty oath than was required for any other state employees. The fact that the regents then went beyond this, to threaten and eventually dismiss tenured faculty, constituted a further violation of "the cornerstone of academic freedom." He refrained from detailing any of his own experiences, reviewing instead the positions taken by the university president, the Academic Senate, and various factions within the Board of Regents. The controversy had been fanned in part, Chew explained, by what he called "fundamentalists," people who struck principled stands on questions like academic freedom even though they had lived uncomplainingly since 1942 with an official university policy excluding Communists from teaching there. The "moral of this very sad story," Chew concluded, was that more explicit procedures needed to be defined for tenure. The rights and roles of the faculty, the Academic Senate, and the Board of Regents needed similar attention and explication, to guarantee that faculty would not be singled out for special treatment again in the future. The procedures of due process might then guard against a repeat of "the present sad and humiliating situation."[43]

Over the course of the 1950s, while teaching in Illinois, Chew became more and more active in what has been called "the atomic scientists' movement." Soon after arriving on campus, Chew founded Urbana's local branch of the Federation of American Scientists (FAS), a national organization dedicated to moderate liberal causes. As Jessica Wang has detailed, by the late 1940s the FAS had become largely a bureaucratic organization, collecting information about some of the more severe abuses of McCarthyism and lobbying certain legislators for reform. In part because of pressure from HUAC and the FBI, the FAS had begun to refrain from its earlier pattern of public demonstrations by the time Chew joined the group, choosing instead the route of "quiet diplomacy."[44]

Chew participated directly in this FAS diplomacy, both on the Illinois campus and, soon, as a visible leader within the national organization. After founding the local FAS branch, Chew immediately began inducting friends and colleagues, such as his fellow physicist Francis Low. As Low recalls, Chew strode up to him soon after he arrived in Urbana and asked simply, "Okay, are you ready to join FAS?" "I was happy to do it," Low continues. "I thought it was a good organization. Geoff's position was very good, and I was happy to take part in it. It was a serious time." Under Chew's direction, the group organized monthly meetings on campus and hosted speakers on topics like the Fifth Amendment. On Chew's initiative, they also became a clearinghouse for campus-wide complaints about unfair treatment, such as problems in obtaining passports. Soon Chew's activities extended

---

Sproul found it ironic that Chew would agree to go to Illinois, but Chew explained that in Illinois this was "the same oath required of all state employees . . . [and] no one feels it to be a restriction on his political activity. . . . The intent of the trustees, therefore, does not seem inimicable [*sic*] to academic freedom." In other words, as far as Chew was concerned, the Illinois oath did not single out faculty for special treatment or unfair scrutiny. Chew to Birge, 24 July 1950.

[43] Geoffrey Chew, "Academic Freedom on Trial at the University of California," *Bull. Atom. Sci.,* Nov. 1950, 6:333–336, on p. 336. The objection that faculty were singled out for closer scrutiny than other people was a common one among Berkeley faculty. Chew noted this in passing on p. 334 of his article; he gave it a more extended discussion in his letter to Birge of 24 July 1950. See also Birge, "History," Vol. 5, Ch. 19; Gardner, *California Oath Controversy* (cit. n. 26), Ch. 3; and Schrecker, *No Ivory Tower* (cit. n. 24), pp. 122–123.

[44] See esp. Wang, *American Science in an Age of Anxiety* (cit. n. 24). On the founding and early years of the FAS see also Alice Kimball Smith, *A Peril and a Hope: The Scientists' Movement in America, 1945–7* (Chicago: Univ. Chicago Press, 1965).

beyond the Urbana campus. In November 1955 he testified before a U.S. Senate subcommittee as chair of the FAS Passport Committee. The FAS objected to the State Department's unwritten policy of denying passports to scientists for political reasons and of further denying the applicants any rights of due process or means of appeal. Usually passports were denied with no reasons given, and appeals met delays lasting months and even years. One of the most famous cases at the time concerned Linus Pauling, who finally received a passport in 1954—after more than two years of attempts—when he applied to go to Sweden to receive his Nobel Prize in chemistry.[45]

In his testimony before the Senate Subcommittee on Constitutional Rights, Chew used several lesser-known cases to lobby for fairer treatment. A passport, he urged, must be "recognized as a right of the U.S. citizen, not merely a privilege." Due process must attend all dealings with passport applications, and only problems with demonstrated relevance to national security issues should result in denials. Applicants denied passports should be supplied with an explicit list of charges against them and given the opportunity for a prompt appeals hearing, at which "confrontation of witnesses and no concealment of evidence, should apply." All appeals hearings should be transcribed and copies made available to all parties. Most important of all, Chew argued, a channel outside of the State Department should be set up to handle further appeals: "We should like to see a well-defined channel" established, "so that applicants will have no uncertainty as to what to do." Both the loyalty oath and the passport situation convinced Chew that only "well-defined channels," operating under due process, could protect the equality and rights of academics. Just as he had explained in his report about the California loyalty oath, Chew labored to make it clear for the committee of senators during his 1955 testimony that academics, and scientists in particular, should neither be singled out for "special privileges" nor subject to special scrutiny or bias.[46] Clear and unambiguous procedures needed to be established so that disagreements would be settled fairly, providing equal treatment to all those affected. With these safeguards in place, Chew believed, scientists could participate in a democratic America as citizens, each equal under the law.

III. PEDAGOGICAL REFORMS: "SECRET SEMINARS" AND "WILD MERRYMAKING"

Chew's pedagogical efforts in the years following his congressional testimony resonated with his more explicitly political activities. Demonstrating the same attitudes as in his lobbying with the FAS, Chew endeavored to make certain that graduate students could work in such a way that none was singled out unfairly and all were encouraged to participate equally. His activities with his own graduate students shaped his approach to

[45] Francis Low, interview with the author, MIT, 11 Apr. 2001; David Kaiser, "Francis E. Low: Coming of Age as a Physicist in Postwar America," *Physics @ MIT*, 2001, *14*:24–31, 70–77, on pp. 71–72; and "Summary of Testimony of Linus Pauling," *Bull. Atom. Sci.*, Jan. 1956, *12*:28.

[46] Geoffrey Chew, "Passport Problems," *Bull. Atom. Sci.*, Jan. 1956, *12*:26–28, on p. 28. This article includes Chew's testimony from 15 Nov. 1955. See also "FAS Congressional Activity in 1955," *ibid.*, p. 45. The specific items Chew lobbied for were conspicuously absent in all kinds of hearings from this period, having been denied to witnesses in HUAC hearings, local security-clearance boards, and often even university committees. See Wang, *American Science in an Age of Anxiety* (cit. n. 24); Schrecker, *No Ivory Tower* (cit. n. 24); and Ellen Schrecker, *Many Are the Crimes: McCarthyism in America* (New York: Little, Brown, 1998). The FAS was quite active during the mid 1950s on the issue of passports and visas for scientists. The entire issue of the *Bulletin of the Atomic Scientists* for Oct. 1952 was dedicated to the topic. See also E.R. [Eugene Rabinowitch], "How to Lose Friends," *Bull. Atom. Sci.*, Jan. 1952, *8*:2–5; Victor Weisskopf, "Visas for Foreign Scientists," *ibid.*, Mar. 1954, *10*:68–69, 112; "American Visa Policy: A Report," *ibid.*, Dec. 1955, *11*:367–373; and John Toll, "Scientists Urge Lifting Travel Restrictions," *ibid.*, Oct. 1958, *14*:326–328.

enlisting collaborators for his autonomous *S*-matrix program. Years later, when his program lay largely abandoned by most particle physicists, Chew continued to assess the turnaround using the language of democratic participation.

Chew's "Little Red Schoolhouse" in Berkeley

While neglecting several large issues concerning tenure, academic freedom, and the legality of state-imposed loyalty oaths, the California Supreme Court ruled in favor of the dismissed nonsigners in October 1952, ordering that the regents reappoint them. This court decision, however, left unresolved the question of back pay, and so for several people the oath controversy lumbered on. This last issue was settled by the court, again in the nonsigners' favor, only in the spring of 1956. Yet as early as 1951, certain senior professors in Berkeley's physics department began to consider how best to lure Chew back to Berkeley. Soon after the first court decision was handed down, Birge tried to entice Chew to return. Reluctantly, Chew decided to stay in Illinois, which had made him a very generous counteroffer on hearing of Berkeley's actions. Still excited by the prospect of returning to Berkeley's stimulating campus, however, Chew spent the spring semester of 1957 there as a visiting professor. Eager to keep Chew in Berkeley, the new department chair, Carl Helmholz, performed some impressive financial gymnastics to convince the administration that it could afford to hire Chew as a full professor. Helmholz's schemes worked, and Chew accepted an appointment as a full professor, beginning in the 1957/1958 academic year.[47]

Immediately Chew began advising a large and growing group of graduate students within the department. Its size was especially notable in that Chew and all of his students were theoretical physicists, for whom working in large groups was still unusual. Often ten or more students would be under Chew's wing at a time, and Chew himself would be engaged in collaborative work with four or five of them; postdocs and research associates made the group even larger. A steady stream of Chew's students completed their Berkeley dissertations beginning in 1959, often with four or five finishing each year.[48] In choosing to train his students in this manner, Chew followed a pattern similar to that set by Oppen-

---

[47] With regard to Chew's return to Berkeley see the handwritten notes between Francis Jenkins, Robert Brode, and Raymond Birge, undated, ca. Apr. 1951, Dept. Physics, Berkeley, Records, ca. 1920–1962, Folder 5:25; on Chew's 1953 offer to return to Berkeley see Chew to Birge, 21 Apr. 1953, Birge Papers, Box 5, Folder "Chew, Geoffrey Foucar, 1924–"; and on his 1957/1958 appointment see Helmholz to Constance, 25 Mar. 1957, Birge Papers, Box 40, Folder "Letters written by Birge, January–May 1957." Helmholz's financial jockeying becomes clear in *ibid.* and Helmholz to Chancellor Clark Kerr, 5 Mar. 1957, Dept. Physics, Berkeley, Records, ca. 1920–1962, Folder 1:26. The reappointment of the nonsigners was conditional on their signing a new statewide loyalty oath, the so-called Levering oath, which was even more explicitly anti-Communist than the original university oath had been. The key difference was that the Levering oath was imposed on all state employees, so that university faculty were no longer singled out for special treatment. See Gardner, *California Oath Controversy* (cit. n. 26), pp. 250, 253–254; and Schrecker, *No Ivory Tower,* pp. 123–125.

[48] Many of Chew's former colleagues and students recalled that his group was unusually large and that he still made time to work carefully with each of them. See Birge, "History," Vol. 5, Ch. 19, p. 51; Frazer, "Analytic and Unitary S Matrix" (cit. n. 15), p. 7; Georgella Perry, "My Years with Professor Chew," in *Passion for Physics,* ed. DeTar *et al.,* pp. 14–16, on p. 15; Steven Frautschi, "My Experiences with the S-Matrix Program," *ibid.,* pp. 44–48, on p. 44; Carleton DeTar, "What Are the Quark and Gluon Poles?" *ibid.,* pp. 71–78, on p. 77; David Gross, "On the Uniqueness of Physical Theories," *ibid.,* pp. 128–136, on p. 128; C. Edward Jones, "Deducing T, C, and P Invariance for Strong Interactions in Topological Particle Theory," *ibid.,* pp. 189–194, on p. 189; William Frazer, interview with the author, 7 July 1998; Jerome Finkelstein, interview with the author, 24 July 1998; Eyvind Wichmann, interview with the author, 13 Aug. 1998; and Henry Stapp, interview with the author, 21 Aug. 1998 (all interviews were conducted in Berkeley). A list of Chew's former graduate students, together with their years of graduation, appears in Frazer, "Analytic and Unitary S Matrix," pp. 7–8.

heimer at Berkeley in the 1930s: the students worked collectively, discussing their research projects regularly with the entire group. The large, close-knit group format contrasted starkly with the approach of Julian Schwinger, for example, who famously advised ten or more Harvard graduate students at a time during the 1950s and 1960s but met with any of them individually only rarely—and never with the whole group.[49]

Whereas Oppenheimer could intimidate students and colleagues alike with his notoriously sharp tongue, Chew's students uniformly recall a much more encouraging advisor, one who, in the words of a former student, "treat[ed] us as full partners in a common effort." In a further gesture of equality, Chew regularly joined the group for informal lunches in the Rad Lab cafeteria. Chew took Oppenheimer's pedagogical model a step further when he instituted what came to be known as the "secret seminar." His entire group of students met weekly to hear presentations from one another; often the meetings took place at Chew's house. The seminar sessions were "secret" because faculty members (other than Chew) were actively discouraged from attending: the goal was to make certain that no graduate students were too intimidated to participate equally with their peers. From deep within Lawrence's sprawling Radiation Laboratory, the original site of American "big science," Chew carved out what one of his former students described as a "little red schoolhouse."[50]

This "little red schoolhouse" approach also shaped how Chew and some of his Berkeley colleagues organized a special conference on the strong interactions, held in Berkeley in December 1960. As handwritten notes from an early planning meeting reveal, Chew, Carl Helmholz, Donald Glaser, and the other members of the committee wanted their conference to bring "new people up to date" on the status of strong-coupling particle physics. As emphasized in these notes, the conference was to be "non-exclusive." Minutes from this planning meeting likewise noted that graduate students' research, as part of the work of Berkeley's department, "should be strongly represented" at the conference.[51]

Meeting these goals would not be easy: physics conferences on special topics were rarely aimed at bringing nonspecialists up to speed, much less highlighting the contribu-

---

[49] On Oppenheimer's pedagogical approach see Robert Serber, "The Early Years," *Phys. Today,* Oct. 1967, 20:35–39; Serber with Crease, *Peace and War* (cit. n. 34), Ch. 2; Alice Kimball Smith and Charles Weiner, eds., *Robert Oppenheimer: Letters and Recollections* (Cambridge, Mass.: Harvard Univ. Press, 1980), Ch. 3; and Kevles, *Physicists* (cit. n. 27), pp. 216–219. In 1958 Schwinger was technically advising sixteen Harvard graduate students; see "1958–59 Department Lists," in Department of Physics, Harvard University, Correspondence, 1958–60, Box A–P, Folder "1958–59 Department Lists," call number UAV 691.10, Harvard University Archives, Pusey Library, Cambridge, Massachusetts. Bryce DeWitt, who completed his dissertation under Schwinger in 1949, talked about Schwinger's style with me during several discussions; William Frazer raised the contrast between Chew and Schwinger during our interview.

[50] Gross, "Uniqueness of Physical Theories" (cit. n. 48), p. 128 ("full partners"). Both Carleton DeTar and Steven Frautschi recalled the lunches during their interviews with Stephen Gordon; see Gordon, "Strong Interactions," pp. 27–28. Details on Chew's "secret seminars" come from Frautschi, "My Experiences with the S-Matrix Program" (cit. n. 48), p. 44; A. Capella *et al.,* "The Pomeron Story," in *Passion for Physics,* ed. DeTar *et al.,* pp. 79–87, on pp. 86–87; and my interviews with Frazer, Finkelstein, Wichmann, and Mandelstam. The term "little red schoolhouse" comes from an interview between Carleton DeTar and Stephen Gordon, conducted May 1997, and quoted in Gordon, "Strong Interactions," p. 29. Several more of Chew's former students with whom Gordon spoke also recalled Chew's "secret seminar." In the 1930s Berkeley's physics department held informal weekly seminars, attended by faculty and graduate students alike, though this single department-wide meeting disappeared after World War II. Carl Helmholz and Howard Shugart discussed these older seminars during my interviews with them in Berkeley on 14 July 1998 and 29 July 1998, respectively.

[51] See the handwritten notes, dated Mar. 1960, on "Special Meeting APS," in Dept. Physics, Berkeley, Records, ca. 1920–1962, Folder 1:39, and the typed minutes from a planning meeting held on 4 Mar. 1960, in the same folder. The handwritten notes are probably by either Carl Helmholz, chair of the department at this time and head of the conference-planning committee, or Howard Shugart, who was a secretary to the conference-planning committee; the notes appear to match Helmholz's handwriting.

tions of graduate students. The Berkeley committee gained some help from the well-known MIT theorist Victor Weisskopf, who had worked in the FAS with Chew throughout the 1950s. As a member of the planning committee for the APS-sponsored special meeting in Berkeley, Weisskopf lobbied hard to obtain additional funding from the National Science Foundation so that the Berkeley conference could indeed include these younger participants. "You can well understand and I am sure you agree," Weisskopf urged, "that such conferences with open attendance are very important for the stimulation of young people or other people who are new in the field." Such openness was especially needed in particle physics, he continued: "The field of high-energy physics is, as you know, very strongly in the hands of a clique and it is hard for an outsider to enter. The Rochester conferences were the only conferences that dealt with that subject and they limited it to invited people only. The Berkeley conference is supposed to break this custom." "Open" meetings, intended for newcomers and students as much as for members of a "clique," were unusual in 1960. It took work on the part of Chew, Helmholz, Weisskopf, and the others to break the mold and keep their meeting "non-exclusive." The "open" meeting attracted about three hundred physicists.[52]

The theoretical portion of this special conference focused almost exclusively on recent work by Chew, Stanley Mandelstam, and Richard Cutkosky on a new framework for approaching particle physics. Just six months before Chew's more outspoken break with quantum field theory at the La Jolla meeting, the material discussed at the "open" Berkeley meeting helped to form the core of Chew's emerging *S*-matrix program. As Chew came to articulate more and more explicitly, the *S*-matrix approach relied on several general principles but eschewed much of the specific formalism of quantum field theory.[53]

The S Matrix and a Democracy of Practitioners

The independence of the *S*-matrix program from many of the esoteric niceties of field theory was given a doubly "democratic" spin by Chew as he championed the new approach and quickly spread its gospel far and wide. First, Chew began to argue for his concept of "nuclear democracy"—that all nuclear particles, subject to the strong nuclear force, should be treated equally, without dividing them into "elementary" and "composite" factions. As we have seen in Section I, Chew argued for this democratic treatment, and his related notion of the bootstrap, largely on the basis of his unusual interpretation of Feynman diagrams. In the rotated diagrams of Figure 3, for example, Chew and his students saw

---

[52] Victor Weisskopf to J. Howard McMillen, 14 Mar. 1960, Birge Papers, Box 29, Folder "Weisskopf, Victor Frederick, 1908–." As it turned out, the NSF refused any financial aid because the Berkeley meeting was under the auspices of the APS; additional funding for the meeting was provided by the AEC and United States Air Force. See the typed report "Conference on Strong Interactions," undated, Dept. Physics, Berkeley, Records, ca. 1920–1962, Folder 1:39; an unsigned, undated postconference report in this folder gives the attendance figure. The Rochester conferences on nuclear and particle physics began in 1950 at the University of Rochester; when they began to move to different venues during the mid 1950s, they still retained the name "Rochester conference." See, e.g., John Polkinghorne, *Rochester Roundabout: The Story of High-Energy Physics* (New York: Freeman, 1989). Like Chew, Weisskopf became quite active with the FAS during the 1950s; while Chew chaired the Passport Committee, Weisskopf headed the Visa Committee. See Victor Weisskopf, "Report on the Visa Situation," *Bull. Atom. Sci.,* Oct. 1952, *8:*221–222; and Weisskopf, "Visas for Foreign Scientists," *ibid.,* Mar. 1954, *10:*68–69, 112.

[53] Schedules and reports on the presentations at the 1960 Berkeley meeting may be found in Dept. Physics, Berkeley, Records, ca. 1920–1962, Folder 1:39. The general principles on which Chew and his collaborators hoped to build their non-field-theoretic *S*-matrix theory included analyticity, unitarity, Lorentz invariance, and crossing symmetry, not all of which are independent from each other. See esp. Chew, *S-Matrix Theory of Strong Interactions* (cit. n. 3); and Cushing, *Theory Construction,* Chs. 5–7.

the ρ meson move from force-carrier to bound-state composite to seemingly elementary particle. Arguing from the structure of these Feynman diagrams, Chew taught his many students to treat all nuclear particles the same way. But his "democratic" sentiment did not end with the new interpretation of Feynman diagrams; Chew fashioned his S-matrix work as "democratic" in a second sense as well. In addition to "democratic" diagrams, Chew championed a "nuclear democracy" among practitioners. Consider his remarks near the close of an ebullient invited lecture at the 1962 New York meeting of the American Physical Society: "I am convinced that a wild period of merrymaking lies before us. All the physicists who never learned field theory can get in the game, and experimenters are just as likely to come up with important ideas as are theorists. They may even have an advantage over us." Chew returned to this theme of nonexperts' advantage over field-theory experts in the S-matrix realm in a lecture at Cambridge University in 1963, reporting that "the less experienced physicists [have] an advantage in working with a new framework. (The inverse correlation of productivity with experience in a situation like this is remarkable.)" Meanwhile, Chew made good on his pledge to bring outsiders into the fold, delivering special lectures and seminars on the new material especially for experimentalists at Berkeley. (See Figure 5.) The 1963 S-matrix textbook by Chew's colleagues Roland Omnès and Marcel Froissart, *Mandelstam Theory and Regge Poles*, similarly carried the subtitle "An Introduction for Experimentalists."[54]

As Chew traveled around the country and beyond, feverishly working his campaign of "wild merrymaking," few could miss his obvious charisma and enthusiasm. As John Polkinghorne later recalled, "We used to call Geoff Chew 'the handsomest man in high energy physics.' I know of at least one senior secretary in a British physics department who kept a photograph of him near her desk. That frank and open face, with just a hint of his one-eighth Burmese ancestry, and his tall commanding figure, made him one of the few theorists in the pin-up class." Rumors spread far and wide that Chew had given up a potential career in professional baseball to work in particle physics. His personal charm and enthusiasm made Chew into an effective "salesman." Polkinghorne attests that "Geoff was definitely a man from whom one would be happy to buy a used car." His talks "were always eagerly awaited," given "their inspirational and encouraging tone."[55]

[54] Geoffrey Chew, "S-Matrix Theory of Strong Interactions without Elementary Particles," *Reviews of Modern Physics*, 1962, *34*:394–401, on p. 400 ("merrymaking"); Chew, "The Dubious Role of the Space-Time Continuum in Microscopic Physics," *Science Progress*, 1963, *51*:529–539, on p. 538 (nonexperts' advantage) (this article contains the text of Chew's 1963 Rouse Ball Lecture at Cambridge); and Roland Omnès and Marcel Froissart, *Mandelstam Theory and Regge Poles: An Introduction for Experimentalists* (New York: Benjamin, 1963) (Froissart spent time working with Chew's group in Berkeley during the early 1960s). Owen Chamberlain in particular recalled Chew's special lectures for experimentalists; see Owen Chamberlain, "Interactions with Geoff Chew," in *Passion for Physics*, ed. DeTar *et al.*, pp. 11–13, on p. 13. Chew's unusual ability and interest in instructing experimentalists was by this time long standing. As a visiting professor at Berkeley in 1957, Chew gave a seminar on the pion-nucleon interaction, which drew an unusual number of experimentalists from the Radiation Laboratory and from Livermore, in addition to the graduate students enrolled in the course. Carl Helmholz reported that the experimentalists "are getting considerable benefit from it even though the subject is quite abstract and mathematical": Helmholz to Constance, 25 Mar. 1957 (cit. n. 47). William Frazer also discussed Chew's informal seminar for experimentalists during our interview.

[55] John Polkinghorne, "Salesman of Ideas," in *Passion for Physics*, ed. DeTar *et al.*, pp. 23–25, on p. 23. Polkinghorne worked on S-matrix theory while based in Cambridge, England. For a similar analysis of the role of charisma in modern physics see Charles Thorpe and Steven Shapin, "Who Was J. Robert Oppenheimer? Charisma and Complex Organization," *Social Studies of Science*, 2000, *30*:545–590. Owen Chamberlain and Georgella Perry (Chew's former secretary) both mentioned Chew's baseball pretensions: Chamberlain, "Interactions with Geoff Chew," pp. 12–13; and Perry, "My Years with Professor Chew" (cit. n. 48), pp. 14–16. Several other physicists recalled these same rumors during interviews with Stephen Gordon: Gordon, "Strong Interactions," p. 15.

Figure 5. *Geoffrey Chew delivering an informal lecture at Berkeley, 1961. Reprinted with kind permission of the Lawrence Berkeley National Laboratory.*

His enthusiasm quickly suffused Berkeley's department and encouraged his graduate students and postdocs to participate as equals in the S-matrix campaign. Louis Balázs, who completed his Ph.D. under Chew's direction in the mid 1960s, reminisced recently that "it was an exciting experience being one of Chew's graduate students at UC-Berkeley in the early 1960s. New ideas were being discussed and developed continually and vigorously, particularly by the postdocs, and it seemed we were on the threshold of a new era in Physics." Those who were pushing the boundaries of the "new era" seemed to be drawn to Berkeley, if they weren't there already: "There was a constant stream of distinguished visitors," Balázs continued, "who seemed to be eager to learn about the new developments in Berkeley." And just as Chew had announced so exuberantly at the 1962 meeting, Balázs too recalled that "even graduate students found that they could make new independent contributions at that time." Another former graduate student, William Frazer, concurs, emphasizing that Chew worked hard to make sure that no students felt intimidated. "He really made a wonderful atmosphere for us to work in."[56] Both students who were still "innocent" of the elaborate quantum field theory formalism, as Chew put it, and experimentalists who had "never learned field theory" stood to contribute as equals in Chew's S-matrix program.

---

[56] Louis Balázs to David Kaiser, 6 Aug. 1998; and Frazer interview. See also Perry, "My Years with Professor Chew," pp. 15–16.

Chew reflected in several places on the best way to train these potential *S*-matrix contributors. As early as 1961, he noted that students who had not learned quantum field
theory, and the usual ways of using and interpreting Feynman diagrams, seemed to fare
best when approaching the new *S*-matrix material. He assured readers of his 1961 lecture
note volume that "it is . . . unnecessary to be conversant with the subleties of field theory,
and a certain innocence in this respect is perhaps even desirable. Experts in field theory
seem to find current trends in *S*-matrix research more baffling than do nonexperts." The
same sentiment appeared five years later, when Chew remarked in the preface to his 1966
textbook: "No background in quantum field theory is required. Indeed, as pointed out in
the preface to my 1961 lecture notes, lengthy experience with Lagrangian field theory
appears to constitute a disadvantage when attempting to learn *S*-matrix theory."[57]

Chew's own students largely followed this prescription. William Frazer, the first student
to complete his dissertation under Chew after his return to Berkeley, worked through
Eyvind Wichmann's "very rigorous field theory course" as a graduate student, though he
found his work with Chew to be much more interesting. "For the student," Frazer recalled
recently, "life was a bit schizophrenic. Either you read on your own the very dry mathematical structure of axiomatic field theory, or you tried to follow the more exciting material." Later students, such as Jerome Finkelstein, recalled that "there were quite a few of
us in Chew's group who really did not spend a lot of time on field theory. I'd taken a
course in it during my first or second year of graduate school, but that was all." Ramamurti
Shankar, another of Chew's students from this time, recently put a humorous turn on this
pedagogical approach:

> I had a choice: either struggle and learn field theory and run the risk of blurting out some four
> letter word like φ^4 in Geoff's presence or simply eliminate all risk by avoiding the subject
> altogether. Being a great believer in the principle of least action, I chose the latter route. Like
> Major Major's father in *Catch 22*, I woke up at the crack of noon and spent eight and even
> twelve hours a day not learning field theory and soon I had not learnt more field theory than
> anyone else in Geoff's group and was quickly moving to the top.

David Gross, who completed his dissertation under Chew's direction in 1966, paraphrased
an often-heard refrain from colleagues who didn't realize that Gross, since the 1970s a
preeminent field theorist, had come from Chew's group: "Funny—you don't look Chewish." One would search in vain to find much usage of specifically field-theoretic techniques
in the sixty dissertations by Chew's graduate students.[58]

So much for Chew's immediate circle in Berkeley. With several well-known texts on
quantum field theory already in print, the next question was how to spread the physics of
"nuclear democracy" beyond Chew's large but geographically limited Berkeley group. Just
as politicians debated a purported "missile gap" with the Soviets, Chew and his collaborators in the early 1960s faced a "textbook gap": given that so many physicists conceivably
could participate in developing the "democratic" *S*-matrix program, the challenge was to
reach these students, experimentalists, and other theorists and deliver the *S*-matrix message.

[57] Chew, *S-Matrix Theory of Strong Interactions* (cit. n. 3), pp. vii–viii; and Geoffrey Chew, *The Analytic S
Matrix: A Basis for Nuclear Democracy* (New York: Benjamin, 1966), p. v.
[58] Frazer interview; Finkelstein interview; Ramamurti Shankar, "Effective Field Theory in Condensed Matter
Physics," in *Conceptual Foundations of Quantum Field Theory,* ed. Cao (cit. n. 4), pp. 47–55, on p. 47; and
Gross, "Uniqueness of Physical Theories" (cit. n. 48), p. 128. Dissertations by Chew's students from 1959 to
1983 may be found in the Berkeley Physics Department Library.

Toward this goal, Chew and many of his *S*-matrix students and collaborators delivered many sets of summer-school lectures—sometimes, like Chew in 1960, giving the same set of lectures at two different schools in the same summer.[59]

Chew and his postdocs also began to publish their lecture notes in inexpensive editions, nearly as soon as the lectures had been delivered. These books themselves reveal much about the quest to attract students to the *S*-matrix team. Most of the important *S*-matrix textbooks were part of the "Frontiers in Physics" series, which began to publish collections of lecture notes and reprints in 1961. Chew's 1961 *S-Matrix Theory of Strong Interactions*, based on his 1960 summer-school lectures, was one of the first books to be included in the new series. By 1964 *S*-matrix tracts constituted nearly one-third of the "Frontiers in Physics" books, even though the series was meant to treat all aspects of physics and not only particle theory.[60]

These books were rushed into print. *Regge Poles and S-Matrix Theory*, by Chew's postdoc Steven Frautschi, stemmed from lectures Frautschi had given once at Cornell University in 1961/1962 and then augmented for delivery at a June 1962 summer school. The lectures presented by Maurice Jacob and Chew in their 1964 *Strong-Interaction Physics* also had been delivered only once, during the academic year 1962/1963. As the series editor David Pines explained, "Frontiers in Physics" was intended to feature just this kind of "rough and informal" lecture notes rather than polished monographs. The very production of the books reflected their mission, as Pines explained: "Photo-offset printing is used throughout, and the books are paperbound, in order to speed publication and reduce costs. It is hoped that the books will thereby be within the financial reach of graduate students in this country and abroad."[61]

The progress of *S*-matrix theorists toward establishing an axiomatic foundation for their new work can also be read immediately from the material form of their books. Chew's early reports on the developing theory in 1961 and 1964 were printed in books that had not been carefully typeset but were printed on inexpensive paper and hurried into print. Chew's first textbook to treat the newly proposed axiomatic version of *S*-matrix theory, *The Analytic S Matrix* (1966), on the other hand, was not published as part of the "Frontiers in Physics" series. Its professional typesetting and glossy pages present a clear contrast with the earlier volumes.[62]

From summer-school lectures to "Frontiers in Physics" volumes, then, the task of cre-

---

[59] Chew's "Double Dispersion Relations and Unitarity as the Basis of a Dynamical Theory of Strong Interactions" (cit. n. 17) appeared in both *Relations de dispersion et particules élémentaires*, ed. de Witt and Omnès (cit. n. 14), pp. 455–514, and in *Dispersion Relations*, ed. Screaton (cit. n. 14), pp. 167–258. See also the "Editor's Note" *ibid.*, p. viii. One might compare Chew and his "missionaries" with Niels Bohr and the spread of the "Copenhagen spirit" as studied in John Heilbron, "The Earliest Missionaries of the Copenhagen Spirit," *Revue d'Histoire des Sciences*, 1985, *30:*195–230.

[60] Lists of the titles published in the "Frontiers of Physics" series were included in the front of each of the books within the series. The *S*-matrix books in the series published between 1961 and 1964 included Chew, *S-Matrix Theory of Strong Interactions* (1961); Omnès and Froissart, *Mandelstam Theory and Regge Poles* (1963); Frautschi, *Regge Poles and S-Matrix Theory* (1963); E. J. Squires, *Complex Angular Momenta and Particle Physics* (1963); and Jacob and Chew, *Strong-Interaction Physics* (1964). During this same period, only two books that focused on quantum field theory for particle physics were included in the series.

[61] See the "Editor's Foreword" by David Pines, dated Aug. 1961, which appears within each of the volumes in the series. Frautschi's book, for example, was reproduced from hand-typed originals. Chew mentioned his friendship with Pines, also a physicist at the University of Illinois at Urbana, as one of the reasons he decided to publish his *S*-matrix lecture notes in this series: Chew to Kaiser, 11 May 1998 (email).

[62] This reading of the material production of Chew's textbooks is inspired by the example of David Cressy, "Books as Totems in Seventeenth-Century England and New England," *Journal of Library History,* 1986, *21:*92–106.

ating and maintaining a community of *S*-matrix theorists involved far more than publishing research articles in the *Physical Review*. Chew was well aware that the dissociation of Feynman diagrams from their circumscribed meaning within quantum field theory would be difficult for students who had already mastered quantum field theory to grasp. His autonomous *S*-matrix program promised opportunities for those students, experimentalists, and other theorists who had not become overly attached to quantum field theory. Inexpensive pedagogical resources, such as lecture notes and reprint collections, could thus serve to broaden the base of "democratic" *S*-matrix practitioners. There is some evidence that this aggressive, democratizing textbook campaign worked: one physicist explained years later that, although he had not been a direct student of Chew's, he had learned particular calculational details of the *S*-matrix program, together with some sense of its special "philosophy," from reading Chew's 1961 *S-Matrix Theory of Strong Interactions*.[63]

The Language of Democracy for a Program in Decline

In addition to emphasizing his democratic team of contributors, Chew drew an increasing contrast during the 1960s and into the early 1970s between the kind of work his *S*-matrix program fostered as compared with that of the field theorists. The field theorists sought to explain the phenomena of particle physics with reference to a small set of basic or unit interactions taking place between a core set of "fundamental" or "elementary" particles. Chew mocked this approach of the "fundamentalists," arguing that such an "aristocratic" arrangement of fundamental particles could not provide an adequate framework for describing the strong interactions.[64] As in his 1950 article describing the loyalty oath controversy at the University of California, Chew reserved the term "fundamentalist" for colleagues who espoused a position at odds with his own "democratic" ideals.

By the late 1960s Chew's program appeared to many physicists to have lost its original focus, becoming mired in details and complexity. Treating the ρ meson within a democratic bootstrap framework was one thing, but moving beyond such a simple system to more complicated calculations had become much more frustrating. Quantum field theory, meanwhile, now augmented with a new emphasis on gauge symmetries and the quark hypothesis, was again attracting the attention of most particle physicists. Chew lamented this turn of events, arguing that it sprang more from physicists' "psychology" than from stubborn experimental data. The trouble, he wrote in the early 1970s, was that the "fundamentalists" dreamed "of the press conference that will announce to the world a dramatic resolution of their quest." Unlike the night thoughts of these fundamentalists, the autonomous *S* matrix was "the cumulative result of many steps stretching out over decades."[65]

Progress for the *S*-matrix camp, Chew explained, was necessarily a gradual game of

---

[63] Al Mueller, "Renormalons and Phenomenology in QCD," in *Passion for Physics*, ed. DeTar *et al.*, pp. 137–142, on p. 137.

[64] Chew used the language of "fundamentalists" throughout his articles "Hadron Bootstrap: Triumph or Frustration?" (cit. n. 18) and "Impasse for the Elementary-Particle Concept" (cit. n. 18). See also Chew, "'Bootstrap': A Scientific Idea?" (cit. n. 18).

[65] Chew, "Hadron Bootstrap," p. 24; and Chew, "Impasse for the Elementary-Particle Concept," p. 124. On the resurgence of quantum field theory see Cushing, *Theory Construction*, Chs. 6–7; Gordon, "Strong Interactions," Chs. 4–5; Andrew Pickering, *Constructing Quarks: A Sociological History of Particle Physics* (Chicago: Univ. Chicago Press, 1984), Chs. 4–8; Abraham Pais, *Inward Bound: Of Matter and Forces in the Physical World* (New York: Oxford Univ. Press, 1986), Ch. 21; Tian Yu Cao, *Conceptual Developments of Twentieth-Century Field Theories* (New York: Cambridge Univ. Press, 1997), Chs. 8–11; and Lillian Hoddeson, Laurie Brown, Michael Riordan, and Max Dresden, eds., *The Rise of the Standard Model: Particle Physics in the 1960s and 1970s* (New York: Cambridge Univ. Press, 1997).

constructing more and more partial models, incorporating the effects of more and more particle exchanges and interactions, all under the rubric of several reigning general principles (such as causality). Chew's program was therefore built on the assumption, he wrote in 1974, "that there will gradually develop a more and more dense coverage of the nuclear world by interlocking models no single model having preeminent status. Such a pattern might be characterized as a 'democracy of models.' " Chew reminded physicists in 1970 that "even though no press conference was called," the stepwise construction of these interlocking models within the *S*-matrix program had already scored several "breakthroughs," a fact he attributed to the "brilliant collective achievement of the high-energy physics community." Reinforcing this point with a history lesson, Chew recalled the "precedent of classical nuclear physics": "This model enjoyed an aristocratic status for almost thirty years, but eventually it was democratized."[66]

This notion of constructing interlocking models through a series of "collective achievements" fit well with Chew's "secret seminar" approach to training graduate students. His students' dissertations reveal this close-knit, mutually buttressing approach: Bipin Desai, completing his dissertation in April 1961, built directly on work by James Ball in his dissertation, filed almost exactly one year earlier; Yongduk Kim's dissertation from June 1961 in turn drew explicitly on the work by Desai and Ball. This pattern continued throughout the 1960s. The abstract from Shu-yuan Chu's May 1966 dissertation, for example, explained that "the method [employed in the dissertation] is an extension of [C. Edward] Jones' proof in the single-channel case, making use of an explicit expression of the determinant D constructed by [David] Gross." Jones completed his dissertation in 1964 with Chew, and Gross submitted his dissertation a few months after Chu.[67] Citations to work, both published and unpublished, by other graduate-student members of Chew's group, and acknowledgments of extended discussions with fellow students, fill nearly every one of Chew's students' dissertations. Chew's pedagogical ideal of equal participation melded seamlessly with his program for the piecewise construction of "interlocking models" by a "collective" of researchers.

By the 1970s, with his program all but abandoned by most physicists, Chew bemoaned the failure of his collective vision to attract those physicists who insisted instead on searching for a single "glamorous-sounding fundamental entity": "Few of the stars of the world of physics are content with the thought that their labors will constitute only a *fraction* of a *vast mosaic* that must be constructed before the complete picture becomes recognizable and understandable." These "stars" of physics, now nearly all within the "fundamentalist" camp, routinely embraced only models based on Lagrangian field theories, Chew noted. They accorded these models "special status," failing to "consider on an equal footing" models not derived from such a field-theoretical basis. The task of the collective-minded *S*-matrix theorist, on the other hand, remained "to view any number of different partially successful models without favoritism."[68]

---

[66] Chew, "Impasse for the Elementary-Particle Concept," p. 124; and Chew, "Hadron Bootstrap," p. 25.

[67] James S. Ball, "The Application of the Mandelstam Representation to Photoproduction of Pions from Nucleons" (Ph.D. diss., Univ. California, Berkeley, 1960); Bipin Desai, "Low-Energy Pion-Photon Interaction: The $(2\pi, 2\gamma)$ Vertex" (Ph.D. diss., Univ. California, Berkeley, 1961); Yongduk Kim, "Production of Pion Pairs by a Photon in the Coulomb Field of a Nucleus" (Ph.D. diss., Univ. California, Berkeley, 1961); and Shu-yuan Chu, "A Study of Multi-Channel Dynamics in the New Strip Approximation" (Ph.D. diss., Univ. California, Berkeley, 1966). Ball's, Desai's, and Kim's dissertations drew heavily on the work by William Frazer and José Fulco, in particular "Effect of a Pion-Pion Scattering Resonance on Nucleon Structure" (cit. n. 15) and "Partial-Wave Dispersion Relations for the Process $\pi + \pi \rightarrow N + \underline{N}$" (cit. n. 15).

[68] Chew, "Impasse for the Elementary-Particle Co̲ncept" (cit. n. 18), p. 124 (emphasis added); and Chew, "Hadron Bootstrap" (cit. n. 18), p. 27.

The late 1960s and early 1970s were a difficult time to be a young physicist in the United States, over and above Chew's frustrations with the fate of "nuclear democracy." With the onset of détente and dramatic cuts in defense spending, U.S. physicists rapidly slid into the worst job shortage the profession had ever witnessed. Enrollments in the placement service registries of the American Physical Society tell the grim tale: in 1968, nearly 1,000 applicants fought for 253 jobs; the next year, almost 1,300 competed for 234 jobs. Only 63 jobs were on offer at a 1970 American Physical Society meeting at which 1,010 young physicists were looking for work; 1,053 competed for 53 jobs at the 1971 meeting. Anecdotal evidence suggests that students who were steeped too heavily in *S*-matrix methods, to the exclusion of field-theoretic techniques, felt the crunch disproportionately. Although many of Chew's students from this later period have gone on to academic careers as theoretical physicists, several *S*-matrix students left the field after earning their Ph.D.'s to become medical doctors or lawyers; others were denied tenure at places like MIT. Although it is difficult to disentangle the root causes of these few theorists' difficulties from the overwhelming across-the-board cutbacks, several physicists to this day continue to associate *S*-matrix training with job-placement difficulties during the late 1960s and early 1970s.[69]

In frustration as in triumph, Chew spoke of his program in distinctly "democratic" terms. In the early 1960s it had seemed open to all: everyone could participate equally, and no one was singled out for special privileges. Later, when the program fell into neglect, the language of Chew's complaints was likewise laden with the tropes of democratic participation. Field theorists improperly elevated an "aristocracy" of particles and granted "special status" only to certain kinds of models. *S*-matrix theorists, on the other hand, strove for an equality of particles, models, and practitioners, all judged "without favoritism" as members of a collective. "Patience" should be the order of the day, Chew wrote, not the yearning for press conferences and the special privileges (such as increased government funding) such singular attention could foster.[70]

In between his many speeches and lectures, Chew built an approach to training students and colleagues that emphasized equal participation and the collective strivings of the group over the "cliquish" dreams of the "fundamentalists." In establishing his "secret seminar," designing new lecture note volumes and textbooks, and delivering special lectures and seminars for experimentalists, Chew sought to make theoretical particle physics a particularly democratic activity. Just as he lobbied for fair treatment of academics and scientists under a controlling state in the Cold War, Chew tried to produce within physics a community of peers, neither singled out for special treatment nor splintered between idea-producing theorists and fact-checking experimentalists. Each contributor to the "vast mosaic" of *S*-matrix theory was to be an equal partner under the law.

IV. THE VIEW FROM PRINCETON

Traces of Chew's outspoken stance on the failure of quantum field theory and on the need to treat all nuclear particles democratically may be found throughout his students' dissertations. Peter Cziffra, writing a year before Chew's famous La Jolla talk, echoed his ad-

---

[69] Several physicists drew these connections during their interviews with Stephen Gordon: Gordon, "Strong Interactions," pp. 50, 53–54. The statistics come from Kaiser, "Putting the 'Big' in 'Big Science'" (cit. n. 28), pp. 33–35.

[70] Chew, "Impasse for the Elementary-Particle Concept" (cit. n. 18), p. 125.

visor's attitude when he opened his dissertation by noting how "stymied" ordinary per-turbative quantum field theory remained when treating the strong interactions. Well into the campaign for nuclear democracy, Akbar Ahmadzadeh reminded readers of his dissertation that insisting with the field theorists on a strict division between "elementary" and "composite" particles often leads to "absurd conclusions" when studying the strong interactions; instead, all particles should be treated as bound states, "on an equal footing." Henry Stapp, a research associate at the Rad Lab when Chew returned there in the late 1950s, pursued an axiomatic foundation for S-matrix theory in the early 1960s. "Early on," Stapp has recalled recently, "I was not really in close touch with Chew; I picked up the S-matrix ideas by osmosis, since Chew's ideas were permeating the area."[71]

Still, despite the group's successes in spreading the word via summer-school lectures and "Frontiers in Physics" volumes, their ideas did not "permeate" all departments of physics in the same way. Princeton's department, in particular, provides a telling contrast with Chew's Berkeley. By the early 1960s Princeton boasted a large and active group of theorists working on many aspects of particle physics. In fact, the Princeton group, like Chew's group in Berkeley, championed and extended many of the nonperturbative, diagram-based tools with which Chew was tinkering. Though a large group of theorists at Princeton pursued topics that fell under the S-matrix rubric, they did not share Chew's zeal for a "nuclear democracy." Comparing the work by these Princeton theorists with that of Chew's group may help, therefore, to highlight which elements of Chew's many-faceted "democracy" remained particular to his Berkeley group.

One of the leaders of Princeton's group had shared many early stops with Chew along a common trajectory. Marvin "Murph" Goldberger had been a graduate student with Chew in Chicago immediately after the war. The two became fast friends, sharing office space, arranging social outings together, completing their dissertations at the same time, and moving together to Berkeley's Rad Lab as postdocs in 1948. Following his postdoc Goldberger took a job back at the University of Chicago, while Chew taught for one year at Berkeley and then resigned over the loyalty oath. Once Chew landed in Urbana, he and Goldberger were again in close proximity; they struck up an active collaboration during the mid 1950s, together with Francis Low and Murray Gell-Mann, also both in the Midwest at that time. Just when Chew left Urbana to return to Berkeley, Goldberger left Chicago to take a position at Princeton, starting in February 1957. Though now separated by the length of the continent, Goldberger and Chew continued to correspond.[72]

[71] Peter Cziffra, "The Two-Pion Exchange Contribution to the Higher Partial Waves of Nucleon-Nucleon Scattering" (Ph.D. diss., Univ. California, Berkeley, 1960), p. 4; Akbar Ahmadzadeh, "A Numerical Study of the Regge Parameters in Potential Scattering" (Ph.D. diss., Univ. California, Berkeley, 1963), pp. 2–3; and Stapp interview. See Henry Stapp, "Derivation of the CPT Theorem and the Connection between Spin and Statistics from Postulates of the S-Matrix Theory," *Phys. Rev.,* 1962, *125:*2139–2162; Stapp, "Axiomatic S-Matrix Theory," *Rev. Mod. Phys.,* 1962, *34:*390–394; Stapp, "Analytic S-Matrix Theory," in *High-Energy Physics and Elementary Particles,* ed. Abdus Salam (Vienna: International Atomic Energy Agency, 1965), pp. 3–54; and Stapp, "Space and Time in S-Matrix Theory," *Phys. Rev.,* 1965, *139:*B257–270. Stanley Mandelstam, a close collaborator of Chew's and an architect of many of the S-matrix techniques, resisted following Chew and his group in renouncing field theory. As early as the Dec. 1960 Berkeley conference, a conference report noted that Chew's presentation based on Mandelstam's work "did not evoke Mandelstam's full assent": "Conference on Strong Interactions," p. 10, Dept. Physics, Berkeley, Records, ca. 1920–1962, Folder 1:39. See also Cushing, *Theory Construction,* pp. 131–132, 145.

[72] Annual Report 1956–1957, pp. 1–2, in Department of Physics, Princeton University, Annual Reports to the [University] President, Seeley G. Mudd Manuscript Library, Princeton Univ., Princeton, N.J. (hereafter cited as **Dept. Physics, Princeton, Annual Reports**); Marvin Goldberger, "Fifteen Years in the Life of Dispersion Relations," in *Subnuclear Phenomena,* ed. A. Zichichi (New York: Academic, 1970), pp. 685–693; Goldberger, "Francis E. Low—A Sixtieth Birthday Tribute," in *Asymptotic Realms of Physics,* ed. Alan Guth, Kerson Huang,

At Princeton Goldberger joined Sam Treiman, who had earned his Ph.D. from Chicago four years after Goldberger and Chew; a year and a half later, Richard Blankenbecler joined the group as a postdoc and later became a regular faculty member. Goldberger, Treiman, Blankenbecler, and their many graduate students spent much of their time during the late 1950s and 1960s on topics that Chew would have called *S*-matrix theory—just the sort of "interlocking models" that he hoped would bring clarity to the strong interaction. Goldberger and Treiman investigated the decay of unstable particles without resorting to field-theoretic Lagrangians; students completed dissertations on the analytic structure of scattering amplitudes and on how to incorporate unitarity, some of the key general principles from which Chew aimed to construct his autonomous *S*-matrix theory. On the surface, these projects all sound as if they could have been completed by Chew's students in Berkeley. Yet when Goldberger reported in 1961 on the Princeton group's many accomplishments, he categorized all of their work, with his characteristic sense of humor, as "the engineering applications of quantum field theory." "This work," Goldberger continued, "is complementary to the purer aspects of quantum field theory." Just at the time when Chew was announcing his clear and decisive break with quantum field theory in La Jolla, the Princeton group celebrated the close fit between their research and Chew's nemesis.[73]

The Princeton group's research was no closer to "engineering" than any of Chew's work; it only seemed more "applied" when compared with the work streaming out from Princeton's other group of theoretical particle physicists, headed by Arthur Wightman. Wightman championed an axiomatic approach to quantum field theory; his research remained at that time, unlike Chew's or Goldberger's, far removed from the details of recent experiments. No doubt with reference to Chew's standing-room-only invited lectures before the American Physical Society, the Princeton group reported in 1966 that "there are presently two approaches to relativistic quantum theory. These are axiomatic field theory and dispersion or S-matrix theory. Notwithstanding some passionate claims, there is not yet any evidence that the two are really different. . . . Both theoretical approaches have been and will continue to be pursued actively at Princeton."[74] Together with Wightman, then, Goldberger and Treiman erected a "big-tent" approach to quantum field theory: "engineering applications" based on *S*-matrix calculational techniques would coexist peacefully with the more "pure" investigations into the structure of quantum field theory. At Berkeley, meanwhile, there were no longer any senior field theorists in town to respond to Chew's challenge; the loyalty oath controversy had ensured that, when figures such as Gian Carlo Wick were fired as nonsigners. Years later, Chew mused on how he might have been more "intimi-

---

and Robert Jaffe (Cambridge, Mass.: MIT Press, 1983), pp. xi–xv; and Goldberger, "A Passion for Physics," in *Passion for Physics,* ed. DeTar *et al.,* pp. 241–245. Andrew Pickering has examined the active collaboration of Chew, Low, Goldberger, and Gell-Mann, emphasizing the importance of their geographical proximity: Andrew Pickering, "From Field to Phenomenology: The History of Dispersion Relations," in *Pions to Quarks,* ed. Brown *et al.* (cit. n. 13), pp. 579–599.

[73] M. L. Goldberger and S. B. Treiman, "Decay of the Pi Meson," *Phys. Rev.,* 1958, *110*:1178–1184; on the fit between the Princeton work and Chew's see Goldberger, "An Outline of Some Accomplishments in Theoretical Physics," Annual Report 1960–1961, pp. 12–13, Dept. Physics, Princeton, Annual Reports. On other work in the department see Annual Report 1958–1959, pp. 2, 7; Annual Report 1960–1961, pp. 12–13; Annual Report 1962–1963, p. 4; Annual Report 1963–1964, p. 66; Annual Report 1964–1965, pp. 79–81; Annual Report 1965–1966, pp. 4–5; and Annual Report 1967–1968, pp. 23–24: Dept. Physics, Princeton, Annual Reports. See also Treiman, "A Connection between the Strong and Weak Interactions," in *Pions to Quarks,* ed. Brown *et al.,* pp. 384–389; and Treiman, "A Life in Particle Physics," *Ann. Rev. Nuclear Particle Sci.,* 1996, *46*:1–30.

[74] Arthur Wightman, "The General Theory of Quantized Fields in the 1950s," in *Pions to Quarks,* ed. Brown *et al.,* pp. 608–629; and Annual Report 1965–1966, pp. 4–5, Dept. Physics, Princeton, Annual Reports (quotation).

dated," and less likely to dismiss quantum field theory outright, had Wick still been in Berkeley. A similar restraint might have been exercised by Francis Low, with whom Chew worked in Urbana in the mid 1950s. Although Low credits Chew with having provided most of the original ideas during their fruitful collaboration, it is likely that if they had remained together in Urbana, Chew's later flamboyant pronouncements about the death of quantum field theory would have been muted by Low's impressive and authoritative grasp of field theory's tenets.[75]

In keeping with the double-barreled approach to field theory at Princeton, and in clear contrast to Chew's group in Berkeley, Goldberger's and Treiman's graduate students actively studied quantum field theory as an essential part of their training. Stephen Adler, who completed his Ph.D. under Treiman's direction in 1964, remembers auditing Wightman's course, since Wightman's work "was seen as undergirding" the calculational techniques of dispersion relations and S-matrix theory. "We were doing an evasive physics," Adler continued: since no one knew how best to treat the strong interactions, "we used whatever methods we could. . . . Dispersion relations were a tool, but we also learned field theory methods because they were useful for treating symmetries." Other Princeton students from this period similarly recall an emphasis on learning quantum field theory.[76] This peaceful coexistence of S-matrix-style calculations with axiomatic quantum field theory also led to a different appraisal of Chew's S-matrix program than the one "permeating" the Berkeley area.

Adler recalls that when Chew came to Princeton to give a talk, "he sounded very messianic." Adler was hardly alone in comparing Chew's active campaigning to religious indoctrination. In fact, many of Chew's colleagues and former collaborators, now working at a distance from him, began to characterize his vigorous pronouncements as religious proselytizing. Goldberger wrote to Murray Gell-Mann in January 1962 that Chew had become "the Billy Graham of physics." He added playfully: "After his talk I nearly declared for Christ. Since I had already changed from Jewish to Regge-ish, it was the only thing I could think of." Years later, John Polkinghorne recalled that Chew proclaimed his ideas "with a fervour" like that "of the impassioned evangelist. There seemed to be a moral edge to the endeavour. It was not so much that it was expedient to be on the mass-shell of the S matrix as that it would have been sinful to be anywhere else." Rather than a ringing democratic political campaign, Chew's efforts often struck his former colleagues as runaway zealotry.[77]

[75] Geoffrey Chew, interview with Gordon, Dec. 1997, quoted in Gordon, "Strong Interactions," p. 33 (see also pp. 32–35). Low credited Chew with the main originality in their collaboration in his interview with me; see also Kaiser, "Francis E. Low" (cit. n. 45), pp. 71–72.

[76] Stephen Adler, telephone interview with the author, 16 Feb. 1999. This emphasis on semiphenomenological tools rather than overarching theory construction became a hallmark of Treiman's group and helped shape Adler's later work on current algebras. See Stephen Adler and Roger Dashen, *Current Algebras and Applications to Particle Physics* (New York: Benjamin, 1968); and Sam Treiman, Roman Jackiw, and David Gross, *Lectures on Current Algebra and Its Applications* (Princeton, N.J.: Princeton Univ. Press, 1972). See also Pickering, *Constructing Quarks* (cit. n. 65), pp. 108–114; and Cao, *Conceptual Developments of Twentieth-Century Field Theories* (cit. n. 65), pp. 229–246. The views of other former Princeton students are expressed in John Bronzan to Kaiser, 15 May 1997 (email), and E. E. Bergmann to Kaiser, 16 May 1997 (email). The same attitude is drawn out in the course notes taken by Kip Thorne when he was a graduate student at Princeton in the early 1960s. See, in particular, Thorne's notes from "Properties of Elementary Particles," a course given by Val Fitch (Spring 1963); "Elementary Particle Physics," taught by Sam Treiman (Spring 1963); "Intermediate Quantum Mechanics and Applications," taught by Goldberger (Fall 1963); and "Elementary Particle Theory," taught by Blankenbecler (Spring 1964). All notes in the possession of Professor Kip Thorne; my thanks to him for sharing copies of these notes.

[77] Adler interview; Marvin Goldberger to Murray Gell-Mann, 27 Jan. 1962, quoted in Johnson, *Strange Beauty*

As Adler recalled, Chew announced in his Princeton lecture "a great hope; but at the same time, Treiman was always a bit skeptical of any grand theory." Indeed, Treiman was skeptical. Whereas Chew's January 1962 lecture before the American Physical Society predicted a "wild period of merrymaking," Treiman's own lecture on *S*-matrix material, delivered ten months later, focused instead on how even the most promising-looking "partial results and insights" remained "all tangled up with approximations which have inevitably to be introduced and which vary in style and severity from one application to another, one author to another." It is important to note that Treiman was no critic of approximations, even "severe" ones, per se. His 1958 work with Goldberger on pion decay relied, in another reviewer's words, on "drastic assumptions" and "feeble arguments." Even in Treiman's own estimation, these were "hair-raising approximations" that he and Goldberger were "quite unable—apart from hand waving—to justify."[78] It wasn't the recourse to approximations that irked Treiman about Chew's program; it was the vehemence with which Chew pitted his work against field theory.

Treiman went on to dismiss the bootstrap hypothesis, trumpeted by Chew as the logical conclusion of nuclear democracy, as "amusing." As we have seen in Section I, the goal of the bootstrap work was to find a single self-consistent solution that would show that strongly interacting particles might each produce the very forces that led to their own production by other particles; each strongly interacting particle, then, might be said to "pull itself up by its own bootstraps." What had captured the imagination of the Berkeley group left Treiman unimpressed. Throughout his 1962 lectures, for example, Treiman emphasized the "conjectural" basis of the bootstrap work and "indulg[ed]" in what he characterized as "pessimistic remarks" regarding the bootstrap program.[79] When it came to relations between *S*-matrix theory and quantum field theory, Treiman was not Chew—and Princeton was not Berkeley.

These differences led to some subtle reinterpretations of the meaning of *S*-matrix work, not only of its relative importance. Princeton's L. F. Cook, another faculty member in the Goldberger-Treiman-Blankenbecler group, reported on his research on Chew's beloved bootstrap mechanism. Despite Treiman's deflating judgment, the bootstrap remained to Chew, Frautschi, and most other members of the Berkeley group the sought-for culmination of nuclear democracy, a kind of holy grail for the equal treatment of all nuclear particles: it *meant* that there were no "elementary" particles; each was a bound-state composite of others. Yet for Cook, surrounded by Princeton's field theorists, elementarity remained central. In his 1965 gloss, "The bootstrap embodies a philosophy which supposedly enables one to calculate" various parameters for "*elementary* particles." Cook's work went on to emphasize the lack of agreement between this favorite topic of Chew's

---

(cit. n. 6), p. 211; and Polkinghorne, "Salesman of Ideas" (cit. n. 55), pp. 24–25. Francis Low similarly remarked that Chew's later efforts seemed "religious" in character: Low interview.

[78] Adler interview; Sam Treiman, "Analyticity in Particle Physics," in *Proceedings of the Eastern Theoretical Physics Conference, October 26–27, 1962,* ed. M. E. Rose (New York: Gordon & Breach, 1963), pp. 127–174, on p. 149; Jackson, "Introduction to Dispersion Relation Techniques" (cit. n. 14), p. 50; and Treiman, "Life in Particle Physics" (cit. n. 73), p. 16. As Treiman himself later remarked of this calculation, "No one but Goldberger and I would have had the effrontery to do what Goldberger and I did": Treiman, "Connection between the Strong and Weak Interactions" (cit. n. 73), p. 388.

[79] Treiman, "Analyticity in Particle Physics," pp. 163, 143. Blankenbecler and Goldberger similarly characterized Chew's bootstrap work as a collection of "interesting speculations" lacking a "physical basis." Making explicit reference to Chew's 1961 La Jolla talk, they dismissed the entire discussion as having merely a "religious nature." R. Blankenbecler and M. L. Goldberger, "Behavior of Scattering Amplitudes at High Energies, Bound States, and Resonances," *Phys. Rev.,* 1962, *126:*766–786, on p. 784. This article was based on Blankenbecler and Goldberger's own talk at the 1961 La Jolla meeting, as indicated in a footnote on p. 766.

and existing experimental data.[80] Even when they turned to the topics most central to Chew's "democratic" campaign, then, Princeton's field theorists clung to the language of "elementary" particles rather than following Chew's group in their vocal break with the "fundamentalists." No matter how completely Chew's democratic vision might have permeated the Berkeley area, that vision did not command a single, unchanging interpretation from physicists further and further removed from Berkeley.

V. CONCLUSIONS: CONDITIONS OF DEMOCRATIC POSSIBILITIES

Perhaps it bears emphasizing that we needn't agree that Chew's many activities were inherently democratic. He maintained Q clearance for several years after his wartime Los Alamos work; such clearance already implied unequal access to research and resources.[81] His "secret seminar" actively discouraged other faculty members from "participating equally." The 1960 Berkeley conference served as a stepping stone for his own S-matrix work, featuring primarily the research of his collaborators. Several friends and former collaborators saw not a democratically minded enrollment campaign but, rather, quasi-religious zealotry in Chew's efforts to interest others in his "democratic" S-matrix program. It is not clear what his thoughts or actions were during the 1964 Free Speech Movement. And so on. Nor is it clear whether Chew was working with a fully articulated or consistent political ideology. While Chew was growing up in Washington, D.C., his father had worked in the Department of Agriculture, which was hardly a politically neutral bureaucracy during the New Deal; perhaps Chew had an ingrained interest, based on his early years, in political issues. Yet whether or not we agree on how "democratic" Chew's efforts ultimately were, or on whether they stemmed from a clear and consistent ideology, one thing remains crucial: Chew and many of his students and colleagues saw his program for strong-interaction particle physics as specifically "democratic"—and as special for that reason.

How, then, are we to interpret the self-proclaimed "democratic" work of Geoffrey Chew? Hard on the heels of his most overtly political activities, Chew began to teach graduate students in a manner consistent with his ideas about democracy. Springboarding from his "open," "noncliquish" conference at Berkeley in 1960, Chew proclaimed at every opportunity that his new physics invited all kinds of participation, from "innocent" students to experimentalists who had "never learned" the rival field-theory formalism. In the midst of these activities, which took Chew from Senate subcommittee hearings to "secret seminars," he began to articulate a new and unprecedented vision of how particles behave and how their interactions should be studied: a successful theory of the strong interactions—unlike quantum field theory, with its "aristocratic" elements—must make no distinctions between the many types of particles. None should be singled out, either for special privileges or for special neglect; all must receive, diagrammatically and mathematically, "equal treatment under the law."

[80] L. F. Cook, in Annual Report 1964–1965, pp. 79–80, Dept. Physics, Princeton, Annual Reports (emphasis added). Interestingly, Cook conducted this research with C. Edward Jones, a new Princeton postdoc who had just completed his dissertation under Chew's direction in Berkeley. Each of the courses on particle theory that Kip Thorne attended as a graduate student at Princeton during this period was labeled "elementary particle physics" or "elementary particle theory," further reinforcing this nonbootstrap view of the field. A second center on which it would be interesting to focus, to extend the analysis of the heterogeneity among S-matrix groups, would be Cambridge, England, which hosted a group centered on R. J. Eden, P. J. Landshoff, D. Olive, and J. Polkinghorne; see their textbook, *The Analytic S-Matrix* (New York: Cambridge Univ. Press, 1966), and references therein. Paul Matthews delivered his inaugural lecture at Imperial College in Nov. 1962 with the title "Some Particles Are More Elementary than Others" (London: Imperial College, Nov. 1962).

[81] On this point see esp. Oreskes and Rainger, "Science and Security before the Atomic Bomb" (cit. n. 24).

Were Geoffrey Chew's ideas about particle physics determined by these particular cultural and political ideas, dug up and exposed by the loyalty oath and McCarthyism? The unidirectional, causal story falls short here.[82] To begin with, not every physicist who experienced the rapid postwar political transitions in Berkeley went on to embrace a "democratic" physics in the way that Chew did. More important, Arthur Wightman—a quintessential "fundamentalist" in Geoffrey Chew's eyes for his insistence on retaining quantum field theory—became, with Chew, an active member of the Federation of American Scientists during the 1950s. No crude equations between voting behavior and theory choice will suffice here.

But the failure of such crude equations is just the beginning of our work as historians, not the disappointing end. Just as the straw-man story of sociopolitical determinism fails, so too does the contention that there was simply no connection between Chew's political engagement, pedagogical reforms, and particle physics. At the biographical level, perhaps Chew's frustration with unfair, anti-Communist practices at Berkeley helped to strengthen an iconoclastic resistance to unquestioned authority and a desire to follow out otherwise-unexplored options—he was, after all, already a self-proclaimed "non-conformist," as he wrote in his letter of resignation to Birge in July 1950. Speaking out against university regents and State Department officials, despite the prevailing orthodoxy, could well have prepped Chew for his similarly outspoken challenges to the field-theory orthodoxy some years later.[83]

Even so, iconoclasm alone cannot explain the specific details of Chew's nuclear democracy—its particular elements and the interpretation he gave to them. Substantive links—not just a consistent underdog stance—appear between the three realms of his postwar activities. In particular, the same language recurs again and again ("fundamentalists," "special status," "without favoritism," "equal partners"); models, particles, and collaborators were all "democratized." This was a vocabulary Chew had honed over a decade filled with angst and activism, in front of regents and senators, years before he began to apply it to Feynman diagrams and ρ mesons. Rather than asking "How much did politics affect Chew's physics?" or "What complicated admixture of politics and culture and society interacted in which complicated ways to produce Chew's ideas in physics?" we can

---

[82] Marcello Cini, who worked as a dispersion-relations theorist during the 1950s, has offered a somewhat strained argument that dispersion relations, which promised "utilitarian" correlations among the newly acquired reams of experimental data, took hold because of its fit with "the dominant ideology in the U.S." Marcello Cini, "The History and Ideology of Dispersion Relations: The Pattern of Internal and External Factors in a Paradigmatic Shift," *Fundamenta Scientiae*, 1980, *1*:157–172, on p. 157.

[83] On the question of iconoclasm, Chew's case warrants comparison with that of David Bohm. Just as Bohm was losing his Princeton job for not cooperating with HUAC, he published two long articles questioning the dominance of the standard interpretation of quantum mechanics and offering a new one in its place. Much as Chew would do in 1961, Bohm thus challenged the accepted physics orthodoxy in a manner consistent with his postwar political convictions. Unlike Chew, however, Bohm (at least later, starting ca. 1960) proclaimed that his ideas in physics were actually inspired by his political thinking. See David Bohm, "A Suggested Interpretation of the Quantum Theory in Terms of 'Hidden' Variables: I and II," *Phys. Rev.*, 1952, *85*:166–179, 180–193; Olwell, "Physical Isolation and Marginalization in Physics" (cit. n. 9); James Cushing, *Quantum Mechanics: Historical Contingency and the Copenhagen Hegemony* (Chicago: Univ. Chicago Press, 1994); F. David Peat, *Infinite Potential: The Life and Times of David Bohm* (Reading, Pa.: Addison-Wesley, 1997), Chs. 5, 6, 8; Mullet, "Political Science" (cit. n. 9); and Kojevnikov, "David Bohm and Collective Movement" (cit. n. 9). Wolfgang Pauli, for one, saw fit to merge Bohm's work in physics with his political troubles, describing Bohm's "younger fellow-travellers (mostly 'deterministic' fanaticists, more or less marxistically coloured)." Pauli believed that Bohm had blurred the line between political engagement and physical theorizing. See Pauli to Léon Rosenfeld, 16 Mar. 1952, in Pauli, *Wissenschaftlicher Briefwechsel*, ed. von Meyenn (cit. n. 39), Vol. 4, Pt. 1, pp. 582–583. Cf. Pauli to Abraham Pais, 7 Mar. 1952, in which he labels Bohm a "*Sektenpfaff*," or, roughly, "cult leader" (*ibid.*, pp. 626–627).

thus build on Chew's curious continuity of language to turn the question around: Why did Chew's work emerge in the form that it did, at the time and in the place that it did? What were the conditions, in other words, that made "nuclear democracy" an intellectual possibility—and indeed not just a "possibility" but the dominant set of techniques for strong-interaction particle physics throughout the 1960s? Why, moreover, was the work subject to so many different, competing interpretations by physicists further and further removed from Chew's immediate group in Berkeley?

When we phrase the question this way, we are no longer driven to squabble over competing ledger sheets, trying to count up "how much political factors determined Chew's physics." Instead, when we ask, "Why then? Why there?" certain plausible connections relate Chew's choices of what to work on and what to lobby for. Feynman diagrams, those staple tools of quantum field theory, appealed to Chew with a usefulness and immediacy far beyond their narrow field-theoretic definitions—an appeal noted by scores of other theorists throughout the 1950s and 1960s as well. Deciding that the diagrams' "true" meaning was not dictated by field theory alone, Chew had some choice in how he would use and interpret them—just as scores of other theorists chose to read and interpret the diagrams in still different ways, toward different calculational ends. As Chew devoted more and more time to political and pedagogical alternatives to what he saw as infringements on equal treatment, perhaps a similarly democratic reading of Feynman diagrams, and of the particles they purported to describe, seemed particularly salient. In other words, perhaps this particular notion of "democracy" and "equal treatment" comprised what Chew mentioned in passing within his 1961 lecture notes as his "general philosophical convictions." That association, at least, would certainly help to explain why it was Chew who produced this fervently "democratic" reading of Feynman diagrams, amid the many other interpretations they received from other theorists. In this sense, "nuclear democracy" seems thoroughly enmeshed with Chew's time and place—it bears the marks of McCarthy-era Berkeley.[84]

These specific resonances, no doubt aided but not uniquely determined by the particular environment of late-1940s and 1960s Berkeley, further help to explain why many other physicists who worked on S-matrix theory at this time did different things with it. To Chew and his many postdocs and students, nuclear democracy and the bootstrap meant that quantum fields and virtual particles simply did not exist and that there was no such thing as an "elementary" particle. Yet many young theorists, such as Sam Treiman's graduate students at Princeton, completed dissertations that treated S-matrix theory neither as a self-consciously "democratic" pursuit nor, much less, as a raised-fist competitor to "aristocratic" quantum field theory—indeed, the bootstrap, at best, seemed to them to offer just one more technique for detailing the properties of the truly "elementary" particles. Pieces of Chew's new calculational machinery were picked up and taught at various places, often bundled with still different theoretical tools and deployed toward different calculational

---

[84] Regarding possible ties between Berkeley's political culture and his physics program, Chew responded recently that he "had never thought about it," though such connections are "a possibility worth considering": Chew, interview with Gordon, Dec. 1997, quoted in Gordon, "Strong Interactions," p. 37. Links between his specific pedagogical efforts and his theoretical approach to particle physics strike a similar chord these days from Chew, who recalls only that "I might have had some idea of that," though "it's hard to recapture the way one was thinking in an earlier period"—which, after all, occurred four decades ago: Geoffrey Chew interview with the author, Berkeley, 10 Feb. 1998. More recently, Chew wrote that an earlier draft of this essay, which made the case for substantive intellectual links between his politics, pedagogy, and physics, was "perceptive and accurate": Chew to Kaiser, 19 Aug. 1999.

ends. Just as Chew appropriated Feynman's diagrams, so too many theorists outside of Berkeley converted Chew's program into their own.

With this tale of democracy in postwar America, we may thus scrutinize and interrogate some complicated, attenuated connections between an intellectual legacy of McCarthyism and reactions to it, and certain ideas and practices within theoretical particle physics. Geoffrey Chew's repeated refrain of "equal participation" and "equal treatment under the law" marked his work—in its various guises and arenas—as a product of his specific time and place. Chew's postwar work thus provides a fertile test case to explore the politico-cultural changes of postwar America and physicists' changing place within it.

Bootstrap Physics: A Conversation with Geoffrey Chew

Fritjof Capra

Introduction (2021)

Geoffrey Chew was one of the deepest and most radical thinkers of twentieth-century physics. His bootstrap theory, technically known as S-Matrix theory, is based on the idea that nature cannot be reduced to fundamental entities, like fundamental constituents of matter, but has to be understood entirely through self-consistency. According to Chew, all of physics has to follow uniquely from the requirement that its components be consistent with one another and with themselves.

This idea constitutes a radical departure from the traditional spirit of basic research in physics, which has always concentrated on finding the fundamental constituents of matter. At the same time, it can be seen as the culmination of the conception of particles as interconnections in an inseparable web of relationships, which arose in quantum mechanics and acquired an intrinsically dynamic nature in relativity theory.

The bootstrap philosophy abandons not only the idea of fundamental constituents of matter but accepts no fundamental entities whatsoever — no fundamental laws or equations, and not even a fundamental structure of space and time. The universe is seen as a dynamic web of interrelated events. None of the properties of any part of this web are fundamental; they all follow from the properties of the other parts, and the overall consistency of their mutual interrelations determines the structure of the entire web.

My conversations with Geoff Chew, reprinted in these pages, were recorded in 1983. During the subsequent years, the bootstrap theory was eclipsed by the success of the standard model, which is very different, as it postulates the existence of fundamental fields and their corresponding particles; and today, bootstrap physics has virtually disappeared from the scene.

However, the standard model does not include gravity, and hence fails to integrate all known particles and forces into a single mathematical framework. The currently most popular candidate for such a framework is string theory, which pictures all particles as different vibrations of mathematical strings in an abstract 10-dimensional space. The mathematical elegance of string theory is compelling, but the theory has serious deficiencies. To begin with, there are several versions with different numbers of spatial dimensions, and the process of reducing these dimensions to the 4 dimensions of our actual space-time is not unambiguous. Even more serious is the problem that the theory has not been verified experimentally. There is a large number of different string theories, none of them capable of explaining the values of the basic parameters of the standard model in a unique way.

Some have argued that a more serious concern of string theory as a theory of quantum gravity lies in the fact that it is currently formulated in terms of vibrating strings moving against a fixed background of geometries of space that do not evolve in time. A consistent theory should be formulated in a "background-independent" way, i.e., in such a way that the structure of space-time emerges from the theory rather than being assumed as the arena in which the physical phenomena take place. In other words, as Chew used to say, space-time will have to be "bootstrapped" — derived from overall self-consistency. If current attempts in formulating a theory of quantum gravity continue to remain elusive, the bootstrap idea may well be revived someday, in some mathematical formulation or other.

1. Introduction (1984)*

For Geoff Chew, the year 1984 is a double jubilee. It is the year of his sixtieth birthday and, at the same time, the twenty-fifth anniversary of his celebrated bootstrap hypothesis. It seemed therefore appropriate to review the history, present status, future potential, and philosophical implications of the bootstrap approach to particle physics. The present paper is the transcript of a wide-ranging conversation about these topics between Chew and the author, which took place in July 1983. The text of the transcript has been edited only minimally in order to preserve the spontaneity of the conversation, and it has been organized as follows.

Contents

1. Introduction (1984)* 179
2. The Bootstrap Idea 179
3. History of the Bootstrap 183
    3.1. Fermi's Influence 184
    3.2. Collaboration with Low; Analyticity and Pole-Particle Correspondence 186
    3.3. Collaboration with Mandelstam; Origin of Bootstrap 188
    3.4. Regge Poles; Chew-Frautschi Plot 190
    3.5. Recognition of S Matrix 191
    3.6. Emergence of Bootstrap Philosophy 192
    3.7. Break with Convention 194
    3.8. Decline of Bootstrap in Late Sixties 195
4. Philosophical Influences 196
5. Recent Advances in the Bootstrap Program 199
    5.1. Topological Expansion; Ordered S Matrix 199
    5.2. Topology — The Language for a New Science? 201

*Reprinted from *A Passion for Physics: Essays in Honour of Geoffrey Chew, including an Interview with Chew*, eds. C. DeTar, J. Finkelstein and C.-I. Tan (World Scientific, 1985).

5.3.   Achievements of Topological Bootstrap Theory 202
5.4.   Outstanding Problems 204
5.5.   Skepticism of Orthodox Physicists 205
5.6.   QCD and Weinberg-Salam Theory 205
6.   Outlook 208
6.1.   Space-Time Continuum and Electromagnetism; Gravity 208
6.2.   Extending the S-Matrix Framework 213
References 215

2. The Bootstrap Idea

CAPRA: Geoff, I would like to begin by asking you: What is the bootstrap? What does it say?

CHEW: The idea tends to evolve in time, and the way I describe it now will probably be different from how I would have described it five or six years ago, or from how I would describe it five or six years in the future. The key notion certainly is consistency; the idea that the laws of nature are controlled by consistency and are not arbitrary. From a bootstrap standpoint, one is not content to see any arbitrary aspect in a theory.

CAPRA: Arbitrary meaning what? Do you mean by that, for example, that the mass of the proton has its reasons?

CHEW: Yes, that is certainly the sense in which I mean it. However, when you actually apply the bootstrap idea, you always accept certain ideas, you have to, and within the context of these accepted ideas you then try to remove arbitrariness as much as possible. For example, in the beginning we simply accepted the idea of particles, the idea of an S matrix, which brings with it the ideas of energy, momentum, superposition, and also the Poincaré group and ideas of analyticity. All these ideas were accepted as the basis of our thinking about consistency.

CAPRA: They were part of your scientific framework, part of your language.

CHEW: Yes, and within that framework an effort was made to understand the properties of the hadrons, just the strongly interacting particles. That was what we operationally called the bootstrap. But then, as time went on, we became more demanding, and we asked: could we also understand the electroweak particles? And this evolved to the feeling that one needs also to understand the origin of space-time, the Poincaré group, probably; ultimately also superposition, the presence of complex numbers, analyticity, all these things.

CAPRA: The presence of complex numbers?

CHEW: Yes, why complex numbers are appropriate to understanding physics. There are lots of other formalisms you might think of. No matter how far you go, because of human limitations you will always have to accept, at any stage, a certain set of ideas.

CAPRA: But apart from those ideas, which are provisionally accepted as fundamental, you don't accept anything arbitrary in your theory. You want to derive everything from overall self-consistency.

CHEW: Yes, that's the idea.

CAPRA: Now, what would you say about the nature of the bootstrap idea? Is it a scientific hypothesis, which has now, maybe, turned into a theory? Is it a philosophy? How would you characterize it?

CHEW: Well it is certainly a philosophy, and I think operationally it has turned into a scientific program. I suppose this scientific program has now enough substance to call it a theory. It is very hard to say when you make these transitions from one category to another.

CAPRA: It is also something like a principle, something like Occam's razor, for example.

CHEW: Yes, or like Mach's principle.

CAPRA: However, there seems to be one problem with this notion of self-consistency. One could say that it is a fundamental principle of an approach which does not accept fundamental principles.

CHEW: That's right. That's the ultimate paradox.

CAPRA: Does that bother you?

CHEW: Well, it disturbs me vaguely, but I don't expect to get to the bottom of the whole thing in my lifetime anyway... All is relative; it's a matter of making a certain amount of progress.

CAPRA: I have sometimes worried about this problem, and I have thought that one could maybe put this principle of self-consistency together with the scientific framework, with the scientific language. It is certainly an important aspect of science that you don't want to be inconsistent, which is again related to the way human beings think and the way we observe.

CHEW: Yes, that would be a way to talk about it.

CAPRA: The notion of self-consistency brings to mind the celebrated paradoxes of quantum mechanics. How do you see the role of these paradoxes?

CHEW: I think that this is one of the most puzzling aspects of physics, and I can only state my own point of view, which I don't think is shared by anybody else. My feeling is that the principles of quantum mechanics, as they are stated, are not satisfactory and that the pursuit of the bootstrap program is going to lead to a different statement. I think that the form of this statement will include such things as: you should not try to express the principles of quantum mechanics in an *a priori* accepted space-time. That is the flaw in the present situation. Quantum mechanics has something intrinsically discrete about it, whereas the idea of space-time is continuous, I believe that if you try to state the principles of quantum mechanics after having accepted space-time as an absolute truth, then you will get into difficulties. My feeling is that the bootstrap approach is going to eventually give us simultaneous explanations for space-time, quantum mechanics, and the meaning of reality. All these will come together, somehow, but you will not be able to begin with space-time as a clear, unambiguous basis and then put these other ideas on top of it.

CAPRA: By the way, I know that some people are confused when they hear you use the term "reality". Whenever you say "reality," you mean Cartesian reality, right?

CHEW: Oh yes, I'm sorry, I should always use "objective reality" or "Cartesian reality".

CAPRA: You don't mean to say that the quantum reality, or the reality of emotions, or of the spiritual realm are any less real?

CHEW: No, no, no! I am just being careless. I mean objective reality, the explicate order, as David Bohm calls it.

CAPRA: Coming back to quantum mechanics, you are saying, in fact, that these paradoxes exist, as Bohr and Heisenberg already emphasized, because we are talking about atomic phenomena in a language which is inappropriate. They were referring to the Cartesian language of classical physics, and what you seem to be saying is that space-time is still a remnant of the classical way of thinking.

CHEW: Absolutely. I don't think the meaning of space-time has ever been separate from classical notions, Somehow, we are trying to grasp the connection between the real world and the quantum principles, and we have to understand that space-time is part of the real world and not something that pre-exists before quantum principles are stated.

CAPRA: I would now like to discuss with you the general significance of the bootstrap idea. I have recently been more and more impressed by the idea that the major shift and the deepest change in our thinking may be the shift from an architectural metaphor of a building, with firm foundations upon which one builds, to the metaphor of a network, which has no foundations but represents a web of interrelated events and, correspondingly, consists of a web of concepts to describe these events. That seems to be the major shift. The notion of a basis, of fundamental concepts, building on strong foundations, etc. — all that runs through Western science and philosophy. Descartes wrote that the knowledge of his time was built on sand and mud and that he was going to build new firm foundations for a new science; and three hundred years later Einstein wrote that the foundations of classical physics, that is of this very edifice of Descartes, were shifting and that he did not see any firm ground upon which he could build a theory. I think, maybe, since the bootstrap, it is now the first time in Western science that we are not looking for firm ground and solid foundations anymore.

CHEW: I think that is true, and it is also true that because of the long tradition of Western science the bootstrap approach has not become reputable yet among scientists. It is not recognized as science precisely because of its lack of a firm foundation. The whole idea of science is, in a sense, in conflict with the bootstrap approach, because science wants questions which are clearly stated and which can have unambiguous experimental verification. Part of the bootstrap scheme, however, is that no concepts are regarded as absolute and you are always expecting to find weaknesses in your old concepts. The bridge, however, between standard science and the bootstrap approach lies in the commonly shared awareness of the approximate nature of all experiments. Even people who are dedicated traditional scientists recognize that no measurement can be completely precise.

CAPRA: But these are two things, the approximate nature of measurement and the approximate nature of concepts.

CHEW: Right, and both are recognized.

CAPRA: By the way, do you have any idea when the appreciation of the approximate nature of scientific theories emerged in the history of science?

CHEW: I don't know for sure, but I suspect that it came along with quantum mechanics. I suspect that in the 19th century people might well have believed that theories like Newton's could have absolute validity.

CAPRA: Anyway, now the approximate nature of science is generally accepted.

CHEW: Yes, but in spite of that, the traditional point of view in science is that at any given stage of activity, there is supposed to be a consensus about certain principles whose validity has not yet been disproved or even challenged. All scientists are supposed to conduct their activities within this framework of accepted principles until some measurement comes along which is accurate enough to show that some principle has to be abandoned. The bootstrap approach recognizes from the start that the principles used are not going to be absolute, that everything is approximate. Nevertheless, it is incumbent upon a bootstrap theorist to get an understanding of the degree of approximation.

CAPRA: But now you have said more or less the same thing about bootstrap science and orthodox science.

CHEW: Well, that's why it is possible for them to coexist. Psychologically, however, there is a difference which causes great misunderstanding. Let me give you an example. At the present time, the overwhelming majority of the theorists working in high energy physics accept an absolute notion of local fields, They do this because it is to them the only known way of combining the quantum principles with the space-time continuum. They accept the space-time continuum as an absolute and accept quantum superposition as an absolute, and they only know one way to put these two things together, which is through the local quantum field, and so they take for granted that whatever the description of natural phenomena is going to be, it will be done through local fields. Now, if you get them in a philosophical discussion such as this one, and if you push them, the more talented ones will agree that probably local quantum fields do not represent absolute truth. But they would say, "So far that has not been shown."

CAPRA: So they would think it might be the absolute truth?

CHEW: I suspect that if you took the most talented ones — people like 't Hooft, Gell-Mann, Weinberg, or Salam — when they are in a philosophical mood they would probably agree that local fields are not the ultimate truth. But they are guessing that within their own lifetime nobody is going to go beyond the capacity of the local field to describe high-energy phenomena. Somehow or other I have come to the belief that it is not too soon to go beyond local fields. What that means is that in trying to develop a theory I don't start with a local quantum field. I start with other ideas, and all these people find this incomprehensible. They say, "Why don't you use local fields? They have never been shown to be wrong." Now, the reason why I don't like them is because they bring in an inherent arbitrariness. Nobody has ever found a way to use local quantum fields without introducing an unpleasant arbitrariness.

3.   History of the Bootstrap

CAPRA: Geoff, I would now like to turn to the history of the bootstrap idea. A little while ago you said, "Somehow or other I came to the belief..." Since the shift from orthodox physics to bootstrap physics is so radical, I am extremely curious to know how you developed these ideas and to what extent you appreciated their radical nature.

3.1. *Fermi's Influence*

CHEW: I made an attempt not long ago to reconstruct some of these developments, and I believe that the beginning came right at the time of my Ph.D. thesis with Fermi. Now there is an irony here, because Fermi was an extreme pragmatist who was not really interested in philosophy at all. He simply wanted to know the rules that would allow him to predict the results of experiments. I remember him talking about quantum mechanics and laughing scornfully at people who spent their time worrying about the interpretation of the theory, because he knew how to use those equations to make predictions, But Fermi suggested as a thesis problem for me an extension of an approximation which he had discovered in connection with the scattering of slow neutrons by molecular systems. He had realized that the molecular binding was important in this process but that nevertheless the interaction of the neutron with the nuclei was overwhelmingly strong compared to its interaction with the rest of the system. While the neutron was interacting with the nucleus you could ignore the molecular forces. It was a very subtle thing which, eventually, became called the impulse approximation. Fermi's idea was that the nuclei behave in response to the atomic forces until the neutron arrives; then, when the neutron is in contact with a particular nucleus, the nucleus forgets that it has any other things around it, until the neutron departs when, once again, it responds to its environment. Now, all of this is done quantum-mechanically, so it's not trivial. But it led Fermi to a certain set of formulas, a recipe of how to know the molecular wave functions, and then all you had to know in addition was the scattering amplitude of the neutron by the nucleus, as if the nucleus were free. And then you could put these two things together to do your computations.

CAPRA: So that was what Fermi had done.

CHEW: Fermi had done that and he suggested that I extend the same idea to scattering of neutrons by nuclei, where you think of the nucleus as being made up of neutrons and protons. The point of the idea was that if the neutron was moving very fast, there might again be something like a neutron interacting with a single nucleon. What Fermi had done here was really to make a practical application of an S-matrix idea. He did not recognize that, but he intuitively understood that there was a complex number which characterizes the scattering of the neutron, which you can measure, and you can use that number in computations. You don't have to say that there is a potential between the neutron and the nucleus; you don't have to go through the apparatus of the Schrödinger equation. All you have to know is that one number, which is an S-matrix element. So that idea got into my head.

CAPRA: And you worked it out?

CHEW: I worked it out for the case of scattering of neutrons by deuterons and various other things.

CAPRA: And it worked?

CHEW: It worked, and it also persuaded me that it was not necessary to have a Schrödinger equation and a potential. Previously, people had always thought that when you computed something you had to have a detailed microscopic interaction between the particles together with a Schrödinger equation.

CAPRA: In other words, you had to have a temporal sequence for the wave function.

CHEW: That's right. Fermi simply produced formulas. You saw no time; you saw no Schrödinger equation. He simply worked directly with amplitudes.

CAPRA: And he did this because he was a pragmatist.

CHEW: That's right. He somehow knew intuitively what he had to do. Now, he did not describe it that way. He described it in ways that very much obscured the S-matrix interpretation. But nevertheless I began to think that a large part of what we normally associate with the Schrödinger equation is simply done by S-matrix principles, and you don't need all this microscopic space-time.

CAPRA: But these S-matrix principles were not formulated at that time.

CHEW: No.

CAPRA: Was the S matrix itself known?

CHEW: Yes. John Wheeler had identified it, 1 think, in 1939. Heisenberg had written papers about certain of its properties in the mid-forties, and he had actually called it the S matrix. Then Christian Møller wrote some review papers which propagated Heisenberg's thinking.

CAPRA: Were you familiar with these papers at that time?

CHEW: Well, that's a very funny thing. I had Fermi's idea, and I knew about the S matrix abstractly, but I did not connect the two; not for a long time. It's very strange. I found the S-matrix theory at that time kind of forbidding. It used an apparatus that was difficult, and I simply did not connect it to those other ideas. But 1 did become aware of the S matrix while I was a graduate student.

After I left Chicago, I continued to work on this impulse approximation for a couple of years, but it was done within sort of a bastard framework. It wasn't S-matrix theory, it was something in between. I was picking up Fermi's intuition and trying to generalize it, and I produced a series of papers in which Murph Goldberger and Giancarlo Wick were also involved. Then I went to the University of Illinois in 1950 and started thinking about π-mesons which had been discovered not long before that. For some reason — I wish I could recall that precisely — I was completely persuaded that the idea of local fields was inappropriate for describing π-mesons. Up until then, people had been dominated by the idea of Yukawa, which was that π-mesons were the analogue of photons. Yukawa had said that the electromagnetic force, which is due to the exchange of photons, was the analogue to the nuclear force due to the exchange of π-mesons. So people were writing down equations just like electromagnetic equations, except that they would have fields associated with the π-mesons.

Now I had been in contact with the early experiments on π-mesons, and it was clear to me that these were particles just like any other nuclear particle, like neutrons or protons, and it seemed silly to me to use fields to describe them. The kinds of experiments you were trying to describe were just like any other nuclear reaction. You didn't use fields in connection with nuclear physics before that; why should you use fields for the π-mesons when they were just another kind of nuclear particle? But people said, π-mesons are not nuclear particles; they are field quanta; they are like photons. It is very strange when you look back now to understand that psychology. So in 1950, when I went to the University of Illinois, I decided to try to make a little model to describe scattering of π-mesons by protons, based on the same idea that Fermi had. I said to myself, suppose the proton is some kind of a structure that contains π-mesons within it and then we shoot π-mesons at it from the outside... Although I did not know that the word S matrix was relevant at that time, it was a model in the spirit of S-matrix theory. It was a model in which you did not use the Schrödinger equation; you just used the superposition of amplitudes. Looking back now I can see that it had much of the Feynman idea that you can build amplitudes by superposition.

3.2. *Collaboration with Low; Analyticity and Pole-Particle Correspondence*

CHEW: The model had a certain amount of success, and then Francis Low came to the University of Illinois, and after a year or so we started to work together. He had made a certain discovery in axiomatic field theory and for some reason either he, or I, or both of us, recognized that his discovery might be relevant to this model that I had developed. So we started to work together on it, and I was so pleased to have somebody of Low's talent to work with that I put aside my feelings about the nature of my model and tried to re-express the content of it in field-theory language. It turned out that, to a large extent, this was possible. Then we wrote a paper together, which many more people could understand. Not so many people could understand the thing I had written at first, but when it was re-expressed in the language of field theory it could be appreciated by many more people. The mathematical structure that came along with it was, in fact, much improved, so that we could see a lot more things.

Now Francis understood that the additional content was, in fact, of a general nature associated with analyticity. It was at that point that I began to be aware of analyticity as a principle. Francis and I had, somehow, come upon the notion that analytic continuation is very powerful. We still did not think of it as S matrix; we thought of it as analyticity suggested by field theory. But, in fact, what we did was to sort of forget the field theory at a certain point and start working with analytic functions. Most of the content of what we did was just based on analytic functions. We started to recognize the complex plane explicitly at that point.

CAPRA: Did the S-matrix framework, as it existed at that time, have analyticity in it, or was this your discovery?

CHEW: I am not quite sure what the honest answer to that is, but I'll tell you what Landau said to me. He was a very dramatic person, very outspoken with no hesitation to express his views about anything. In 1959, at a meeting in Kiev, he expressed annoyance to me about the work that I was doing with Mandelstam on the dynamics. He said we wasted our time with approximations, dealing with a system that was incomplete. He was partly right, but through our effort we discovered general things which we would not have discovered had we not made that effort. In any case, in the course of criticizing me for putting so much effort into this π-π dynamics, Landau said, "You know, you have discovered an absolutely crucial point, which is that particles correspond to poles, that the S matrix is an analytic function, and that the poles of the S-matrix are the particles." He attributed that discovery to me. I didn't think of that as my discovery, but when I look back and ask myself, who was it who first really appreciated that particles correspond to poles, maybe it was me; I am not sure. It was an idea that was floating; it occurred in various special forms here and there, but somehow the generality was not recognized. For example, a few people, such as Wigner, had come to the idea that the notion of an unstable atomic state could be associated with a complex pole in something or other, and that the imaginary part of the pole location was associated with its lifetime.

CAPRA: Was that the Breit-Wigner resonance?

CHEW: The Breit-Wigner resonance formula was an example. It's hard to tell how general Breit and Wigner thought these concepts were. At that time, such ideas were always presented as if they could be derived from perturbation theory, but at least Fermi knew very well that they had to be general; they couldn't rest on perturbation theory.

CAPRA: Now, what about the analyticity of the S matrix?

CHEW: Well, the S matrix itself was not a well-recognized notion.

CAPRA: It seems that what you contributed, then, was the emphasis on analyticity which put the whole notion of a pole in a different context. Without even mentioning the S matrix, it was nevertheless a step in that direction.

CHEW: That's right; that's true. I certainly contributed something, but it's hard to say exactly what it was. I remember being puzzled at the time that there weren't lots of people recognizing these points, and I felt there must be something the matter with me, because it seemed so evident to me that we were dealing with an analytic function which has poles. But nobody else... and then Landau! That was a tremendous thing. Here I go to Russia, and here comes Landau and congratulates me for exactly recognizing this. Then there was one other place that I know of where I was given credit, This was in a paper by a Berkeley mathematician who was studying the abstract mathematical problem of how to extrapolate a function which you only know incompletely. He was focusing on the use of the idea that the function had poles with known locations, and he attributed the use of this information for the extrapolation of an incompletely known function to me. In presenting the history of this problem, he referred to a paper of mine, which might be the first paper in which a definite statement about the association of particles with poles is made. This was a paper on the problem of deducing the pion-nucleon coupling constant from nucleon-nucleon scattering data, which I had written in 1958.[1]

Just as Francis Low and I were doing that work, Gell-Mann and Goldberger had started to develop their dispersion relations, which had a big influence on me. They believed the relationships that they employed were all based on field theory, but I remember I was quite convinced that it was analyticity that counted and that field theory was not really necessary. I still did not make the connection with the term S matrix. It's very strange; that was already in 1955-56.

3.3. *Collaboration with Mandelstam; Origin of Bootstrap*

CHEW: I left Illinois in 1957 and came here to Berkeley, and then I met Stanley Mandelstam, who had discovered double dispersion relations and had thereby solved a problem I had been struggling with for a long time. I had become aware of the fact that analytic continuations in energy needed to be extended to an angle — that you had to continue both in energy and angle. I could not figure out quite how to do it, but Mandelstam did.

CAPRA: He developed that whole framework of the s and t variables, didn't he?

CHEW: Yes, that's right. Well, I got Mandelstam to come to Berkeley and we worked together. He was at Columbia and had gotten his Ph.D. at Birmingham with Peierls. At Columbia nobody knew what he was doing; nobody paid any attention to him. I heard him give a talk at the Washington meeting of the American Physical Society, and I remember that I said to myself when I heard his talk, "Oh, this young guy, he doesn't know how hard his problem is. He thinks he solved it, but I'm sure he hasn't solved it because he doesn't know this difficulty and that difficulty, and so on." And I thought, in kindness to him I'll point out some of these difficulties after his talk. But when I started to ask him questions he just answered every question. I was totally overwhelmed; he had really solved the problem! So I persuaded him to come out to Berkeley, which he happily did, and we collaborated on two papers extending the whole idea to π-π scattering. Up until then it had always been πN scattering, but now that Mandelstam had extended the analytic continuation, it was possible to think about π-π scattering.

We did not get a satisfactory theory, of course, because the pion is not the end of the story, although at that time we thought it was. It is funny to look back at this now. We thought, somehow, that the π-meson was the key, and if you could understand how pions interacted with pions, you really got it. But we discovered that something was loose; the system did not close.

CAPRA: In all this there was no S matrix yet, and of course no bootstrap?

CHEW: That's right, but this was where the idea came into my head, and in 1959 the word "bootstrap" appeared in print for the first time, although rather casually. Mandelstam and I had pushed our π-π analysis to the point where we could see that there might be a solution of the following character. The pions would interact to produce either a bound state or a pseudo-bound state, and that bound state by crossing would then constitute a force which would be the agent for making a bound state in the first place. We could see this possibility quite clearly in the way the coefficients of the equations arose.

CAPRA: Had the crossing property been identified at that time?

CHEW: Yes, crossing had been discovered by Gell-Mann and Goldberger, but it had not been used in the sense of dynamics to make a theory of forces. It was understood that ingoing particles became outgoing antiparticles by crossing, but the crossing property had not been applied to talk about forces in the cross channel. Somehow it needed Mandelstam's representation to do that. Well, Mandelstam and I figured those things out, and in 1959 there was that conference in Kiev at which I was a rapporteur. When I reported on our work, together with the work of others, I used the term "bootstrap" for the first time in the text of this report, referring to that possibility that we had noticed. Now, you have to realize that the ϱ-meson had not been discovered at that time; Mandelstam and I thought of a bound state of two pions being simultaneously, through its exchange, the force that holds the pions together. A corresponding particle was not yet experimentally known. So all this was very tentative. Nevertheless, we had noticed this possibility and we described this as a kind of bootstrap dynamics.

CAPRA: "We" meaning Mandelstam and you?

CHEW: I said it. I don't think Mandelstam would have used the term, but it certainly came out of our joint work, I think in discussions with Mandelstam I had used the term "bootstrap". Stanley never endorsed it but, being a very mild person, he did not fight it, and the term also appeared in one of our papers.[2] What happened then was that a number of other people, in particular Zachariasen and Zemach, and some others, used the term once they had grasped the idea. Going along with it was an approximation called the N/D approximation, in which you use S-matrix principles to do computations within the framework of scattering amplitudes, which you analytically continue without using the Schrödinger equation. If you think of Mandelstam's subsequent interests, he never picked up the bootstrap idea. He allowed me to use it in one of our joint papers, but he never felt comfortable with it. Stanley is a beautiful example of the kind of physicist I was talking about earlier. He feels a need for something fundamental. I think he always believed that he had firm ground under his feet.

CAPRA: Even though you could say that his double dispersion relations were really the first tool that pulled out the firm ground from under you.

CHEW: That's right, that's right; that's exactly right!

CAPRA: If you disperse in one channel, and then you turn everything around and disperse in the other channel, that is very much connected with that whole network idea that was later to emerge.

CHEW: That's right, but Stanley thought that it was based on field theory. By the way, Francis Low's earlier work was of somewhat similar status. Low did the same thing in one variable, and then Mandelstam did it in two variables.

3.4. *Regge Poles; Chew-Frautschi Plot*

CAPRA: Where did Frautschi come in?

CHEW: Frautschi came a year or so after that, in 1960. Mandelstam and I had been frustrated in our N/D calculations by a certain divergence that appeared in these equations. It turned out to be impossible to avoid this with the methods that we were aware of. This was associated with a power behavior that goes along with the spin 1 of the ϱ-meson. We had to use the spin 1 in order for anything interesting to happen, and then we got into this difficulty in connection with asymptotic behavior. I was furious because I felt intuitively that there should not be any divergence. The Q-meson was not an elementary particle, it was coming out as a composite, and it was ridiculous that it should produce a divergence just like in field theory. It was most irritating to have a difficulty characteristic of field theory just because the Q-meson had spin 1. When Frautschi came — now this is a very important historical question — somehow or other we became aware of a paper by Tullio Regge. I forgot who told us about that paper, maybe it was Mandelstam. I am not sure, but it was probably Mandelstam. Anyway, somebody told us about the paper by Regge, which seemed to have something to do with our difficulty of the spin-I asymptotic behavior. So we tried to read Regge's paper. In the beginning we did not understand it very well, but we did grasp the idea of an angular momentum which depends on energy. Regge somehow made an analytic continuation away from the integer angular momentum so as to make it smooth.

CAPRA: Was this one of his basic papers on complex angular momentum?

CHEW: It was practically the only one. As far as I know, he just wrote one paper, and he did it in the context of potential scattering. Frautschi and I, in frequent consultation with Mandelstam, came to the belief that this kind of behavior was general, that it would apply to the relativistic problem. I remember that Mandelstam was not very keen on this at first. We had long, long arguments about this question, which went on into 1961, I think. Frautschi was enthusiastic, and he and I worked together and developed some phenomenological applications before we understood clearly what was going on. Strangely, Murray Gell-Mann played a big role in this. He got interested, and it was Gell-Mann, I think, who said you should call these things Regge poles. He thought it was a big joke, because Regge himself did not have a clue as to what we were doing with them. He had simply written that one paper, and from his point of view it was mathematics based on the Schrödinger equation, and there was no connection with a more general problem.

CAPRA: Did you see the relevance of crossing to Regge's formalism at that time?

CHEW: Yes, we certainly did. There was a confused period there, in which we were sure we had come upon something of generality and importance, but we weren't clever enough to get it really straight. We kept talking to other people about it and getting their advice, and gradually a number of other people became interested in the development. Gell-Mann certainly did; he was very enthusiastic about it; also Goldberger and some other people. Then Frautschi and I wrote a paper applying the idea. There were just enough baryon masses that had been measured at that point, so you could begin to see a Regge trajectory developing. But my real interest, and I suppose also Frautschi's, was to apply this to the bootstrap idea. We wanted to take the equations that Mandelstam and I had developed and apply this Regge boundary condition to them, so that we would get away from that divergence. Well, there was progress made in that respect, but in retrospect you can see that the understanding of Regge behavior still did not close the problem. During that period we worked up a lot of enthusiasm for the bootstrap notion, and other people picked it up and started to work on it from a variety of standpoints. So in the early sixties the term "bootstrap" was very widely spread, and a lot of different approaches to it were developed.

CAPRA: It seems that there were a number of simultaneous developments that generated great interest and enthusiasm at that time: the bootstrap, Regge poles, S-matrix theory, and all that.

3.5. *Recognition of S Matrix*

CHEW: That reminds me of the time when the S matrix finally became recognized. It was not until I tried to write a book in 1961. I wrote a little book for the Benjamin series, *S-Matrix Theory of Strong Interactions*.[3] When I prepared that book and a talk for a conference that was held in La Jolla I said to myself: After all these years of pretending that what I was doing was field theory, I finally want to be honest and say that I don't really believe in it; that what is important is analytic continuation. I kept looking for a word to contrast it with field theory, and suddenly I became aware that the S matrix was the point. It was not until 1961 that I really grasped that this was the concept Wheeler and Heisenberg had discovered twenty years before. I think at that talk in La Jolla I used the term "S matrix", and I certainly did in the book, and from then on I kept using it. Henry Stapp provided amplification very quickly by extending the idea to the description of spin.

CAPRA: When was the axiomatic work by Stapp, Iagolnitzer, and others done?

CHEW: That was somewhat later, but in 1962, and probably starting in 1960/61, Henry worked on the problem of spin.

CAPRA: What about Polkinghorne? He and some others wrote a book about S-matrix theory.[4]

CHEW: That's a tricky business. There was a team of four authors: Polkinghorne, Eden, Olive, and Landshoff. They had been working on dispersion relations and analytic continuation, and their book, which finally appeared around 1966, comes actually in two parts. Part of it is sort of straight S-matrix theory, which was mainly written by David Olive, and the other part is Feynman diagrams. I remember, at the time I didn't like that. I thought the book ought to be just on straight S-matrix theory, that they should not spend all this effort on Feynman diagrams. Well, subsequently I've changed my mind; what they were doing was very relevant.

CAPRA: Around the mid sixties, then, the S-matrix framework was more or less established.

CHEW: Yes.

CAPRA: This was the time when you wrote *The Analytic S Matrix*.[5] So that must have represented the culmination of your ideas at that time.

CHEW: At that time, it did. Yes.

CAPRA: Did you have any new insights while working on this book? You said that while working on the first book you really recognized the S matrix for the first time. Was there anything like that connected with the second book?

CHEW: I think there was a good deal less. As I remember, the book was a disappointment to me, because it was not able to move past two-particle channels. When Mandelstam and I developed our techniques we knew how to discuss poles and we knew how to discuss the two-particle branch points, but that was it. We did not know how to discuss anything higher — and we still don't! This impasse has never really been overcome. Essentially; what the second book did was to add the Regge theory in a good deal more detail.

CAPRA: And I suppose you also presented things in a more systematic way.

CHEW: Yes.

3.6. *Emergence of Bootstrap Philosophy*

CAPRA: Now what about the philosophical side of the bootstrap idea? First of all, it seems that the bootstrap idea was always tied to S-matrix theory, even before you knew that you were dealing with S-matrix theory.

CHEW: Yes, that's right.

CAPRA: It seems that the whole idea emerged out of a pragmatic position and in a sort of technical way.

CHEW: Right.

CAPRA: You did not sit down to think how the world was built; you did not entertain general philosophical thoughts?

CHEW: No.

CAPRA: So when did the whole bootstrap philosophy emerge?

CHEW: I think it was during my collaboration with Mandelstam. I particularly remember one item in those discussions, which focused the issue. In developing the equations for the π-π system with Mandelstam, we not only encountered the possibility of the ϱ-meson being generated as a bound state and also producing the force necessary to sustain the bound state; we encountered a parameter in connection with the s-wave scattering, which Mandelstam wanted to associate with a standard field theoretical parameter. In standard scalar field theories there is a $\Lambda\phi^4$ term in the Lagrangian, which corresponds to an s-wave interaction for spin-0 particles. The parameter which showed up in our equations could be interpreted as such a coefficient in a Lagrangian. That's how Mandelstam wanted to interpret it. We had long arguments about this, and I don't remember exactly how the paper was finally written, but I didn't like that at all. I couldn't believe this system was going to admit a fundamental parameter of that character. It seemed absurd to me that it would. Mandelstam felt this was just a representation of field theory, and he thought X was the fundamental parameter and that the ϱ-meson would then, somehow, emerge driven by that parameter. Our equations did not indicate that; they indicated that the ϱ-meson was driving itself. There was no real connection between that parameter and the ϱ-meson. The parameter was just dangling out there, and subsequently it-has been understood that it is no more fundamental than anything else. Nowadays one would not dream of referring to Λ as a fundamental parameter. Because of that particular aspect of our theory I was forced to think hard about fundamental parameters. In this example it seemed clear that Λ could not be a fundamental parameter. And I said to myself: But here is Mandelstam who believes that it's fundamental. Why? Because he believes in a Lagrangian. So, at that point I said: there is something sick about the whole Lagrangian idea that causes people to think there have to be parameters sitting there, which you are not going to be able to understand. I think in trying to defend myself, in trying to find a language that would express my idea, somehow the term "bootstrap" was helpful. I tried to explain to people why I felt that way. That parameter should not be there, because this was a bootstrap system which would not allow such things. I am pretty sure that up until then I had been vague; I had been rather unclear in my own head as to what I believed concerning fundamental parameters, elementary particles, and the like. At that point I had somehow crossed the bridge.

CAPRA: The fact that you should be able to derive the masses, or rather the mass ratios of particles seems to be much more intrinsic to the bootstrap framework, crossing, and all that. Didn't you feel that?

CHEW: Yes, but you know it is remarkably hard to put aside ideas you grew up with. When 1 was a student it was accepted that neutrons and protons were fundamental particles, and nobody dreamed of explaining their masses. I was aware that the logic of the bootstrap said you had to be able to determine them, but still, because nobody believed that you could determine them, the possibility was hard to accept. But then Frautschi and I were led to make our Regge plot, and — my God! — there was the proton sitting on the same curve with these other things. That was, somehow, a real punch — to see the mass of the proton in what was clearly a dynamical context.

CAPRA: So the Regge formalism really helped you to work out your philosophy.

CHEW: Oh, yes, tremendously! When you see the mass of the proton sitting on the same curve with a lot of other things, that tends to dissolve prejudices.

3.7. *Break with Convention*

CAPRA: Now the bootstrap idea is extremely radical compared to the whole scientific tradition. Was this radical nature apparent to you, and was there a struggle? You know, when you read Heisenberg and Bohr, you realize that they struggled like hell. Did you go through a similar phase?

CHEW: I remember going through something like that before that conference in La Jolla, asking myself: do I really believe this? Am I really prepared to back up this position? I went over all the developments that I had been exposed to until then, and I could only come to one conclusion, and that was the idea of nuclear democracy. None of the nuclear particles really could be said to have a fundamental status; they all had to be bound states of each other. It's true I was aware this was a radical idea, but nothing else made sense to me at that point. I couldn't see any alternative. The fact that the Schrödinger equation was not to be taken as a fundamental statement of dynamics had been working gradually on me over these years. I had seen how much the S matrix could do. Mandelstam, by the way, drove the final nail into that coffin, because he explicitly showed, using his double dispersion relations, how you could recapture the Schrödinger equation as an approximation. So I had no feelings any longer that one needed an equation of motion.

CAPRA: Several years ago you told me that this was very crucial in your thinking, because in giving up the equation of motion you also give up the notion of "things". When you have an equation of motion there is a thing that moves.

CHEW: That's correct.

CAPRA: I have heard people refer to a talk of yours where you were very enthusiastic and very radical. You said, "From now on you can forget about Lagrangians", and things like that. Was that the La Jolla talk?

CHEW: Yes, that was the La Jolla talk in 1961. I suspect there was a lot of emotional stress associated with preparing that talk and giving it. Up until then I had continued to operate as if I accepted field theory, even though I didn't believe it. I felt torn and dishonest, but I was so anxious to get problems solved that I didn't want to let arguments with my colleagues get in the way of solving these technical problems. But for some reason I decided...

CAPRA: ...to come out of the closet, as they say these days.

CHEW: That's right (laughs), and having done that I was probably more inclined to think about philosophical questions. Once you have said to yourself: conventional wisdom does not have to be accepted, that's a big psychological break. I guess I was sort of expecting that as a result of that talk somebody would come and give me an overwhelming argument to the contrary. But they didn't. They didn't like what I said; they were furious, but there was no counter statement that was substantial. 1 remember a story about Arthur Wightman, who was doing axiomatic field theory at that time. He was furious at my La Jolla talk, but he also had a sense of humor. He put up a sign at his office door in Princeton, which said "Closed by order of G. F. Chew."

3.8. *Decline of Bootstrap in Late Sixties*

CAPRA: During the late sixties, there was a decline of the bootstrap idea, probably because of the difficulties you mentioned before, the inability to go beyond two-particle channels and to find the right kind of approximation.

CHEW: Yes, right!

CAPRA: At the same time the quark idea gained momentum. How did you feel in those years? You must have been disappointed, of course, but did you actually have doubts as to whether the bootstrap program could be carried out?

CHEW: I did not have doubts about the ultimate story; I certainly had doubts about the time scale, about whether I was going to see any significant part of it. I resisted the quark business very strongly at the beginning, because I felt that it was abandoning the whole bootstrap idea.

CAPRA: In those days people thought of quarks as particles, I suppose.

CHEW: Well, there was a confused period at the beginning, from 1962 to 1966 or so, in which Gell-Mann played a very big role. He did not call them particles.

CAPRA: He was talking about mathematical quarks, I remember.

CHEW: That's right; he didn't think they were particles. Then gradually naive but phenomenologically successful models were developed, by Dalitz for example, and, I guess, when people discovered the color concept to resolve the difficulty of the symmetry of the baryon wave function, they started to be less inhibited about calling the quarks particles. Then, finally, when QCD was invented, they lost all their inhibition.

CAPRA: So what was your attitude in those years?

CHEW: I resisted the quark idea for quite a number of years, but I began to be more receptive when the dual models began to show up.

CAPRA: So that was quite late.

CHEW: That's right; 1968–69.

CAPRA: Did you sense then that there was something behind the quark idea other than quarks as particles?

CHEW: I certainly resisted the idea of quarks as particles. I have never been able to swallow that, and I couldn't fit the quark idea into anything that made sense to me until the dual models appeared.

CAPRA: There is an interesting coincidence here. I remember you giving a talk at a conference in Irvine in 1969. I thought that this was a very pessimistic talk, very subdued. Actually, it was the only pessimistic talk that I have ever heard you give. At that time you must have been at the end of a long stretch of years where there did not seem to be much hope for the bootstrap.

CHEW: Yes.

CAPRA: And yet, it was at that very conference that the Harari-Rosner diagrams were also discussed. So that was the lowest point, and from then on it went uphill.

CHEW: Yes, I think that's about right.

CAPRA: However, at the same time you wrote two general, more philosophical articles about the bootstrap,[6] one of them called *Hadron Bootstrap — Triumph or Frustration?* In these two articles you expressed, basically, a positive outlook. Now, what made you keep your faith?

CHEW: I think, by that time there were so many philosophical elements in the picture which seemed to be stronger than the difficulties. always felt that the difficulties were just lack of imagination. It wasn't that the boot strap idea itself was wrong; it was just that we were without a good technique for pursuing it. I never really changed from that attitude. But, you know, when you are speaking at a meeting of physicists, and you haven't got anything to present...

CAPRA: That's not much fun.

CHEW: That's right. It is much easier to write a philosophical article and express your enthusiasm.

CAPRA: It is interesting that by that time, by the end of the sixties, the philosophy had become so strong that you could actually do that. In 1961, say, you couldn't have done it.

4. Philosophical Influences

CAPRA: Given the radical nature of this bootstrap philosophy, I have always been very curious about your philosophical background. You are obviously a very philosophical person in the way you do science. Were you always interested in philosophy?

CHEW: No, I was not aware of being interested in philosophy. I tended to model myself after Fermi. I find this paradoxical in retrospect, but for a long time I tried to think that I was going to behave as much as possible in the spirit of Fermi. As a matter of fact, I recall that during the period of collaboration with Francis Low, one day we were riding in a car back from a conference, and Francis brought up the question of whether quantum mechanics was really understood, and we began discussing some of the crazy things about quantum mechanics. And I remember feeling what a waste of time this was to think about such things. I couldn't respond to Francis' arguments, but I was still very much a student of Fermi at that point, and I just didn't believe that scientists should spend their time worrying about issues like that.

CAPRA: You know, it's interesting that the S-matrix approach does have this pragmatic aspect, and it also has a very deep philosophical aspect. It's a very curious mixture.

CHEW: That's absolutely right. The S-matrix idea is the clearest expression of the Copenhagen interpretation.

CAPRA: You must have been interested, though, in the whole mystery of quantum mechanics, the Bohr-Einstein debates, and so on.

CHEW: No, I wasn't. I think I appreciated that there was a difference between quantum mechanics and the Schrödinger equation. Fairly early on I knew that quantum mechanics really meant the S matrix. So my war at that point was with the Schrödinger equation or, if you like, with the use of the space-time continuum as the underpinning, I felt that the S matrix was completely capable of doing everything that needed to be done, and I didn't worry very much about the philosophical significance of that position. It really was only a good deal later, when I had to write and give talks, that I started to think about that.

CAPRA: This is very difficult for me to imagine. I met you in 1969, when I was at UC Santa Cruz, and I remember that you came and gave a talk about the significance of small parameters from the bootstrap point of view. I was very impressed, already then, by your way of presenting things and by the depth of your thinking. So I have always known you as a very deep thinker and a very philosophical person. Did you turn into that at some stage?

CHEW: Hmm! (smiles)

CAPRA: You see, this is really a surprise to me that you say you weren't interested in philosophy, not even in the philosophical aspects of physics,

CHEW: Well, somewhere in the 1960s, I guess, there must have been... Well, okay, I can remember, when I was in England in 1963, that I was asked to give a lecture. I was beginning to become more philosophical during that year. I remember that in addition to this big lecture, which had a certain amount of philosophy in it, 1 also gave a small lecture to a Cambridge college, in which I tried to persuade them that there was no absolute truth in science. I remember that this was a pretty radical thing to do, at that point, in Cambridge.

CAPRA: You see, I always had the idea that there must be something in your interest — some philosophical tradition, some religious tradition, or something in the world of art — something that influenced your thinking. We know that Niels Bohr was influenced by Kierkegaard and by William James, that Heisenberg was reading Plato. Some of the ideas from these traditions influenced them and helped them in their conceptual crisis. But there doesn't seem to be anything of that kind in your

CHEW: I can't identify anything like that. That's quite true.

CAPRA: Maybe that just means that you are really an original thinker.

CHEW: No, I don't think so. The influence was there, coming in various ways that were not so obvious. Let's see, maybe I can identify a few roots. You know, Edward Teller was somebody who had a substantial influence on me in addition to Fermi. Fermi was not interested in philosophical questions, but Teller really was. When I was a student at the University of Chicago, Teller made me aware, either in formal lectures or in private conversations, of some of the great philosophical issues associated with quantum mechanics. In particular he told me a few things which somehow stuck. I haven't thought about this for a long time, but either Teller or somebody else made the point that the quantum theory of electromagnetism, which had been analyzed by Bohr and Rosenfeld and which implied some extremely puzzling aspects in connection with electric charge, measurement, and so on, that all this only made sense because of the zero mass of the photon. You could understand the known facts about electromagnetism and also the presence of quantum principles only because of the zero mass of the photon. There was an approximation involved that had to do with the dimensions of the measuring apparatus, and you couldn't expect the notion of a local quantum field to have any final, definite meaning. I guess I've never forgotten that, So I got this idea early on that we really depend on approximations. And not only do we depend on approximations, but the nature of the theories that we construct depends on certain physical parameters. There is a remark, along the same line, attributed to Bohr, which I recall. If the fine-structure constant were not small, our whole way of looking at quantum mechanics and the real world would be totally different. It is very, very important that the fine-structure constant be a small number in order for matter to be involved in a way that allows us to think about it the way we do. We depend on the smallness of 1/137 very, very much. I found it troubling that most physicists, when they carry out their activities, ignore those considerations, that they never stop to think about the significance of the parameters. Also, George Gamow had an influence on me. I met Gamow extremely early, when I was only 18 years old, at George Washington University. His courses were anecdotal and not very systematic; he picked out the spectacular and glamorous aspects of physics. You know that he also wrote a series of popular books.

CAPRA: Yes, of course.

CHEW: From him, I guess, I must have learned some of these peculiarities of quantum mechanics very early. And those were things you could not grasp within the traditional view. That's right. When you put your finger on it, the fact that quantum mechanics makes sense has a strong bootstrap implication. If the parameters were not right, it wouldn't make sense. I believe that Gamow might have gotten that across in his discussions of some of the paradoxes that arise when you suppose that quantum principles govern the phenomena of our ordinary world. In discussing these examples he must have taken the parameters of the real world and shown that some very good approximation was involved. So that idea that approximation was crucial and that parameters were always important must have come very early.

CAPRA: So all these philosophical influences on you really came from scientists. There was no parallel influence, apparently, from any school of philosophy.

CHEW: Well, I am certainly not aware of any. I realize, when I talk to philosophers, that I know so little about philosophy, it is embarrassing.

5. Recent Advances in the Bootstrap Program

5.1. *Topological Expansion; Ordered S Matrix*

CAPRA: Now I would like to come to the recent history of the bootstrap. What was the decisive advance, and when did you become aware of it?

CHEW: There were many steps which impressed me, and the cumulative effect is a little hard to break up into pieces. I became seriously interested in the new developments in 1974 when I ran into Veneziano at CERN and learned about his notion of a planar approximation. He identified the idea that there was a level in something like a topological expansion, which was topologically planar, where some remarkably simple things happened and the bootstrap became really very much clearer and simpler to understand. Not only that, there were also experimental facts which supported the usefulness of this topological expansion. Shortly thereafter Carl Rosenzweig showed up here in Berkeley. He had been in contact with Veneziano and we started to work together and wrote a few papers which looked quite promising. We were also in touch with a fair number of other people who worked on related things, and then in 1977 we undertook to write a review of those new developments.[7] In the course of getting ready for this review we discovered the concept of what we then called the ordered S matrix. It was a formalization of Veneziano's thinking, but I remember that, when it was presented to Veneziano, he was quite clear in saying that this was something new, that it was something added to his ideas. Again, it is a little hard to say in which way it added, but he did make that statement. Certainly in my own thinking, seeing the concept of the ordered S matrix appear was a big support. It meant that there was a mathematical area which was suitable for bootstrap theory. Up until that point there were a lot of vague statements floating around and we didn't know how to convert them into something that was really a discipline.

CAPRA: Now I want to backtrack a little bit. In the years between 1969 and 1974, in those five years, there were a lot of ideas which were precursors of the new development — duality, the Veneziano model, etc. Did you recognize those as being relevant to the bootstrap?

CHEW: That's a good question. Between 1969 and 1974 1 was aware of these dual models and very interested in them, but I didn't know how to take them and do something with them in terms of the S-matrix framework.

CAPRA: These models were very much associated with quarks.

CHEW: That's right, but it was apparent to me that they were not field theory. There was something else that was going on. Unfortunately, from my standpoint, people succeeded in translating a lot of that into something that was called a string model. The string model has a funny in-between status, which just threw a fog over everything. It is not field theory, but it is Lagrangian theory with arbitrary parameters and uses the space-time continuum as a base. I was quite confused and put off by all the activity that went into string models; it did not seem the right thing one ought to be doing.

CAPRA: In a sense this seems to be a parallel to the early history of S-matrix theory, where people were doing something that would later turn out to be relevant to a new development in S-matrix theory but were doing it from a field-theory perspective.

CHEW: Yes, that's right, So I listened to what people were saying about the string model, but I didn't work on it. What I did during those years was work related to the concept of the pomeron. There were several papers with Pignotti and Snyder on trying to clarify the status of the pomeron, and that set me up for the influence of Veneziano in 1974. What I became sensitive to in those years was the fact that this phenomenon. that was called the pomeron had a lot of simplicity to it, but that there was a mysterious weakness associated with it. This was perplexing to me because it seemed to contradict one of the assumptions of bootstrap theory, which was that strong interactions were self-generating. Here was a piece that had clearly strong interactions with some very simple properties but with a strength that was very weak. I was puzzled; where could this small number come from? I think I wrote several papers in which i made an effort to understand what it was that made the pomeron weak. Well, in 1974, when I went to CERN and talked to Veneziano, that was the thing that really hit me. His topological expansion didn't have the pomeron at the planar level. The pomeron was a correction, and that immediately appealed to me as a natural explanation for its weakness. You see, that was kind of symbolic of the whole new development. Even though strong interactions are strong, nevertheless it is profitable, via the topological expansion, to make some sort of a classification of different levels of strength. I think that was for me a very, very big step; to get over the idea that all of strong interactions need to be understood at the same time. Maybe we could do bootstrap but nevertheless have hierarchies.

CAPRA: At that time you must have felt a tremendous surge of enthusiasm, after these ten years or so of "crossing the desert", as it were.

CHEW: Yes, that's absolutely right. Of course, the enthusiasm did not suddenly come in 1974, It began, and then, as Rosenzweig and I started to work, it built up as we saw more and more things that wanted to emerge.

5.2. *Topology — The Language for a New Science?*

CAPRA: So the really new development was the recognition of order as a new ingredient in particle physics; and topology, of course, is very closely related to order.

CHEW: Right.

CAPRA: From the most general point of view you could say that, when you have that philosophy of "no foundation", you deal with relationships and topology seems to be the language of relationships par excellence. Therefore, it would seem to be the language most appropriate for this whole web philosophy and for the bootstrap idea, I really see a tremendous. potential here. Topology could really be the mathematical language for a new science.

CHEW: Yes, I agree. That's my feeling about it. I tried to say that while David Bohm was here, Bohm has emphasized the importance of language, and I suggested to him that in this problem of getting beyond explicate order, as he calls it, the order of the ordinary real world, our language is extremely prejudicial, because so much of the language that we use is based on explicate order. Bohm knows this and he tries hard to get around it. So what I was proposing to him is that the language of topology, and in particular of graphs, seems very suitable. It is my feeling that in the future some extremely deep questions are going to be approached using the language of graphs.

CAPRA: Now, topology and graph theory are two distinct even though closely related languages. Topology seems the more general framework.

CHEW: I have asked myself that question many times, and I am not sure. I have also asked Poénaru and he is not very definite on this subject. Certainly, graphs without "thickening" are not sufficient, but what I don't grasp is whether this thickening of a graph is all of topology or whether it is only a teeny bit of topology.

At the moment I find it still a puzzling feature of the topological bootstrap theory that there is so much redundancy. Sometimes we find it appropriate to talk about graphs, sometimes about surfaces, and the two are so interlocked that it's hard to know ...

CAPRA: Coming back to the problems of language, you have mentioned several times in discussions we had over the years that the question-and-answer framework of ordinary scientific investigation will be found unsuitable when we want to go beyond the Cartesian framework. Have you given that any more thought?

CHEW: Well, I have come to believe that topological language is a very good candidate for going beyond the question-and-answer framework. The way our theory has developed in the last few years, we quite typically don't know what question to ask. We don't get into the posture of saying: Here is the question; let's try to answer that question! We simply, ...

CAPRA: ... sort of muddle along ...

CHEW: ... muddle along and use consistency as the guide. Each increase in the consistency then suggests something that is incomplete, but it rarely takes the form of a well-defined question. We are constantly downgrading concepts that in the recent past would have been considered fundamental and would have been used as the language for questions. You see, when you formulate a question you have to have some basic concepts that you are accepting in order to formulate the question. This is also related to something that you and I have already discussed, The description of our subject can begin at a great variety of different places. There isn't any clear starting point. So it's very hard to say, when you are writing about this business, that this is the problem that is being attacked and then go from there.

CAPRA: Right. Since the whole system represents a network of relationships without any well-defined foundation, you can start at almost any arbitrary point.

5.3. *Achievements of Topological Bootstrap Theory*

CAPRA: Geoff, I would now like to ask you to give an assessment of the topological bootstrap, sort of a summary of the results achieved so far.[8] Which puzzles, actually, did it really solve? Where were the most important achievements?

CHEW: I am almost sure that what I am going to say will not be what I would say a year hence, because this is a subjective question. As a matter of fact, some of the things that impressed me a number of years ago no longer impress me now, because I have gotten so used to them. Let me see if I can overcome that. One of the important developments has been the explanation of the distinction between strong and nonstrong interactions. The connection of this distinction with the topological expansion has been, I think, a big success. I believe that in the future it will also illuminate the meaning of space-time, because the notion of a continuum, which has some connection with space-time, does not emerge until you get to the electroweak level. The ingredient that characterizes the strong interactions qualitatively is the contraction idea, which goes with zero entropy and which is also very much connected with inaccessible degrees of freedom. We can now see not only how these things cause the strong interactions to be strong; we also understand the order of magnitude of the observed strength. I think that's an enormous step forward. However, it will take further development of the theory to convince other people of this. We also understand why the machinery of quantum electrodynamics with its reliance on space-time should work and why strong interactions can co-exist without having any such underpinning. All interactions share the S matrix — the analyticity properties, unitarity — all these notions co-exist, and yet we get separation between strong and electroweak interactions.

CAPRA: Now let's talk about the strong interactions themselves. I think the understanding of the nature of quarks would have to be counted as one of the big successes there.

CHEW: Oh yes, and going into more detail I would say the meaning of the mysterious color degree of freedom has been exposed and the meaning of quark generation tentatively brought out, although we are still not completely sure about quark generation.

CAPRA: Baryon number conservation has also been understood.

CHEW: Yes, baryon and lepton number conservation, electric charge quantization — all these famous quantization characteristics have been explained.

CAPRA: How would you now characterize the quark concept?

CHEW: The quark concept is an extremely useful way of describing part of the order, which is present in the strong interactions. A more detailed description would be the following one. In contrast to the standard approaches, we do not identify the quarks with the lines of the Feynman graph. When you express the order you thicken the Feynman graph by embedding it in a bounded surface, and then it is the boundary of the surface which houses the quarks.

CAPRA: More generally, can you say that, if particles are relationships then quarks are patterns in these relationships?

CHEW: Yes, you can say that.

CAPRA: And this is what has been understood. The relationships are not arbitrary; there are constraints, which produce patterns, and these patterns are the quarks. Now the confinement is also something that has been understood.

CHEW: Quark confinement is automatic from that standpoint; because quarks don't carry momentum; they are not identifiable as particles.

CAPRA: Finally, the fact that all particles are built of two topological elements seems to be a major result.

CHEW: Yes, the theme of twoness is pervasive in this theory, and the full significance of that feature has yet to be understood. We are finding more and more two-valued quantities to be required by consistency, and their combinatorics control the whole business.

CAPRA: Might that be connected with the two dimensions of the surfaces we are using?

CHEW: I don't think so. It is rather that the idea of an *orientation* is two-valued. We have two kinds of boundary units, and these units come with a variety of two-valued orientations. Out of these patterns of twoness we think we have understood the major characteristics of the strong interactions and quite a number of things about electroweak interactions.

CAPRA: So this pattern of twoness is now more important than the threeness that was emphasized before — the three sides of the triangle, the three colors, etc.

CHEW: I am hesistant to say that. The threeness comes from the triangle or, you can also say, from the cubic vertex, A graph becomes nontrivial as soon as it develops a cubic vertex, and from the cubic vertex you can build everything. So you need threeness in addition to twoness.

CAPRA: As far as particles are concerned, the theory gives a finite number of particles, right?

CHEW: If you speak of *elementary* particles, that's right, a finite but large number. I once counted how many elementary hadrons there are. There are about 18,000 of them.

CAPRA: 18,000?

CHEW: Yes, most of them are particles with 6 topological constituents that we have begun to call "hexons". There are maybe 200–300 elementary mesons, 1000 elementary baryons and all the rest are hexons. We are expecting the majority of physical hexons to have too large a mass to be detected with present accelerators.

5.4. *Outstanding Problems*

CAPRA: Now, Geoff, what would you say are the major outstanding problems in the topological bootstrap? Let's first talk about hadron physics.

CHEW: In hadron physics the identifiable major outstanding problem is the breaking of generation symmetry. We have not yet understood how that comes about. We have some candidate ideas, and they all have to do with the coupling of strong and electroweak interactions. Without going to electroweak interactions, we do understand the breaking down of the hadron supermultiplets, for example the differences between mesons and baryons, or the difference between a ρ and a π-meson. But for distinguishing between the generations, for example between the π and the K-meson, you have to go to electroweak interactions. The infrared phenomenon is another major area to be developed, both from the practical and conceptual standpoints. The fact that we start with zero-mass electroweak particles is of great significance, both quantitatively and qualitatively, and one of the main things for the future is the development of a real theory of measurement.

CAPRA: I want to come to this whole question later, because that is a big question.

CHEW: Right.

CAPRA: In connection with hadrons, can you think of other major outstanding problems?

CHEW: For the hadrons ... the hadrons ...

CAPRA: What about the zero-entropy bootstrap?

CHEW: That's a technical problem. I don't see it as a tremendous puzzle. I think it's a challenge to our ingenuity to figure out better ways of solving that problem. We do need some new ideas, but I wouldn't put that in the category of major outstanding problems.

CAPRA: That really attests to the tremendous success of the topological bootstrap for hadron physics.

CHEW: Yes, I am hard put to think of qualitative questions that have not been explained. All the quantum numbers — the parities, the spins, and all that — all of that seems to be understandable.

CAPRA: With respect to electroweak interactions, of course, the whole theory is still changing.

CHEW: That's right.

5.5. *Skepticism of Orthodox Physicists*

CAPRA: Now, if the topological bootstrap has been so successful in hadron physics, why is there such tremendous skepticism among orthodox physicists? Why do they hesitate so much to accept it and what will convince them?

CHEW: I think there are various ingredients in the answer. One is that QCD was very successfully sold as *the* theory of strong interactions. Although the deficiencies of QCD are recognized by many people, there is this prevailing sense that it is the correct theory. Some of the things that it doesn't explain and which the topological theory explains would be, for example, the origin of quarks, the origin of three colors, the origin of all these quantum numbers which you simply have to put into QCD. But people have gotten used to putting things in, and they have forgotten that one needs to answer questions about why such ingredients are there in the first place. At some point they *will* come around to thinking about that. Part of the problem, Fritjof, is that the topological theory has not been presented in any easily understandable way. You wrote a review and I wrote a review,[8] but in neither of these reviews did we go into any detail about Feynman rules.

CAPRA: In other words, nobody can apply the theory.

CHEW: Nobody can apply it, that's right. I think there are lots of young people around who would do things with it if they could get their hands on it. Maybe I made a mistake in not sticking with the strong interactions; I became deeply involved in this problem of the electroweak interactions.

CAPRA: So what happens is that practically all the people working on the topological theory work on the fundamental development of the theory, and nobody worries too much about applications.

CHEW: Well, there are some who do, like Nicolescu.

CAPRA: Yes, but since the entire number is very small those are just one or two.

CHEW: That's right. On the other hand, given the state of QCD at the moment, I think that people are subconsciously aware that QCD is not doing the job, but without an alternative they are not recognizing this explicitly.

CAPRA: And you think the feeling that QCD is not living up to the expectations is spreading?

CHEW: Well, my best measure is the graduate students here, During this last year I repeatedly found graduate students who were not satisfied with QCD; there was a lot of interest in the topological approach at the level of graduate students.

5.6. *QCD and Weinberg-Salam Theory*

CAPRA: Maybe, at this point, I could ask you to summarize your criticism of quantum field theory, that is both of QCD and of the Weinberg-Salam theory.

CHEW: My position derives from the way the topological theory has been evolving and from the fact that it makes so much sense. What has developed is that you can associate with both weak and strong interactions a Feynman expansion. That goes somewhat beyond what we used to say. We used to think that strong interactions would be pure S matrix and would not admit an off-shell continuation. The way the theory has evolved, there is an off-shell continuation. This was painful to accept; we did not really want it; we thought that it was contrary to the S-matrix spirit, but after long discussions we came to the conclusion that you really had to extend off-shell and that this did not upset the essential ideas. More importantly, the theory tells you how to go off-shell; it isn't something that is left open. But this extension does not imply a field theory. This is the strange thing. For a long time people said: if you go off-shell this is equivalent to field theory. That's not true.

CAPRA: So you don't have local interactions?

CHEW: You don't have local interactions, even though you have Feynman rules. The big difference is that the vertices for strong interactions are not simple polynomials in momentum.

CAPRA: Because of contraction?

CHEW: Because of contraction. Each hadronic vertex represents the contraction of an infinite number of topologies, so that the function associated with the vertex has an infinite sequence of singularities. This is qualitatively different from what you get in ordinary field theory. If you Fourier-transform a topological vertex function you will not get a local space-time interaction. This is the situation for the strong interactions. Now, the extension to electroweak interactions, right from the beginning, has not admitted contractions. The logic of this situation has never been explored as carefully as would be appropriate. Next year I hope to start writing a complete and systematic account of the whole thing, and then I will try to work out why the electroweak topologies do not admit contractions. There are all sorts of indications that they should not admit contractions; therefore the vertices for their Feynman rules are local if you Fourier-transform.

CAPRA: So then you do have local fields.

CHEW: Yes, we think that you can define local fields. There is a tricky point here. The locality really is not a property of a single field; it's a property of their interactions; for example, you have three fields interacting at the same point in space-time with each other. So it's a little bit misleading to talk about just local fields individually; it's a matter of the Lagrangian.

CAPRA: And this locality is a feature of the electroweak topologies?

CHEW: Right. Without having insisted on locality, it seems that the noncontraction associated with the electroweak interactions implies that the interactions are local,

CAPRA: Regarding your criticism of the orthodox theories, there seem to be two key issues involved then, the locality and the arbitrariness.

CHEW: That's right, and the two are connected, because it was contractions that eliminated the arbitrariness for the strong interactions, where the zero-entropy level is completely controlled by contractions. Now, you can then ask: what is it that makes the electroweak theory non-arbitrary if you don't have contractions. Here we are in a fuzzy area, but there are at least two recognized sources of non-arbitrariness for electroweak theory. Firstly, segments of surface boundary are building blocks of *all* elementary particles. They control strong interactions and you use the same objects to build electroweak particles. If you just said that, however, there would still be a fair amount of arbitrariness.

CAPRA: Well, the other link would be via the "naked cylinder".

CHEW: That's right, but I think it's probably true that I am the only one so far who has paid much attention to this. Of course, the argument is not tight; it has been loose. Finkelstein and Poénaru have been sympathetic to my reasoning, but . . .

CAPRA: This is probably because they came in later, after you had worked out the whole cylinder business. So you probably have a better feeling for it.

CHEW: That's probably true. We were talking before about this strange period between 1969 and 1974, when I was working on pomeron theory. Now that was all cylinder! So later, when the electroweak stuff came along, I immediately saw it as cylindrical. It was always cylindrical, right from the beginning.

CAPRA: So the arbitrariness is reduced, or even eliminated, by that link to the naked cylinder.

CHEW: I think that it will eventually be eliminated, but it is reduced already.

CAPRA: So what, then, is your assessment of QCD and the Weinberg-Salam theory on the basis of the topological theory?

CHEW: I can see Weinberg-Salam as basically O.K. It's a question of what the full group is; what they have done is to identify the dominant low-energy subgroup. The idea of using a gauge field theory for describing the electroweak interactions seems to be supported by the topological approach.

CAPRA: So you expect the Weinberg-Salam theory to be recuperated more or less completely by the topological approach.

CHEW: Yes, it almost has been already. We have come close to recapturing the whole thing. What we are doing now is trying to figure out the part of the story that has not already been guessed by Weinberg and Salam.

CAPRA: And what about QCD?

CHEW: QCD is a local field theory, which treats strong interactions exactly the same way as electroweak. It is a naive extension of Weinberg-Salam; you take Weinberg-Salam and you replace SU(2) by SU(3), Otherwise, you do very little to it.

CAPRA: Which means, you ignore the fundamental difference between strong and electroweak interactions.

CHEW: Right,

CAPRA: They would argue, of course, that this is justified by the phenomenon of asymptotic freedom.

CHEW: Yes, they say that there is another scale at which these things are unified, but there are qualitative differences beyond that. The degree of freedom associated with SU(2) is isospin, or electric charge, which is an observable degree of freedom labeling the particles. The analogous degree of freedom for SU(3) is color, which is not observable. Now that is a clear qualitative difference between these two situations, and there isn't any logical explanation why you should get confinement in the one case and not in the other, They set up the two theories in completely parallel fashion, and then they just say you get confinement in one case and not in the other.

CAPRA: Now, in the topological theory, can you relate confinement to contraction?

CHEW: Yes. We are talking here about color confinement, which in topological theory means that color is an inaccessible degree of freedom, similar to the cyclic order of lines around a vertex. It is another kind of order which you sum over. Now, if you don't have contractions, you don't have inaccessible degrees of freedom either. The inaccessible degrees of freedom and the idea of contraction go together. There are no contractions without inaccessible variables and in electroweak interactions all variables are accessible.

CAPRA: That is an interesting connection.

CHEW: Yes, it is tremendously interesting. This is one of the things I am going to try hard to get straight next year.

CAPRA: Coming back to QCD, do you think that it will be thrown away entirely?

CHEW: Yes, as a fundamental theory, because it is fundamentally wrong. As a phenomenological theory for GeV-scale physics, part of QCD will survive. It gets certain things right, such as energy-momentum conservation and Lorentz invariance, and people have put in by hand the right number of degrees of freedom. They looked at experiment and said, we see there are three colors, and there are two spins on the quark, and they inserted these features, so the theory is bound to have phenomenological utility, but it's profoundly wrong.

CAPRA: Now if QCD is wrong, then the grand unification schemes will have no future either.

CHEW: No, they will not. By the way, topological bootstrap theory predicts absolute stability for the proton, in contrast to most grand unification schemes.

6. Outlook

6.1. *Space-Time Continuum and Electromagnetism; Gravity*

CAPRA: I would now like to talk about future developments, and to begin with I would like to discuss the concept of the space-time continuum with you. During the course of our conversation you have repeatedly been critical of the use of a space-time continuum, and you have also said that the paradoxes of quantum mechanics arise, in your view, because in the framework of quantum mechanics atomic phenomena are embedded in a continuous space-time. Now, evidently, atomic phenomena *are* embedded in space-time. You and I are embedded in space-time, and so are the atoms we consist of. Space-time is a concept that is extremely useful, so what do you mean by the statement that one should not embed atomic phenomena in space-time?

CHEW: Well, first 'of all, I take it as obvious that the quantum principles render inevitable the idea that objective Cartesian reality is an approximation. You cannot have the principles of quantum mechanics and, at the same time, say that our ordinary ideas of external reality, the explicate order, are an exact description. You can produce examples showing how a system subject to quantum principles begins to exhibit classical behavior when it becomes sufficiently complex. That is something which people have repeatedly done. You can show how classical behavior emerges as an approximation to quantum behavior. The WKB approximation is a famous example, and there are lots of others.

CAPRA: Well, more generally, you can derive the basic laws of Newtonian physics from quantum physics...

CHEW: ... as approximations, when things become adequately complex. So the classical Cartesian notion of objects, and all of Newtonian physics are approximations, I don't see how they can be exact. They have to depend on the complexity of the phenomena which are being described. If the phenomena become too simple, the classical description won't work.

CAPRA: So you have a quantum level at which there are no solid objects and at which those classical concepts do not hold; and then, as you get to higher and higher complexity, the classical concepts somehow emerge.

CHEW: Yes.

CAPRA: And you are saying, then, that space-time is such a classical concept.

CHEW: That's right. It emerges along with the classical domain and you should not accept it at the beginning.

CAPRA: And now you have also some ideas about how space-time will emerge at high complexity.

CHEW: Right. The idea, actually, was present in a qualitative form long before and probably goes back right to my graduate-student days. When you learn quantum field theory, you learn that the quantum fields are introduced as operators, just like the operators in standard quantum mechanics, but they don't correspond to observables.

CAPRA: Except for the electromagnetic field.

CHEW: Except for the electromagnetic field; that's right! That's the only exception. And that just hit me like a bomb. Why should the electromagnetic field be different? I never forgot that, and I was constantly amazed that my colleagues, who did field theory, never seemed to pay any attention to that. It never fell out of my consciousness that there had to be some very special feature connected with the electromagnetic field. In my student days the idea was put into my head that the possibility of the electromagnetic field being an actual observable, in the same sense, as say, momentum, was associated with the zero photon mass. It is actually associated with more than that, but I remember being told that zero mass was the essential ingredient. And as time went on and I began to think about the S matrix in connection with strong interactions, I was much impressed by the fact that electromagnetic fields which could be described classically would not tolerate an S-matrix description. There was a funny complementarity. The zero photon mass, which allowed the electromagnetic field to have the status of a classical observable, also means that you cannot describe photons by the S matrix.

CAPRA: Is this statement still true in view of the recent successes of topological electroweak theory?

CHEW: It is still true because of the infrared phenomenon. At our present level of understanding, we have to think of the electroweak particles as massless, and that leaves this very important infrared problem still to be faced.

CAPRA: And we might have to go beyond the S matrix to solve it?

CHEW: I am sure. Henry Stapp has already taken the first step in that direction by showing how you can start, basically, with an S-matrix approach but then, when you are faced with these zero-mass particles, have to change the basis from the asymptotic states which are based on counting individual numbers of particles to asymptotic states where you superpose coherently different numbers of particles. Now, when you are superposing states containing differing numbers of particles you are not, strictly speaking, going outside the S-matrix idea, because the S matrix allows superposition, but it is not the way we normally use the S matrix. The idea that these states are then measurable as superpositions, that's where you get into a new area. What is it you are measuring when your apparatus is sensitive to a superposition of states with an indefinite number of particles? But that is exactly what a classical apparatus does. When you get to the classical level, you are using the fact that soft photons are coming in whose number is not well-defined. I want to emphasize again that the idea of the S matrix implies such an extension should be possible. The statement that the S matrix is unitary is based on the idea that you can superpose the basic states. You can always give a meaning to an arbitrary superposition. In fact, you can't prove unitarity unless you assume that an arbitrary superposition of states is possible. That means also states of different numbers of particles. It is not a notion that you normally use in S-matrix theory, but it really is implied by unitarity. The thing which is perhaps qualitatively new is this idea that the superposition should involve states of arbitrarily large numbers of particles.

CAPRA: So the very special nature of electromagnetism stayed in the back of your mind during all those years while you were developing the S-matrix framework?

CHEW: Oh yes. Because the zero mass kept electromagnetism out of the standard framework and because it allowed the status of a classical observable for the electromagnetic field, I have always been very much impressed by the importance of electromagnetism, and you will find occasional references to it in my papers, here and there.

CAPRA: I know. In fact, I would like to read to you a passage from a paper you wrote in 1971:[9]

> *Electromagnetism is deeply mysterious and its origin unlikely to be explained within our current scientific framework because the unique attributes of this interaction are inextricably enmeshed with the framework itself.*

In view of the recent progress, have your views changed since you wrote this passage?

CHEW: No.

CAPRA: But you expect now to unravel a little more of that mystery don't you?

CHEW: That's right, but we *are* departing from the scientific framework. That's one of our problems of communication, because in challenging the standards of space-time we are off into the unknown, to some extent. Now we can conceal it when we publish papers and talk to people, and people won't know that we are really...off the deep end (laughs). But the fact is, you are, as soon as you say that the meaning of space-time is coming from the graphs and not vice versa.

CAPRA: So your statement still stands; and you are saying the more we learn about electromagnetism the more we are forced to give up the standard scientific framework. It is at that expense that we can learn about electromagnetism.

CHEW: That's correct.

CAPRA: Now we got into this discussion of electromagnetism by talking about space-time.

CHEW: I know. You see the understanding of what a measurement is, of what space-time is, of what an observer is, of what electromagnetism is — all these are tied together. The language that is useful for discussing this is the language of graphs, and the key notion is the idea of gentle events.[10] This idea is uniquely associated with photons. What it means is that in the Feynman graphs there is a special kind of vertex where two of the lines are very closely connected — what is carried in on one line is almost exactly carried out on the other line — and the third line is the photon, which is bringing almost nothing in or out.

CAPRA: That is, a very slight disturbance.

CHEW: Very slight disturbance, and it is in the analysis of why this is possible that you understand the special features of the photon. It has to have zero mass, and it also turns out that it has to have spin 1. It has to flip the chirality, that's the strange thing. Connected with spin 1 , furthermore, is the attraction of unlike charges and the repulsion of like charges, which produces the effect of large clumps of matter that are almost electrically neutral. That is necessary for the appearance of classical reality. A classical object is recognizable as such, partly, because it doesn't carry an enormous electric charge.

CAPRA: So there are these gentle events, and they pile up.

CHEW: Yes, and they pile up coherently, and these coherent superpositions of photons generate the classical fields.

CAPRA: And how is this connected with space-time?

CHEW: Well, from the point of view of a quantum starting point, everything is discrete in the beginning; there is no continuum, and the discreteness is represented by the vertices of graphs. The graphs have no metric associated with them; there is no meaning to the distance between two vertices. These vertices I would call the "hard" vertices, and they are interspersed with all the "gentle" or "soft" photon vertices. There will be an infinite number of superpositions of these gentle vertices, whereas the hard vertices will remain finite and discrete. What Stapp showed, in order to solve the infrared problem — the specific mathematical problem associated with the zero mass of the photon — is that you are led to certain particular superpositions which, in effect, approximately localize the hard vertices. So after you have added this infinite coherent superposition of soft photons a hard vertex, which to begin with had no sense of space-time localization, acquires an approximate localization.

CAPRA: And the more soft vertices you have, the more precise the localization?

CHEW: Well, it doesn't ever localize beyond a certain limit.

CAPRA: You mean the limit given by the uncertainty principle?

CHEW: Yes. You don 't ever get arbitrary localization, but you discover that you get as much localization as you expect in order to have a classical interpretation.

CAPRA: Now, you are saying that you could derive the uncertainty principle from that?

CHEW: Oh sure, You can derive the uncertainty principle and you can derive space-time. The idea of deriving space-time, of course, is really the more striking.

CAPRA: So what emerges is the notion of continuity and the notion of approximate localization.

CHEW: Right.

CAPRA: And with it the uncertainty principle.

CHEW: Yes. And, I would say, with- it a meaning for measurement.

CAPRA: I see. All these steps are still tentative, though.

CHEW: That's right, But out of this will also come, at the same time, the capacity for recognizing certain patterns of events as representing an observer looking at something. Once you have the gentle-photon idea in the picture, you can begin to do that. In this sense, I would say, you can hope to make a theory of objective reality, But the meaning of space-time will come at the same moment. You will not start with space-time and then try to develop a theory of objective reality.

CAPRA: Now what about the dimensions of space-time? That does not follow from the gentle events.

CHEW: No. I think that comes out of the topology, and we are getting some very, very strong hints about that right now. The fact that momentum has four components is associated with a 2×2 matrix, and we suspect very much that this matrix is related to these pairs of two-valued orientations, that is to the apparent twoness inherent in the nature of elementary particles. So it seems that these topological notions can be translated into the statement that momentum must be a four-component object.

CAPRA: What about the conservation of energy and momentum?

CHEW: Well, that is something we don't have such a specific idea about, but I have a conjecture about it. Momentum appears to be built from indices which are closely connected to spin indices, and the same kind of topological symmetry that gives rise to $SU(2)$ isospin invariance also seems to give rise to $SU(2)$ spin invariance, which is rotational invariance and means angular momentum conservation. So the topology does definitely promise to explain why angular momentum is conserved. Now, if spin and momentum are built basically from the same topological indices and you can get conservation of angular momentum, it seems to be very plausible that you are also going to get conservation of linear momentum and energy. I feel that sooner or later we will understand that.

CAPRA: Now I have another question. Considering that you envisage deriving the basic properties of space-time, what about the constancy of speed of light? Will that be derived too?

CHEW: Well, that somehow just comes along with the Lorentz invariance. If you can get rotational invariance, the analytic continuation of the rotation group will lead you to the Lorentz group, and the Lorentz group effectively contains this notion of the speed of light.

CAPRA: So you foresee that all this could really come out of the topology.

CHEW: Oh yes. I don't think that's implausible.

CAPRA: You have often said that you expect gravity, too, to emerge together with continuous space-time in the high-complexity limit.

CHEW: Yes, that's true.

CAPRA: If so, then what about gravitons?

CHEW: From the topological standpoint you don't start with gravitons; you don't have gravitons on the same footing with the elementary particles. Gravitation is expected to emerge as a manifestation of extremely high complexity, *beyond* the level at which space-time becomes recognizable. If- gravitation has meaning only at a classical level, there may be no gravitons.

6.2. *Extending the S-Matrix Framework*

CAPRA: Looking into the future, I am curious about the necessity of either extending or abandoning the S-matrix framework.

CHEW: We have already extended it in two ways, actually. We have extended it off-shell and also by changing the basis in connection with the infrared problem.

CAPRA: It seems now that this extended framework of the S matrix is appropriate for all of particle physics.

CHEW: If it is understood in the sense of being supplemented by a topological expansion, I do think so. However, there is something else. When you say "*the* S matrix is the basis for understanding everything," I would object and say that the S matrix is defined in terms of momentum. But the whole notion of momentum, as we have been discussing in the last ten or fifteen minutes, promises not to be something you have to accept on an *a priori* basis, but will come from somewhere else. Now where does it come from? In the end, the meaning of momentum has to come from large-scale objective reality. The striking point about momentum is that in order to measure it precisely, you need large-scale apparatus. The more accurately you measure momentum, the bigger is the apparatus you are using. And the fact that high complexity is involved is somehow essential. So there is some elusive quality to the conceptual structure. There is another dimension which should also be mentioned, and that is cosmology. It is the aspect of the story which says that the ideas of physical phenomena and physical laws are somehow localized in time. That also has to be an approximation. Any sophisticated cosmological approach will tell you that the conditions which lead to our sense of Cartesian reality don't necessarily always exist.

CAPRA: So there is no abstract, eternal validity of the laws of nature?

CHEW: No, I think we are surely making an approximation.

CAPRA: Now, coming back to the bootstrap program, in the past you have often expressed the idea of a mosaic of interlocking models. This idea has been of tremendous value for people outside physics. I know many people who have been inspired by it. But it seems now that in particle physics it might no longer be necessary.

CHEW: Well, let me think whether I agree with that or not... I think there are two considerations to bear in mind. One of them is that already at the level of the zero-entropy bootstrap one may not be able to find an exact solution. One may always have to use some kind of model. So the notion of a mosaic of models could be relevant right there at zero entropy. For example, Espinosa has been doing just that in his thesis. He is making models which are appropriate to certain portions of the zero-entropy space.

CAPRA: So what you felt would apply to the entire hadron physics applies now to the nonlinear part, which is confined to zero entropy.

CHEW: Could be. It may that we shall be condemned to always understand zero entropy in a piecemeal fashion. The other point is that somewhere consciousness has to enter the picture. So far I have been talking about using soft photons to develop an understanding of space-time, objective reality, and the Cartesian-Newtonian view. But you know that in this domain of high complexity there is something else which, in a vague way, people describe by this term "consciousness," and which has an actual impact. The fact that I am raising my hand to gesticulate...

CAPRA: Of course, but there is much more to it. There is the hand itself, which you will not explain in the S-matrix framework. There is the whole world of living organisms.

CHEW: That's right. But that has to affect the accuracy of our description of the physical world. You can't have a completely accurate description of the physical world which leaves out consciousness, because it is clear that consciousness interacts with the physical world. How can I possibly have a complete description of the physical world if I don't include consciousness in the story? At the moment we work with a model that neglects consciousness; it's an approximation.

CAPRA: It's obvious that the inclusion of consciousness would give a more complete picture, but will this actually be necessary to make progress in our understanding of physical phenomena?

CHEW: Eventually, surely,

CAPRA: At what point do you think that we will be forced to go into this domain?

CHEW: I don't have well-developed views about at what level consciousness will be necessarily the concern of physical scientists, but I feel sure that the time has to come. It is clear that if you push your desire for complete understanding far enough, eventually you will have to bring it in.

References

1. G. F. Chew, *Phys. Rev.* **112**, 1380 (1958).
2. G. F. Chew and S. Mandelstam, *Nuovo Cimento* **19**, 752 (1961).
3. G. F. Chew, *S-Matrix Theory of Strong Interactions* (Benjamin, New York, 1962).
4. R. J. Eden, P. V. Landshoff, D. I. Olive and J. C. Polkinghorne, *The Analytic S Matrix* (Cambridge University Press, 1966).
5. G. F. Chew, *The Analytic S Matrix* (Benjamin, New York, 1966).
6. G. F. Chew, *Science* **161**, 762 (1968); *Physics Today* **23**, 23 (1970).
7. G. F. Chew and C. Rosenzweig, *Phys. Rep.* **41C**, No. 5 (1978).
8. See also G. F. Chew and V. Poenaru, *Z. Phys.* **C11**, 59 (1981); F. Capra, *Surveys in High Energy Phys.* **4**, 127 (1984).
9. G. F. Chew, *Phys. Rev.* **4D**, 2334 (1971).
10. See G. F. Chew, *Zygon* **20**, No. 2, 159 (1985).

The Bootstrap Principle and the Uniqueness of Our World

Basarab Nicolescu

CNRS, Paris, France and
University Babes-Bolyai, Cluj-Napoca, Romania
19 villa Curial, 75019 Paris, France

We present an history of the bootstrap principle. We also investigate its philosophical implications[*]. We conclude with personal recollections about Geoffrey Chew.

1. Eddington and the Epistemological Principles

The English astronomer and physicist Arthur Stanley Eddington (1882–1944) is the first scholar who was able to reformulate the old problem regarding the possible uniqueness of our world in modern terms.

At the origin of Eddington's reflection lays the recognition of certain laws of modern physics, such as Heisenberg's uncertainty relations, regarded as laws with an absolutely novel character, authentic *epistemological principles*. This observation, in itself, is not absolutely original. The uniqueness of Eddington's thinking is manifested in the relationship that he postulates between these epistemological principles and the observer him/herself.

Eddington is convinced that all the fundamental assumptions that determine the development of physical theories can be replaced by epistemological principles. "All the usual laws of nature, considered as fundamental," writes Eddington, "can be entirely provided by epistemological considerations."[1] Epistemological principles are therefore more general than the fundamental laws of nature: *a very limited number of epistemological principles can explain a large number of fundamental laws of nature.*

According to Eddington, epistemological principles themselves can be deduced based on *the sensory and intellectual knowledge of the subject.* To the extent that this structure is unique, reality itself should be unique. The observer is integrated in the reality that he/she observes.

Eddington's ideas are summarized in what is called *Eddington's principle:*[2] all

---

[*]This includes a revision of a section of *From Modernity to Cosmodernity: Science, Culture, and Spirituality* by B. Nicolescu, published by SUNY Press [2014].

the fundamental statements of physics can be deduced using logical reasoning, based on qualitative statements: thus, pure numbers of physics that are *constant* (such as the constant of the fine structure, characterizing electromagnetic interactions) can be derived without recourse to any experimental data.

Of course, this principle is concerned with the most general aspects of nature, related to the *invariance* of physical laws: the radius of the atom or the radius of Earth would never be deduced, taking into account epistemological principles.

The assertion that Eddington's ideas demonstrate an extreme idealism is false. The outside world remains more necessary than ever: "We refuse to admit," writes Eddington, "the probability that the outside world, despite the difficulties we face to get to it, could be disqualified because it would not exist ... The outside world is the one examined by our common experience and, for us, no other could play the same role ..."[3]

It is well known that Eddington's ideas were very poorly received both by philosophers and by physicists, who saw Eddington's approach as a consequence of an intolerable confusion between philosophy, epistemology and science: how could one accept the idea that epistemology becomes the engine of science, taking into consideration the most general and fundamental parts of it?

The implicit question, which is the core of Eddington's reflection, still remains of great current interest: could the arbitrary be tolerated in a theoretical description of reality — whether it is about concepts (why are there three dimensions of space and one dimension of time? why is there a well-defined number of quarks?) or about pure constants (why has the constant of fine structures the value it has?)? The fact that scientific experiments show the validity of a particular scientific concept or the numerical value of a certain constant is not an answer to our question: we want to know the *what for* about things. In fact, the progress of a scientific theory is related to the elimination of arbitrariness from the scientific description: a theory based on a small number of axioms is rightly considered far more interesting than a model tolerating hundreds of concepts and arbitrary parameters, in order to describe the same set of phenomena. Some physicists believe that a certain degree of arbitrariness will always exist though. But there is also the logical possibility of a total elimination of the arbitrary.

2. Unity and Self-Consistency: The Bootstrap Principle

The *bootstrap* hypothesis emerged first as a possible explanation of certain experimental data in particle physics. This hypothesis was formulated for the first time in 1959 by Geoffrey Chew[4] and was immediately used, for detailed physics calculations, by Chew and Mandelstam.[5] The word *bootstrap* itself is untranslatable. Indeed, bootstrap, in the proper sense "laces," also means "to levitate while dragging your boots." The most appropriate term in translation would be that of *self-consistency*.

The bootstrap theory has emerged as a natural reaction against classical realism, which received a death blow, and against the idea, to which it was associated, of a need for equations of motion in space–time, during the formulation of quantum mechanics, around 1930.

According to Newton, we learned about the existence of equations of motion, in order to describe physical reality: Newton's equation regarding macroscopic bodies, Maxwell's equations for electric and magnetic fields, and Schrödinger and Dirac's equations for the movements of atomic systems. The movement described by these equations is that of certain entities considered as fundamental building blocks of physical reality, defined at each point of the space–time continuum. By definition, these equations possess an intrinsic deterministic character (the fact that, in some cases, large ensembles of objects can lead to a chaotic behavior does not alter the deterministic character of the basic equations of motion).

Quantum entities are not subject to classical determinism. The bootstrap theory is just drawing the logical conclusions of this situation by proposing the abdication of any equation of motion. This attitude is consistent with the schedule of the matrix S ("S" is the initial for the English word *scattering*) initiated by Heisenberg in 1943:[6] a realist theory must be expressed in terms of quantities directly related to experimental observation.

The abdication of any equation of motion has an immediate consequence: the absence of any fundamental brick of physical reality. "After all," writes Chew, "if one accepts the need for an equation of motion, this means that there is a certain fundamental entity that is in motion; this cannot be a particle — it must be something different, maybe a field. As I noted, there might be *no* fundamental entity, field, or something else."[7]

Here is a more precise definition of the bootstrap, given by Chew: "The only mechanism that meets the general principles of physics is the mechanism of nature;...the observed particles are the only quantum and relativistic systems that can be designed without internal conflict...Each nuclear particle plays three different roles: (1) *constituent* of the compounds; (2) mediator of the force responsible for the overall cohesion of the compound; (3) *composed* system..."[8]

In this definition, the part appears simultaneously as the whole. Nature is conceived as a global entity, *inseparable* at a fundamental level. The particle plays the triple role of a *system* in the irreducible interaction with other systems, which is a first rapprochement between the bootstrap theory and the current systemic thinking.

What is in question in the bootstrap theory is actually the notion of a precise identity of a particle, substituting the notion of a relationship between events, the *event* signifying the context of the creation or annihilation of some particles. Responsible for the emergence of what is called a particle are the *relationships* between events. There is no object in itself, possessing its own identity, which can be defined as separate or different from other particles. A *particle is what it is because all other particles exist simultaneously*: the attributes of an entity that is physically determined are the result of the interaction with other particles.

The bootstrap is therefore a vision of the world's unity, a principle of nature's self-consistency: *the world built on its own laws through self-consistency*. It represents a break with the scientific thinking based on the tradition of Democritus and Newton: the concept of fundamental (and thus *arbitrary*) entities, characterizing material *substance*, is replaced by a concept of the organization of matter — that of self-consistency. It should be noted that the bootstrap principle is both *organizational* and *structuring*: an *infinite* number of self-consistency conditions determines the existing particles in a *unique* way.

The only compatible world with the laws of nature (which are themselves deducible, in principle, through self-consistency) is, according to the bootstrap principle, the world of nature: it is impossible to find a system without internal contradiction on the logical plane, and which is at the same time, in agreement with everything that is observed or will be observed. In other words, we can only make a series of approximations. It is a confirmation of modesty when facing the rational order of nature, but also a statement of hope.

The bootstrap has important implications for the nature of scientific prediction. The knowledge of the whole claims a long and patient investigation. Therefore, it must be admitted that the ways of explaining the self-consistency of nature are researched, ensuring the fact that behind the approximation hides a fundamental coherence, a rational order without gaps: once obtained partial information on the real world, the knowledge of the rest of the world is not arbitrary — it is obtained by self-consistency.

Obviously, there are different degrees of generality in the formulation of the bootstrap's principle.

Thus, the conception of a postulation of a very general form of the bootstrap principle is very logical, including not only particles, but also macroscopic bodies, life and even consciousness: the self-consistency of the whole requires including all aspects of nature. Under this very general form, the bootstrap principle has, with the current state of knowledge, an *unscientific* character. The fact was emphasized by Chew himself in an article published in 1968: "Bootstrap: A Scientific Idea?"[9] Thus, we have to make a difference between the entire, unscientific bootstrap, and the partial, truncated *bootstrap*, which can be scientifically productive and effective. Such a partial *bootstrap* — the *hadron bootstrap* — is the one which enjoyed the interest of physicists in the decade 1960–1970.

3. Is There a Nuclear Democracy?

The partial bootstrap proposed by Chew in 1959 refers to the world of hadrons (like the proton or the neutron) which have a strong interaction, which is exercised at the atomic nucleus scale. Most of the currently known particles are included in this category. Other particles such as leptons or photons are excluded from *the hadron bootstrap*, which is already an inherent limitation of the approach.

The hadron bootstrap is formulated within the general framework of the S matrix theory.[10] The essential idea of this theory is that of focusing efforts on the scientific description of the hadrons, as they appear *before* and *after* their interaction. An element of the S matrix describes a specific reaction between particles, depending on the quantity of motion and on the initial and final hadrons' spins (*the spin* is the intrinsic angular momentum of the particle). It is not described what happens at the very moment of the interaction. Thus, it is not necessary to introduce a space–time at a microscopical scale: macroscopical space–time is sufficient for the experimental definition of the quantities of motion and of the hadrons' spins. In particular, it is not necessary to define the position of a hadron.

These elements of the S matrix describe all the reactions that can be designed between hadrons. They represent the amplitudes of probability, regarding the probability that a given set of measures in the final state comes next after a given set of baseline measures. These amplitudes of the probability of transition directly correspond to measurable quantities.

A true theory must therefore be able to predict the ensemble of these elements of a matrix that is the S matrix itself. The hadron partial bootstrap postulates that it is possible that a single S matrix could be predicted based on some very general restrictions. The axiomatic meaning of these constraints determines the partial character of the hadron bootstrap: in turn, those restrictions should, in principle, be deductible by self-consistency.

The hadron bootstrap considers four general restrictions, all based on extensive experimental evidence.

In the hadron bootstrap theory, all the restrictions imposed play the role performed by the equations of motion inside the theories that are part of the Newtonian tradition. In fact, a set of coupled, nonlinear equations, infinite in number, is generated. It is possible for such a system of equations, highly restrictive, to generate a *unique solution* of the hadrons' bootstrap. But the problem is extremely complex on the mathematical plane and it has not been possible to solve this system of equations; therefore, it is not even known if the system has a solution (not necessarily unique). Subtle and refined mathematical methods, adapted to completely solve the bootstrap problem, have simply not yet been invented.

We had to satisfy ourselves with the formulation of models that tried to take into account the restrictions of the hadron bootstrap as much as possible. These models have been important for both theory and experimental designs. Thus, one of them, Veneziano's model, has generated a whole line of research — that of superstrings. The agreement of certain models of the hadron bootstrap with experimental data was impressive. Yet, physicists have moved away from the theory of the hadron bootstrap.

The mathematical complications inherent in the bootstrap program do not represent the only explanation for this decline. Apparently, an irreconcilable opposition was separating the hadron bootstrap, together with its nuclear democracy, from the theory of quarks. On the other hand, encouraged by the experimental success of unified theories, scholars were moving their center of interest towards the theories of the unification of all physical interactions. Or, by definition, the hadron bootstrap included only the strong interactions.

The bootstrap had to leave the physics scene, as a means of calculating physical interactions; however, the bootstrap principle still remained alive. It is referred to in the most diverse contexts, whenever the self-consistency of a theory is required: "With regard to dynamics, when everything else fails, our Supreme Court of Appeal is the bootstrap mechanism, the principle of the self-consistency of the Universe..." — writes Abdus Salam.[11]

4. The Bootstrap and the Anthropic Principle

Apparently, there is no relationship between the bootstrap theory, formulated in particle physics, and the *anthropic principle*, postulated in cosmology. A subtle and significant connection can nevertheless be established between the two approaches.

Let us simply recall some important aspects of the anthropic principle.

The anthropic principle ("anthropic" comes from the Greek word *anthropos* meaning "man") was introduced by Robert H. Dicke in 1961, two years after the formulation of the hadron bootstrap. Its usefulness was largely demonstrated by the works of many scientists.

Today there are different formulations of the anthropic principle. Despite their diversity, there could be recognized a common idea that crosses all these formulations: the existence of a *correlation* between the occurrence of the human, of intelligent life in outer space (and therefore on Earth, the only place where we could identify this intelligent life) and the physical conditions governing the evolution of our universe. This correlation seems to be subject to strong restrictions: if you change the value of certain physical constants or that of the parameters occurring in some places a little, then the physical, chemical and biological conditions, which allow the appearance of humanity on Earth, are *no longer* met.

The existence of these strong restrictions could lead to the hasty conclusion (actually, already drawn by certain scientists or philosophers) that our universe was designed in order to allow the emergence of intelligent observers — us — according to a predetermined project. This conclusion is too simplistic and crumbles in front of a finer analysis concerning the anthropic principle in cosmology.

First of all, Dicke takes some numerical coincidences observed around 1930 as a starting point for the formulation of the anthropic principle. In particular, Dirac noticed the existence of certain empirical relationships between some important constants when they are expressed in a dimensionless form (that is in the form of pure numbers that are not linked to some units of measure). Thus, for example, the age of the universe expressed in atomic units, the number of non-zero mass particles (like protons) in the visible region of the universe or a certain dimensionless shape of the gravitational coupling constant may be taken into account. Dirac noticed that all these constants are expressed as integer powers (positive or negative) of the same number (10 to the power 40), suggesting that these coincidences can be the manifestation of an unknown law of that age. It is also interesting to note that, for reasons that have to do with the logical consistency of his suggestion, Dirac was forced to postulate that certain constants, such as that of gravitational coupling, must *vary over time*. Referring to Dirac's suggestion, Dicke made the fundamental observation that *the age of the universe cannot have an arbitrary numeric value* when taking into account human existence, in its physical, chemical and biological constitution: it must be that those non-uniformities, which are the galaxies and which give rise to stars and planets, take the time (and the necessary conditions) to compose themselves; it is also required that the heavy elements that make up our bodies and are needed for life, have time to compose themselves through stellar nucleosynthesis.

Some strong constraints seem to be exerted not only on the age of the universe, but also on other physical and astrophysical quantities. Brandon Carter underlined the role of the gravitational coupling constant, which has to be close to the experimentally observed value, so as the planets could survive long enough, and could appear. A gravity that is too strong or too weak leads to the impossibility of the formation of planets. The coupling constant characterizing strong interactions is also very constrained. "If the strong force had been slightly stronger than the way it is," says Frank Tipler, "then it would have been possible for helium to form with only two protons... Thus, they could maintain themselves together in a normal way, but the electromagnetic force would be unable to reject two foreign protons and it would therefore become very easy, for this reason, to trigger a nuclear response by only lighting a match;... in the early universe, all the hydrogen would have been converted to helium and there would have been no hydrogen available in order to form the stars of first importance... If, on the contrary, the strong force had been weaker, then complex atoms, such as the carbon, could no longer exist. The rejecting electromagnetic force would tear the nucleus."[12]

In fact, a vast *self-consistency*, that concerns both physical interactions and the phenomena of life, seems to govern the evolution of the universe. Galaxies, stars, planets, humanity, and the atomic and quantum world seem united through one and the same self-consistency. The relationships emphasized by the anthropic principle are a sign of this self-consistency. Again, they raise the problem of the uniqueness of the observed world.

It is precisely through this self-consistency that a link between the general bootstrap principle and the anthropic principle might be determined. It might be even argued that the anthropic principle is a special case of the bootstrap principle.

Certain formulations of the anthropic principle have a strong flavor of finalism, which is unacceptable to the modern scientific spirit. As Stephen Hawking noted, despite his contribution to the deepening of the anthropic principle, "the anthropic principle is not completely satisfactory and we cannot stop thinking that there should be a deeper explanation. In addition, this principle could not give account of all the regions of the universe . . . Perhaps our entire galaxy has been composed this way, but it does not seem necessary that the same happens with other existing galaxies, not to mention a few billions of them, that we are spotting, spread almost uniformly throughout the observable universe. This large-scale homogeneity of the universe makes it very difficult to adopt an anthropocentric point of view or to believe that the structure of the universe is determined by something as peripheral as a complicated molecular structure, present on a single planet, spinning on the orbit of an average planet, in the outer suburbs of a fairly typical spiral galaxy."[13]

Even if the human being is, in a sense, a peripheral phenomenon, he/she seems to be necessary for the self-consistency of the Whole. The human being is not the center, but rather a link of reality that includes him/her, a participant in the dynamic, evolving structure of the universe: the center is everywhere. It is the interpretation that we could offer if the anthropic principle is placed inside the perspective of the general bootstrap's principle. The finality (which passes through the human being, but which is not limited to him/her) is not pre-existing: it is built by the universe itself.

It is also instructive to put in parallel the bootstrap theory in physics and the anthropic principle, as it is applied to cosmology, from the same perspective of a general self-consistency. The anthropic principle could only be strengthened by an increasingly more detailed dynamic modeling. It is not unlikely that the next decades could witness the birth of a new formulation of the bootstraps principle that would take into account the phenomena of life. Thus, it would be gradually understood whether life is a necessary, non-accidental phenomenon, required by the self-consistency of our universe.

5. Methodological Considerations

Ever since its formulation, the bootstrap principle has had bitter detractors and passionate advocates.

Thus, in 1965, three Japanese physicists belonging to the materialist-dialectical school in Nagoya (formed around Sakata and Taketani), wrote; ". . . such trials will lead us to Leibniz's philosophy, which conceives the universe as having a predetermined harmony. This view will introduce religious elements in science and will stop scientific thinking at this level."[14] While giving full credence to neither the ideas of Nagoya's school, nor the idea of the bootstrap, the Israeli physicist Yuval Ne'eman wrote in 1975: "we see now, at Berkeley and elsewhere, another attempt at describing the hadron matter, followed by an almost equal dogmatic narrowness, and where the fundamental motivation comes from the belief that we have reached the end of the road. This is the bootstrap current . . ."[15] Echoing these words, in 1977 the journalist Nigel Calder wrote: "The bootstrap . . . involves a conscious rejection of the objective of traditional physics which is to explain events in terms of forces that act between well defined particles . . . The particles themselves cease to be an object of analysis; they are rather relationships between events. If Chew was right, this would mean very bad news for Western philosophy and science, with their aim to banish the redundant mystery of the universe, revealing its fundamental entities and laws."[16] A similar, very plausible controversy could be produced, also with regard to superstrings.

The first novelty of the bootstrap principle is, as physicist James Cushing rightly notices, of methodological nature.[17] The hypothetical-deductive method has become increasingly more prevalent in physics over the last three centuries: it presupposes the existence of certain fundamental laws, raised to the rank of axioms, all their consequences being deducted. A scientific explanation should follow this path. Usually, the assumption that these laws shall be exercised within certain fundamental entities, contingent and independent, but in interaction, is also formulated.

By proclaiming the absence of laws and fundamental entities, the bootstrap principle introduces a radically new methodology into science. It is true that, in practice, it is limited to a partial bootstrap, which tolerates some axioms as a starting point, and so there is a certain resemblance to the standard, hypothetical-deductive methodology. But the refusal to introduce fundamental entities and equations of motion that are associated with them, persists in various schemes of the partial bootstrap. Therefore, the bootstrap, as it is applied in physics, represents a deviant methodology, which is only accepted when the standard methodology is in crisis.

This is precisely what happened with the hadron bootstrap — formulated in an era characterized by a tremendous proliferation of various hadrons: in such circumstances, it was impossible to maintain the assumption of fundamental entities among these hadrons. Moreover, if the idea of fundamental entities is old, prior to our century, then the idea that particles can be *concurrently* elementary and composed could not have a meaning without the contributions of the quantum theory and those of the relativity theory. In fact, this is reflected in some important developments of the quantum chromodynamics.

The aesthetic attraction of the bootstrap approach, associated with its successes regarding experimental data, has made such a deviant methodology become dominant in the physics of strong interactions in the decade 1960–1970. Meanwhile, the standard methodology has seen a dramatic reversal by inventing the quarks model and by developing the quantum field theory.

Beyond physics and its methodology, the bootstrap principle has also attracted the attention of philosophers. An immediate temptation is to relate the bootstrap to Leibniz's monadology. Scholarly studies were published in this direction by George Gale.[18] It is still difficult to advance anything other than a few analogies. As a simple substance, without parts, the monad, even if it is a mirror of everything, seems very different from a particle, which consists of all other particles. The relationship between monads is static, while the one between particles is dynamic, based on a continuous exchange of information. In addition, the bootstrap's world is not "the best of all possible worlds," it is neither the worst, nor the best, and it is simply the only one that can exist by self-consistency. The bootstrap principle seems in this respect, closer to Anaxagoras (c. 500–c. 428 BC), which proclaims that everything is in everything, and that an object is what it is because all the other objects exist at the same time (an object simply corresponds to its predominant component): "In reality, nothing is neither made, nor destroyed, but is made and separated, based on beings that exist."[19] His doctrine about the *homeomeries* is not foreign to the notion of unobservable sub-constituents. Other ideas of Anaxagoras, such as the unity of contradictories (e.g. "snow is white, but water is black"), the unity of matter in the universe, the purpose that is created through the progress of rationality, deserve also to be mentioned in this context.[20]

Another temptation is to approach the bootstrap and Taoism.[21] However, the bootstrap is closer to Anaxagoras, Boehme, and Peirce or to the current systemic thinking, than to Leibniz or Eastern philosophy.

Finally, one last attempt is to see the *bootstrap* principle as an illustration of Hegel's dialectics.[22] But the way the bootstrap realizes the unity of contradictions is closer to Stéphane Lupasco's view[22] than to Hegel's.

Whatever the fascination exerted by the bootstrap principle on the philosophical plan its interest lies in something else. More than a new theme in physics, it is rather a *symbol*, causing a view on the world's unity. This symbol is inexhaustible, still always remaining the same. Its richness includes the manifestation in the field of natural systems. Certainly, there is a *total bootstrap*, which is a view of the world, and a *partial bootstrap*, which corresponds to a scientific theory. One without the other is poor and, ultimately, sterile. The view of the world is nourished and enriched with information derived from the natural world and, in turn, the scientific theory gets a *human* dimension through the existence of a vision. The major interest of the bootstrap principle is that of unifying a view on the world with a scientific theory in one and the same approach.

As a vision of the world's unity and through its consequences on the nature of Reality, the *bootstrap* principle seems close to Tradition. Was it not Boehme who said, in the *Mysterium pansophicum*: "all together is only one being..."[24]? And in *Aurora,* was not he formulating a true principle of the *cosmic bootstrap*, when he wrote: "The sun is born and produced by all stars. It is the light extracted from universal nature and, in turn, it shines in the universal nature of this world, where it is linked to other stars, as if all of them together were one single star"[25]?

In his article published in 1968, provocatively titled "Bootstrap: A Scientific Idea?,"[26] Chew noted that the idea of the *bootstrap*, in its most general formulation, is "much older than particle physics." And he continued: "The number of *a priori* concepts decreases together with the progress of physics, but it seems that science, as we know it, still requires a language based on a number of *a priori* accepted concepts. So, semantically, the attempt to explain *all* concepts can hardly be called 'scientific...'" And Chew concludes: "If at its logical extremes, the *bootstrap* hypothesis implies that the existence of consciousness, considered together with all other aspects of nature, is necessary to the auto-consistency of everything, such a notion, even if not completely meaningless, is clearly unscientific..." Here we find a precise definition of the inexhaustible, irreducible nature of the idea of the *bootstrap*, which gives it the property of being a symbol.

6. Bootstrap, Superstring Theory and the Cosmic Bootstrap

The history of superstring theory is surprising, because its origin can be found in a theory that is completely outside unification theories: the *bootstrap* theory. "The superstring theory has resulted from a concept called the principle of the *bootstrap* — Murray Gell-Mann writes —...a nice and simple principle of auto-coherence..."[27] Physicists, like me, who have lived through the period of 1960–1985 are aware of this fact, but new generations of physicists completely ignore it.

The disappearance of the *bootstrap* from the physics scene coincides paradoxically with the attainment of a significant result, which would constitute the basis of the superstring theory: the use, in 1968 by Gabriele Veneziano, of a simple formula, invented long before Euler, describing, in a compact, unified manner several major features of the hadron world. A fundamental property of duality characterizes this approach: the changes on one reaction path are related to the changes on another reaction path. Two years after Veneziano's discovery, Yoichiro Nambu, Holger Nielsen and Leonard Susskind showed that Veneziano's formula can be interpreted in terms of the interaction between very low, one-dimensional vibrating *strings*. In 1971, Pierre Ramond, André Neveu and John Schwarz elaborated an extension of the theory including the fermions: this approach is the recognized predecessor of the superstring theory. Finally, in 1974 a crucial step was made: John Schwarz and Joël Scherk revealed that the theory in question predicted the existence of the graviton, and therefore general relativity. The two researchers had the genius to turn a defeat into victory: a particle was raising serious questions about the validity of the whole approach, but Schwarz and Scherk were able to recognize here the graviton.

In *De nature rerum,* Lucretius (c. 98–55 BC) ridiculed the doctrine of Anaxagoras (c. 500–c. 428 BC) on behalf of common sense, which claimed that Everything is in everything. In the modern era, the opposition between Anaxagoras' doctrines and those of Lucretius is found in the dispute, which marked the decade 1960–1970, between the bootstrap supporters and the supporters of the quantum field theory, and, also, in the current discussions on the status of the superstring theory and about the cosmic bootstrap.

How does the universe function? Is it just sort of a machine, of course a spectacular one, but after all, a machine, composed of practically independent systems, mechanically linked to each other? Or rather, does it have an underlying unity, provided by a dynamic intelligence, continuously evolving, functional at every level of nature? Are there laws that run across all scales of nature (particles, atoms, people, planets, etc.), invariable laws, which still have different effects, depending on the scale on which they function?[28] In other words, is there a kind of mutual packing between the different scales of nature, or is the universe just a sad machine, each scale being destined to destruction and death, by the continuous growth of disorder, of entropy?

The ensemble of self-creation and self-organization processes of the universe has been named the *cosmic bootstrap,* by the English physicist Paul Davies: "the universe fills itself, exclusively from within, thanks to its physical nature, with all the energy needed to create and animate matter, thus supplying its own explosive origin. This is the cosmic *bootstrap.* We owe our existence to its amazing power."[29]

But how far can we go with the self-organization of the universe? Does humanity have a place in the cosmic order?

The great American philosopher, logician and mathematician, Charles Sanders Peirce (1839–1914) thought about *indeterminacy* in the formulation of laws: "The only possible way of understanding the laws of nature ... is to involve them as being the result of evolution. This means that they are not absolute, that they cannot be accurately respected. This implies an element of indeterminacy, of spontaneity ... in nature."[30] Therefore, there is always "a certain deflection of facts from any well-defined formula."[31] The analogy with the contemporary hypothesis of a Gödelian structure of knowledge is striking. "Things that are indeterminate in themselves have the function to determine each other ...,"[32] wrote Peirce at the beginning of this century, in a prophetic presentation of the bootstrap principle.

7. Personal Recollections

I met Geoffrey Chew due to my colleague Denyse, experimental physicist at the University of Paris 6, who married Geoff in 1972.

In September 1972, Geoff was member of my PhD (Doctorat d'État) jury at the University of Paris.

In 1976–1977 I spent an one-year post-doctoral stay at Lawrence Berkeley Laboratory as "Guest scientist." This was a fabulous opportunity to begin my collaboration with Geoffrey Chew in the new field of "topological bootstrap." After this period, Geoff came several times in Paris (Orsay), at Institute of Nuclear Physics and I made regular one-month visits at Berkeley.

We published several papers together in the period 1979–1984.[33–35]

In 1984 I had the joy to participate at "Chew Jubilee" in Berkeley, marking the 60th anniversary of Geoff, where I gave a talk (based on a text written in collaboration with Valentin Poenaru) in which I made a summary of the theoretical and phenomenological status of the topological bootstrap.[36]

The interaction I had with Geoffrey Chew and the great French topologist Valentin Poenaru was one of the most exciting periods of my life as a physicist. We were all convinced that topology infuses new life to the bootstrap approach. In the period 1979–1985 we had an incredible number of long letters exchanged between us (sometimes even two letters per day).

The most impressive moments, for me, were the so-called "secret seminars," which occurred every Friday in the Geoff's office and at which a small number of people (like Henry Stapp, Jerry Finkelstein and Fritjof Capra) were convened. I was fascinated to see, during these "secret seminars," how the mind of Geoff works. Geoff was traversed continuously by new ideas and spoke for hours. When he met a contradiction, he abandoned everything and started again from the beginning.

In our private encounters, I questioned him about philosophical aspects of bootstrap. It was obvious for me that Geoff had no philosophical background and no great philosopher influenced him. However, he had a very deep philosophical intuition.

In 1981, the known French writer Michel Random (1933–2008) had the idea of taking several interviews of Geoffrey Chew during his stay in Paris, in order to publish a book. When the manuscript of the book was finished, Geoff considered that the level of the book, from physics point of view, is not good and the project was abandoned.

The most important testimony about philosophical matters comes from Geoff himself, when I asked him to become founding member of the International Center for Transdisciplinary Research and Studies (CIRET) from Paris.[37] In his letter of acceptance addressed to me at November 24, 1986, Geoff writes: "I was pleased to hear of the prospective creation of CIRET and I am honored to be included in the list of founding members. My own research in recent times has not been describable as "physics" in the traditional sense and has frequently been stimulated by people from other disciplines, such as philosophy, linguistics and computer science."[38]

The last letter I received from Geoff was on February 9, 2019, short time before he died. Previously, William Stranger, Provost the New Paradigm Institute (Lake County, California) mentioned to Geoff how pleased he was when he discovered that bootstrap theory plays a prominent role in my book *From Modernity to Cosmodernity — Science, Society and Spirituality*.[39] In his answer to William Stranger and to me, Geoff writes: "You and Basarab may be interested in the attached paper[40] that I have written but not yet published. The term "bootstrap" appears, but I distinguish "physics" from "quantum cosmology"."[41]

Geoffrey Chew was dominated by an overwhelming passion for physics. He worked until the last moments of his life.

He is alive forever in ourselves.

References

1. A. S. Eddington, *The Philosophy of Physical Science* (Cambridge University Press, Cambridge, 1949), p. 57. The first edition was published in 1939.
2. E. Whittaker, *Eddington's Principle in the Philosophy of Science* (Cambridge University Press, Cambridge, 1951).
3. A. S. Eddington, *La Nature du monde physique* (Payot, Paris, 1929), p. 286.
4. G. F. Chew, Theory of strong coupling of ordinary particles, Rapporteur's Talk in *Proc. 9th Int. Ann. Conf. on High Energy Physics* (Kiev, 1959), p. 332; G. F. Chew and M. Jacob, *Strong Interactions Physics* (W.A. Benjamin, New York, 1964).
5. G. F. Chew and S. Mandelstam, *Nuovo Cim.* **19**, 752 (1961).

6. J. T. Cushing, *Theory Construction and Selection in Modern Physics — The S Matrix* (Cambridge University Press, Cambridge,1990).

7. G. F. Chew, Impasse for the elementary-particle concept, in *The Great Ideas Today 1974* (Encyclopaedia Britannica, London, 1974), p. 114.

8. *Ibid.*, p. 119; see also G. F. Chew, *Phys. Today* **23**(10), 23 (1970).

9. G. F. Chew, *Science* **161**, 762 (1968).

10. G. F. Chew, *S-Matrix Theory of Strong Interaction* (W.A. Benjamin, New York, 1961).

11. A. Salam, The Nature of the "ultimate" explanation in physics, in *Ideals and Realities — Selected Essays of Abdus Salam*, eds. Z. Hassan and C. H. Lai (World Scientific, Singapore, 1983), p. 310, https://doi.org/10.1142/0047

12. J. Barrow, F. Tipler and M.-O. Monchicourt, *L'Homme et le cosmos — Le Principe anthropique en astrophysique modern* (Imago, Paris, 1984), p. 79.

13. S. Hawking, *CERN Courier* **21**, 4 (1981).

14. Z. Maki, Y. Ohnuki and S. Sakata, in *Proc. Int. Conference on Elem. Part.*, eds. Y. Tanikawa and N. G. Kaiji (Kyoto University, 1966), p. 109.

15. Y. Ne'eman, Concrete versus Abstract Theoretical Models, in *Interaction between Science and Philosophy*, ed. Y. Elkana (Humanities Press, Atlantic Heights, 1975), p. 16.

16. N. Calder, *The Key of the Universe* (British Broadcasting Corporation, London, 1977), p. 93.

17. J. T. Cushing, *Stud. Hist. Phil. Sci.* **16**, 31 (1985).

18. G. Gale, *Hist. Id.* **35**, 339 (1974); G. Gale, *Rev. Met.* **29**, 323 (1975).

19. J.-P. Dumont, Anaxagore, in *Dictionnaire des philosophes*, ed. D. Huisman (PUF, Paris, 1984), vol. I, p. 77.

20. J. H. Graisser and T. K. Graisser, *Am. J. Phys.* **45**(5), 439 (1977).

21. F. Capra, *The Tao of Physics: An Exploration of the Parallels Between Modern Physics and Eastern Mysticism* (Shambala, Berkeley, 1975).

22. J. P. Merlo, Physique et philosophie, preprint Centre d'Études Nucléaires de Saclay, 1975 (unpublished).

23. H. Badescu and B. Nicolescu (eds.), *Stéphane Lupasco — L'homme et l'oeuvre* (Editions du Rocher, Monaco, 1999).

24. J. Boehme, *Mysterium pansophicum*, trans. J. R. Earle. (A. A. Knopf, New York, 1920), p. 190.

25. J. Boehme, *L'Aurore naissante*, trans. Le Filosophe Inconnu. (Arché, Milano, 1977) p. 72.

26. G. F. Chew, *Science* **161**, 762 (1968).

27. M. Gell-Mann, *Le Quark et le jaguar* (Flammarion, Paris, 1997), pp. 150–151.

28. B. Nicolescu, *Manifesto of Transdisciplinarity*, trans. K.-C. Voss. (State University of New York (SUNY) Press, New York, 2002).

29. P. Davies, *Superforce — The Search for a Great Unified Theory of Nature* (Simon & Schuster, New York,1984), p. 195.

30. *Ibid.*, p. 148.

31. D. D. Roberts, *The Existential Graphs of Charles S. Peirce* (Mouton, Illinois, 1973), p. 147.

32. *Ibid.*, p. 147.

33. G. F. Chew, B. Nicolescu, J. Uscherson and R. Vinh Mau, Topological quark-gluon structure of elementary hadrons beyond mesons and baryons, CERN report TH.2635. (CERN, 1979), unpublished.

34. G. F. Chew, J. Finkelstein, B. Nicolescu and V. Poenaru, *Zeit. F. Phys. C* **14**, 289 (1982).

35. G. F. Chew, D. Issler, B. Nicolescu and V. Poenaru, GeV partons and TeV hexons from a topological viewpoint, in *Proc. of the XIXth Rencontre de Moriond, Vol. "New Particle Production"*, ed. J. Tran Thanh Van (Frontieres Editions, France, 1984), pp. 143–165.

36. B. Nicolescu and V. Poenaru, From Baryonium to hexons, in *A Passion for Physics. Proc. G. F. Chew Jubilee*, eds. C. DeTar, J. Finkelstein and C. I. Tan (World Scientific, Singapore, 1985), pp. 195–221.

37. See the site of CIRET. Accessed on March 2, 2020.

38. Letter of Geoffrey Chew addressed to Basarab Nicolescu on November 24, 1986. Archives of Basarab Nicolescu.

39. B. Nicolescu, *From Modernity to Cosmodernity — Science, Culture, and Spirituality* (State University of New York (SUNY) Press, New York, USA, 2014). Chapter 7 of this book is entirely devoted to the bootstrap approach.

40. G. Chew, Chiral-Electromagnetic Gravitational Theory of Every "Thing" Evolving Gelfand-Dirac Hamilton-Riemann Quantum Cosmology, LBL preprint (LBL, February 8, 2019). Unpublished.

41. Letter of Geoffrey Chew addressed to William Stranger and Basarab Nicolescu on February 9, 2019.

Part III
Reprints of Selected Articles

PHYSICAL REVIEW   VOLUME 101, NUMBER 5   MARCH 1, 1956

Effective-Range Approach to the Low-Energy p-Wave Pion-Nucleon Interaction

G. F. Chew and F. E. Low

Department of Physics, University of Illinois, Urbana, Illinois

(Received November 28, 1955)

The theory of p-wave pion-nucleon scattering is reexamined using the formalism recently proposed by one of the authors (F.E.L.). On the basis of the cut-off Yukawa theory without nuclear recoil it is found, for not too high values of the coupling constant, that: (a) For each p-wave phase shift a certain function of the cotangent should be approximately linear at low energies and should extrapolate to the Born approximation at zero total energy. The value of the renormalized unrationalized coupling constant determined in this way from experiment is $f^2 = 0.08$. A special feature of the predicted energy dependence of the phase shifts is that δ_{33} is positive and the other p phase shifts are negative. (b) The so-called "crossing theorem" requires a relation between the four p phase shifts, so that in addition to the coupling constant only two further constants are needed to completely specify the low-energy behavior. (c) The direction of the energy variation in the (3,3) state is such that a resonance will occur for a sufficiently large cut-off ω_{max}. Rough estimates indicate that $\omega_{max} \approx 6$ will produce a resonance at the energy required by experiment. It is argued that the results (a) and (b) are very probably also consequences of a relativistic theory but that (c) may not be.

I. INTRODUCTION

THE ability of the Yukawa theory to describe quantitatively the pion-nucleon interaction is still uncertain. The recent discovery of new particles, the hyperons and K-mesons, makes it unlikely that this theory can be valid in the multi-Bev energy region, but there still remains an interesting and important question: Can a theory of the Yukawa type quantitatively correlate experiments in the sub-Bev region, below the threshold for production of "curious" particles? Recently, it has been shown[1] that a crude static model of the pion-nucleon interaction, based on the Yukawa idea, is quite powerful in correlating certain experiments; however, the relation of the model to a true theory has been obscure. One of the main purposes of this paper and the one following (which will be concerned with photomeson production) is to show that the most important predictions of the model are actually independent of its details and thus may also be predictions of a "true" theory.

The results to be presented are based on a new set of equations[2] which can be applied both to the static model and to the relativistic Yukawa theory, and which exhibit the low-energy properties of both in a clear and useful way. Some new and quite general predictions about p-wave pion-nucleon scattering have been achieved without explicit solution of the equations, and the second purpose of this paper is to report these new predictions.

Throughout the first six sections of this paper, arguments and derivations will be given in terms of the static model. The simplification achieved by eliminating antinucleons and recoil is enormous. In Sec. II, a new and simplified derivation of the equations for meson scattering is presented. In Sec. III, those properties of

the scattering are discussed which can be deduced without explicit solution of the equations. Section IV deals with the equations in what might be called the "one-meson approximation." In Sec. V, an effective-range treatment of the scattering problem is presented. Section VI deals with certain total-cross-section sum rules, including a sum rule for the renormalization constants of the theory. Finally, in Sec. VII, we discuss the possible extension of our results to more complicated cases such as the relativistic pseudoscalar theory.

II. DERIVATION OF THE SCATTERING EQUATIONS

We present here a much simpler derivation of the scattering equations than is given in reference 2. We follow a method suggested by Wick[3] which unfortunately applies to the fixed-source theory only. For a derivation of the appropriate expression when nucleon recoil is to be included, see reference 2.

We take as our Hamiltonian[4]

$$H = H_0 + H_I, \tag{1}$$

where

$$H_I = \sum_k V_k^{(0)} a_k + V_k^{(0)\dagger} a_k^\dagger, \tag{2}$$

$$H_0 = \sum_k a_k^\dagger a_k \omega_k, \tag{3}$$

and

$$V_k^{(0)} = i f_{(r)}^{(0)} (\boldsymbol{\sigma} \cdot \mathbf{k}/\sqrt{2\omega_k}) \tau_k v(k). \tag{4}$$

Here $a_k^\dagger$ and a_k are, respectively, creation and annihilation operators for single mesons, $\omega_k = (1 + k^2)^{\frac{1}{2}}$, $\boldsymbol{\sigma}$ is the nucleon spin vector, and τ_k is the kth component of the nucleon isotopic spin operator. In our notation, the meson quantum numbers are all described by a single symbol (k) which includes the three components of momentum and the isotopic spin. Also, $f_{(r)}^{(0)}$ is the rationalized but unrenormalized coupling constant.

The Hamiltonian (1) has a complete set of eigenstates Ψ_n. These states include the four single-nucleon states

[1] G. F. Chew, Phys. Rev. **95**, 1669 (1954).
[2] F. E. Low, Phys. Rev. **97**, 1392 (1955). Closely related equations have also been derived by Lehmann, Symanzik, and Zimmerman, Nuovo cimento **1**, 1 (1955).
[3] G. C. Wick, Revs. Modern Phys. **27**, 339 (1955).
[4] We take $\hbar = c = \mu = 1$; μ is the meson rest mass.

Ψ_0 (we suppress the spin and isotopic spin indices), the one-meson states Ψ_q, two-meson states, etc. We are of course particularly interested in the one-meson states with outgoing waves, $\Psi_q^{(+)}$. These are solutions of the Schrödinger equation

$$(H-E_q)\Psi_q^{(+)}=0,$$

with

$$E_q=E_0+\omega_q, \tag{5}$$

where E_0 is the single-nucleon energy. Following Wick, we set

$$\Psi_q^{(+)}=a_q^\dagger\Psi_0+\chi^{(+)}, \tag{6}$$

where

$$(H-E_q)\chi^{(+)}=-(H-E_q)a_q^\dagger\Psi_0.$$

Now

$$[H_0,a_q^\dagger]=\omega_q a_q^\dagger$$

and

$$[H_I,a_q^\dagger]=V_q^{(0)},$$

so that

$$(H-E_q)a_q^\dagger\Psi_0=V_q^{(0)}\Psi_0$$

and

$$(H-E_q)\chi^{(+)}=-V_q^{(0)}\Psi_0.$$

Dividing through by $(H-E_q)$, we have

$$\chi^{(+)}=-\frac{1}{H-E_q-i\epsilon}V_q^{(0)}\Psi_0, \tag{7}$$

where the $-i\epsilon$ is inserted to produce outgoing waves in $\chi^{(+)}$.[5] The reader will recall that $1/(a-i\epsilon)=\mathrm{P}(1/a)+i\pi\delta(a)$,[6] where $\mathrm{P}(1/a)$ stands for the Cauchy principal value. We rewrite Eq. (7) as:

$$\chi^{(+)}=-\sum_n\frac{1}{E_n-E_q-i\epsilon}\Psi_n^{(-)}\rangle\langle\Psi_n^{(-)},V_q^{(0)}\Psi_0\rangle, \tag{8}$$

where the $\Psi_n^{(-)}$ are the complete orthonormal set of *incoming* wave eigenstates. We assume in writing Eq. (8) that there are no bound states.

The next problem is to relate the result (8) to the scattering matrix. Wick[7] has been able to show that the S-matrix is directly related to the coefficient of $\Psi_n^{(-)}$ in the sum in (8) by the equation

$$\langle n|S|q\rangle=\delta_{nq}-2\pi i\delta(E_q-E_n)T_q(n), \tag{9}$$

where $T_q(n)=(\Psi_n^{(-)},V_q^{(0)}\Psi_0)$.

The following simple proof of (9) was kindly communicated to us by B. S. de Witt. The starting point is a well-known formula for the S-matrix:

$$\langle n|S|q\rangle=(\Psi_n^{(-)},\Psi_q^{(+)}). \tag{10}$$

Now rewrite $\Psi_q^{(+)}$ so that the corresponding incoming

wave solution appears explicitly:

$$\Psi_q^{(+)}=a_q^\dagger\Psi_0-\frac{1}{H-E_q-i\epsilon}V_q^{(0)}\Psi_0$$

$$=\Psi_q^{(-)}+\left\{\frac{1}{H-E_q+i\epsilon}-\frac{1}{H-E_q-i\epsilon}\right\}V_q^{(0)}\Psi_0$$

$$=\Psi_q^{(-)}-2\pi i\delta(H-E_q)V_q^{(0)}\Psi_0. \tag{11}$$

Substitution of (11) into (10) clearly leads to the desired result (9).

The total cross section for mesons of momentum and isotopic spin q is then

$$\sigma_q=\frac{2\pi}{v_q}\sum_n\delta(E_n-E_q)|T_q(n)|^2, \tag{12}$$

where $v_q=q/\omega_q$ is the incident meson velocity.

It will be seen by (9) that for $E_q=E_n$, the quantity $T_q(n)$ is the conventional T matrix of scattering theory, but for $E_q\neq E_n$ this is no longer the case. $T_q(n)$ remains everywhere closely related to the energy-conserving T matrix at energy E_n, since it depends on the variable q only in a trivial way. In contrast, the conventional T matrix usually depends on its two indices in a non-factorable way and its values off and on the energy shell are not simply related. The trivial dependence of $T_q(n)$ on the variable q is an important simplification achieved by the present method of calculation.

In order to investigate the properties of $T_q(p)$, we note that since $\Psi_p^{(-)}$ can be written

$$\Psi_p^{(-)}=a_p^\dagger\Psi_0-\frac{1}{H-E_p+i\epsilon}V_p^{(0)}\Psi_0, \tag{13}$$

it follows that

$$T_q(p)=(a_p^\dagger\Psi_0,V_q^{(0)}\Psi_0)$$

$$-\left(\frac{1}{H-E_p+i\epsilon}V_p^{(0)}\Psi_0,\,V_q^{(0)}\Psi_0\right) \tag{14}$$

$$=(\Psi_0,V_q^{(0)}a_p\Psi_0)$$

$$-\left(\Psi_0,\,V_p^{(0)\dagger}\frac{1}{H-E_p-i\epsilon}V_q^{(0)}\Psi_0\right), \tag{15}$$

where we have made use of the fact that $V_q^{(0)}$ and a_p commute.

Let us normalize H (by subtracting the nucleon self-energy) so that

$$H\Psi_0=0, \tag{16}$$

$$H\Psi_p=\omega_p\Psi_p, \tag{17}$$

etc. The annihilation operator a_p may be eliminated from the first term of (15) by making use of its commutator with the Hamiltonian:

$$[a_p,H]=\omega_p a_p+V_p^{(0)\dagger}.$$

[5] B. Lippmann and J. Schwinger, Phys. Rev. **79**, 469 (1950).
[6] P. A. M. Dirac, *The Principles of Quantum Mechanics* (Oxford University Press, New York, 1947), third edition, p. 198.
[7] G. C. Wick, Revs. Modern Phys. **27**, 339 (1955).

Thus,

$$[a_p H - H a_p - \omega_p a_p - V_p^{(0)\dagger}]\Psi_0 = 0$$

or

$$(H + \omega_p) a_p \Psi_0 = -V_p^{(0)\dagger}\Psi_0, \qquad (18)$$

so that

$$a_p \Psi_0 = -\frac{1}{H + \omega_p} V_p^{(0)\dagger}\Psi_0, \qquad (19)$$

and Eq. (15) becomes

$$T_q(p) = -\left(\Psi_0,\ V_p^{(0)\dagger}\frac{1}{H - \omega_p - i\epsilon} V_q^{(0)}\Psi_0\right) \qquad (20)$$

$$-\left(\Psi_0,\ V_q^{(0)}\frac{1}{H + \omega_p} V_p^{(0)\dagger}\Psi_0\right).$$

Finally, since $V_p^{(0)\dagger} = -V_p^{(0)}$, we have

$$T_q(p) = \left(\Psi_0,\ \left[V_p^{(0)}\frac{1}{H - \omega_p - i\epsilon} V_q^{(0)}\right.\right.$$

$$\left.\left. + V_q^{(0)}\frac{1}{H + \omega_p} V_p^{(0)}\right]\Psi_0\right). \qquad (21)$$

We may write Eq. (21) in a somewhat more familiar form by reintroducing the complete orthonormal set of states $\Psi_n^{(-)}$:

$$T_q(p) = \sum_n (\Psi_0, V_p^{(0)}\Psi_n^{(-)})(\Psi_n^{(-)}, V_q^{(0)}\Psi_0)/(E_n - \omega_p - i\epsilon)$$
$$+ \sum_n (\Psi_0, V_q^{(0)}\Psi_n^{(-)})(\Psi_n^{(-)}, V_p^{(0)}\Psi_0)/(E_n + \omega_p). \qquad (22)$$

Although Eq. (22) is strongly reminiscent of second-order perturbation theory, it is an exact result. The difference from perturbation theory lies, of course, in the use of initial, intermediate, and final states which are exact eigenstates of the total Hamiltonian H rather than of the free Hamiltonian H_0.

A more compact writing of (22) evidently is achieved by the use of the matrix element,

$$T_q(n) = (\Psi_n^{(-)}, V_q^{(0)}\Psi_0), \qquad (23)$$

already introduced in Eq. (9), and the complex conjugate matrix element,

$$T_q^*(n) = (V_q^{(0)}\Psi_0, \Psi_n^{(-)})$$
$$= (\Psi_0, V_q^{(0)\dagger}\Psi_n^{(-)})$$
$$= -(\Psi_0, V_q^{(0)}\Psi_n^{(-)}). \qquad (24)$$

Even more notational simplification results if we consider (22) to be an operator equation with respect to the nucleon spin and isotopic spin variables. To illustrate how this works out, we write these variables explicitly in a typical matrix element. Take a final state β, initial state α, and 4 intermediate states Ψ_n^γ, $\gamma = 1, \cdots 4$. Then

$$\sum_\gamma \langle\beta| V_p^{(0)} |n,\gamma\rangle\langle n,\gamma| V_q^{(0)} |\alpha\rangle$$
$$= -\sum_\gamma \langle\gamma| T_p(n) |\beta\rangle^* \langle\gamma| T_q(n) |\alpha\rangle$$
$$= -\sum_\gamma \langle\beta| T_p^\dagger(n) |\gamma\rangle\langle\gamma| T_q(n) |\alpha\rangle$$
$$= -\langle\beta| T_p^\dagger(n) T_q(n) \alpha\rangle.$$

Thus, if indices corresponding to initial and final nucleon states are as usual suppressed, Eq. (22) may be re-expressed as

$$T_q(p) = -\sum_n \left[\frac{T_p^\dagger(n) T_q(n)}{E_n - \omega_p - i\epsilon} + \frac{T_q^\dagger(n) T_p(n)}{E_n + \omega_p}\right], \qquad (25)$$

a form of the equation which exhibits clearly its most important general properties.

III. GENERAL PROPERTIES OF THE EQUATION

A. Unitarity of the S-Matrix

We begin by examining the well-known requirement that the scattering matrix shall be unitary. From the relation (9), it follows that the unitarity condition $S^\dagger S = 1$ is equivalent to the following statement:

$$T_p^\dagger(q) - T_q(p) = 2\pi i \sum_n \delta(E_n - \omega_q) T_p^\dagger(n) T_q(n), \qquad (26)$$

when $\omega_p = \omega_q = \omega$. If one takes the conjugate transpose of (25) to obtain $T_p^\dagger(q)$, the only change to occur on the right-hand side is the replacement of $-i\epsilon$ in the first denominator by $+i\epsilon$. Then, since

$$\frac{1}{E_n - \omega - i\epsilon} - \frac{1}{E_n - \omega + i\epsilon} = 2\pi i \delta(E_n - \omega),$$

the unitarity condition (26) is evidently satisfied by any matrix function satisfying (25). It would seem, therefore, that one novel feature of (25), the quadratic rather than linear dependence on T of the right-hand side, is largely a reflection of the unitarity requirement. The second term of the right-hand side, however, has nothing to do with unitarity. Its presence has rather to do with a second and a quite different general property of the theory, which we discuss next.

B. Crossing Theorem

Gell-Mann and Goldberger[8] have pointed out an important symmetry possessed by theories of the Yukawa type. In terms of Feynman diagrams for meson-nucleon scattering, one may express this symmetry by saying that for any given diagram, another must exist which is obtained from the first by exchanging the incoming and outgoing meson lines. This exchange is not a simple time reversal because the nucleon line is not inverted.

In the present formulation of the theory this "crossing symmetry," as it is sometimes called, can be simply expressed in terms of a certain matrix function of a complex variable, z, which we define by

$$t_{qp}(z) = -\sum_n \left[\frac{T_p^\dagger(n) T_q(n)}{E_n - z} + \frac{T_q^\dagger(n) T_p(n)}{E_n + z}\right]. \qquad (27)$$

---

[8] M. Gell-Mann and M. L. Goldberger, in *Proceedings of the Fourth Annual Rochester Conference on High Energy Nuclear Physics* (University of Rochester Press, Rochester, 1954).

Note that $t_{qp}(z)$ is a Hermitian matrix function of z in the sense that

$$t_{qp}(z) = t_{pq}{}^{\dagger}(z^*).$$

Note further that the dependence on both p and q is now trivial. Only the dependence on z is unknown.

Clearly, the limit of $t_{qp}(z)$ as z approaches the positive real axis from above $(z \to \omega_p + i\epsilon)$ is $T_q(p)$; but for an expression of the crossing symmetry we must keep the functional dependence on p and z separate. The symmetry is expressed by the relation

$$t_{qp}(z) = t_{pq}(-z), \qquad (28)$$

and it becomes apparent that the reason for the second term in (27) is precisely to satisfy (28).

C. Pole of the Function $t_{q,p}(z)$ at the Origin

It is helpful to get clearly in mind the nature and location of the singularities of the function $t_{qp}(z)$. This is possible from inspection of (27) because the energy eigenvalues of the Hamiltonian, E_n, are known even though the eigenfunctions are not.

The lowest eigenvalue (after the self-energy subtraction) is zero, belonging to the four zero-meson states. These states, then give rise to a simple pole at $z=0$, with residue $R_{qp} = [T_p{}^{\dagger}(0)T_q(0) - T_q{}^{\dagger}(0)T_p(0)]$. This statement is an analog of the Kroll-Ruderman theorem[9] and provides a method for measuring the coupling constant by scattering experiments. The point is that $T_q(0)$ contains the zero-meson wave functions only in a matrix element,

$$(\Psi_0{}^{(\alpha)}, \sigma_q \tau_q \Psi_0{}^{(\beta)}),$$

which for reasons of invariance must be a q-, α-, and β-independent multiple, say Z, of the matrix element

$$(u_\alpha, \sigma_q \tau_q u_\beta),$$

where u_α and u_β are normalized Pauli spinors ("bare-nucleon" wave functions). If we now *define* $f_{(r)} = Z f_{(r)}{}^{(0)}$ and call $f_{(r)}$ the renormalized (rationalized) coupling constant, it follows that

$$T_q(0) = V_q,$$

where V_q is obtained from $V_q{}^{(0)}$ by replacing $f_{(r)}{}^{(0)}$ by $f_{(r)}$. The function $t_{qp}(z)$ in the neighborhood of $z=0$ is therefore completely determined by the renormalized coupling constant and vice versa. Since our theory is a finite one, we are of course free to define the coupling constant in the most convenient way. The present definition coincides with those previously given by Chew[10] and Lee[11]; most important, it is also appropriate (in the sense of the Kroll-Ruderman theorem) to the calculation of threshold photomeson production.

[9] N. M. Kroll and M. A. Ruderman, Phys. Rev. 93, 233 (1954).
[10] G. F. Chew, Phys. Rev. 94, 1748 (1954).
[11] T. D. Lee, Phys. Rev. 95, 1329 (1954).

D. Behavior at Infinity

It is clear from inspection of (27) that as $z \to \infty$, $t_{qp}(z)$ behaves like $1/z$. It is an interesting but not very useful fact that the coefficient of $1/z$ at infinity is a multiple of the residue at the origin which is independent of q and p and of the nucleon variables. The proof of this relation is given by reference back to Eq. (22). As z (or ω_p) approaches infinity, the dependence on the energy eigenvalue E_n in the denominator is removed and closure may be applied to evaluate the sum over states, giving

$$\lim_{z \to \infty} z t_{qp}(z) = -(\Psi_0, [V_p{}^{(0)}, V_q{}^{(0)}]\Psi_0),$$

which is to be compared to the residue at the origin,

$$R_{qp} = -[V_p, V_q].$$

By using the standard commutation properties of the σ's and τ's, these two coefficients are easily shown to differ only by a factor which is independent of p and q as well as of the nucleon spin and isotopic spin, but the factor is not unity. The point is that here one is dealing essentially with matrix elements of σ_q or τ_q rather than the product $\sigma_q \tau_q$ as in the preceding section.

E. Location of Branch Points and Cuts

From the form of (27) it is clear that, in addition to the pole at the origin, all the other singularities of (27) also lie along the real axis. The next lowest eigenvalue of the energy corresponds to a single meson at rest; then there is a continuous distribution in energy of one-meson states up to $+\infty$. At an energy equal to two meson rest masses, a continuous distribution of two-meson states begins, and so on, for all higher numbers of mesons. It follows that, because of the one-meson states, the function $t_{qp}(z)$ has a branch point at $z=1$ and a cut along the real axis for $z>1$. The two-meson states produce a branch point at $z=2$ and a cut for $z>2$, etc. The "crossed" terms in (27) obviously produce similar singularities in the left half-plane.

It will now be shown that the conditions listed under A, B, C, D, E, above, together with the factorability of the T matrix which is implied by Eq. (9), are completely equivalent to Eq. (27). Consider the condition that except for a simple pole of residue R_{pq} at the origin all the singularities of $t_{qp}(z)$ are confined to branch points with cuts running along the real axis for $z > +1$ and $z < -1$. If $t_{qp}(z)$ goes like $1/z$ at infinity, it may be expanded in the form

$$t_{qp}(z) = \frac{R_{qp}}{z} + \int_1^{\infty} dx' \left[\frac{F_{qp}(x')}{x' - z} + \frac{G_{qp}(x')}{x' + z} \right], \qquad (29)$$

where $F_{qp}(x)$ and $G_{qp}(x)$ are weighting functions defined for $x \geqslant 1$. Evidently the functions F_{qp} and G_{qp} are given by the jump in the function t_{qp} going across the real axis in the right and left half-planes, respectively. We have, in fact,

$$2\pi i F_{qp}(x) = \lim_{z \to x+i\epsilon} t_{qp}(z) - \lim_{z \to x-i\epsilon} t_{qp}(z) \qquad (30)$$

for $x \geqslant 1$, and

$$2\pi i G_{qp}(x) = \lim_{z \to -x-i\epsilon} t_{qp}(z) - \lim_{z \to -x+i\epsilon} t_{qp}(z), \qquad (31)$$

where x is still $\geqslant 1$. Now, if we define $T_q(p)$ as the limit of $t_{qp}(z)$ as $z \to \omega_p + i\epsilon$ and further impose the reality condition, $t_{qp}(z) = t_{pq}{}^\dagger(z^*)$, then (30) leads to

$$F_{qp}(\omega_p) = \frac{1}{2\pi i}[T_q(p) - T_p{}^\dagger(q)]_{\omega_p = \omega_q}. \qquad (30')$$

Imposing the crossing relation (28) allows (31) to be written

$$G_{qp}(\omega_p) = \frac{1}{2\pi i}[T_p(q) - T_q{}^\dagger(p)]_{\omega_p = \omega_q}. \qquad (31')$$

Finally the unitarity condition (26) transforms Eq. (29) via (30') and (31') into our original Eq. (27) for the special case $\omega_p = \omega_q$. The latter restriction is of no consequence, however, since we have noted before that one may move off the energy shell at will.

To summarize, if it is possible to find a Hermitian matrix function, $t_{qp}(z)$, which has a simple pole at the origin of residue R_{qp}, goes to zero like $1/z$ at ∞, has otherwise only branch points and cuts along the real axis for $z > 1$ and $z < 1$, and which satisfies unitarity as well as the crossing relation, then one has a solution of Eq. (27). Unfortunately, the formulation of the unitarity condition involves multimeson (two-meson and higher) states and cannot be written down on the basis of *a priori* arguments in terms of $t_{qp}(z)$. However, if multimeson states are neglected, as in the next section, then the above conditions form a practical basis for solving the scattering problem.

IV. ONE-MESON APPROXIMATION

If we assume that the inelastic cross sections are small compared to the elastic ones (for all values of the energy) then, as a first approximation, the contributions of multimeson states to the unitarity condition (26) may be neglected. In this case it is convenient to re-express the conditions on the matrix t_{qp} in terms of phase shifts.

We set

$$t_{qp}(z) = -v(q)v(p)\frac{4\pi}{(4\omega_q\omega_p)^{\frac{1}{2}}} \sum_{\alpha=1}^{4} P_\alpha(p,q)h_\alpha(z), \qquad (32)$$

where

$$\begin{aligned} P_{11} &= \tfrac{1}{3}\tau_p\tau_q(\boldsymbol{\sigma}\cdot\mathbf{p})(\boldsymbol{\sigma}\cdot\mathbf{q}), \\ P_{13} &= \tfrac{1}{3}\tau_p\tau_q[3\mathbf{p}\cdot\mathbf{q} - (\boldsymbol{\sigma}\cdot\mathbf{p})(\boldsymbol{\sigma}\cdot\mathbf{q})], \\ P_{31} &= (\delta_{pq} - \tfrac{1}{3}\tau_p\tau_q)(\boldsymbol{\sigma}\cdot\mathbf{p})(\boldsymbol{\sigma}\cdot\mathbf{q}), \\ P_{33} &= (\delta_{pq} - \tfrac{1}{3}\tau_p\tau_q)[3\mathbf{p}\cdot\mathbf{q} - (\boldsymbol{\sigma}\cdot\mathbf{p})(\boldsymbol{\sigma}\cdot\mathbf{q})]. \end{aligned} \qquad (33)$$

The P_α's are projection operators for the four eigen-

states of total angular momentum and isotopic spin. In the subscript $\alpha = (2I, 2J)$, I is the total isotopic spin and J the total angular momentum.

The new functions $h_\alpha(z)$ are simply related to phase shifts. Application of the unitarity condition shows that

$$\lim_{z \to \omega_p + i\epsilon} h_\alpha(z) = e^{i\delta_\alpha(p)} \sin[\delta_\alpha(p)]/p^3 v^2(p), \qquad (34)$$

where the $\delta_\alpha(p)$ are real and identical with the conventionally defined phase shifts for $\omega_p \leqslant 2$. It is well known that the scattering phase shifts of the (1,3) and (3,1) states are equal in this theory, so we henceforth confine our attention to the three functions $h_1 = h_{11}$, $h_2 = h_{13} = h_{31}$, and $h_3 = h_{33}$.

The conditions on the function $t_{qp}(z)$ given in the preceding section may be translated into conditions on the functions $h_\alpha(z)$. In addition to unitarity, which is expressed by (34), we have the crossing relation

$$h_\alpha(z) = \sum_{\beta=1}^{3} A_{\alpha\beta}h_\beta(-z), \qquad (35)$$

where

$$A = \begin{bmatrix} 1/9 & -8/9 & 16/9 \\ -2/9 & 7/9 & 4/9 \\ 4/9 & 4/9 & 1/9 \end{bmatrix}. \qquad (36)$$

The condition that $t_{qp}(z)$ be a Hermitian matrix function of z implies that $h_\alpha(z)$ is a real function of z in the sense that $h_\alpha(z^*) = h_\alpha^*(z)$. The boundary condition at infinity is that h_α behave like $1/z$, while at the origin h_α should have a simple pole of residue λ_α, where

$$\lambda_\alpha = \tfrac{2}{3}f^2 \begin{bmatrix} -4 \\ -1 \\ 2 \end{bmatrix}. \qquad (37)$$

Here f^2 is the nonrationalized coupling constant. We may at this point mention two useful properties of the crossing matrix $A_{\alpha\beta}$:

(1) $$\sum_\beta A_{\alpha\beta}A_{\beta\gamma} = \delta_{\alpha\gamma}, \qquad (38)$$

(2) $$\sum_\beta A_{\alpha\beta}\lambda_\beta = -\lambda_\alpha. \qquad (39)$$

Finally, we require that all other singularities of h_α be confined to two branch points at $z = \pm 1$, with cuts along the real axis to $\pm\infty$. This set of conditions is equivalent to the following equations for $h_\alpha(\omega)$:

$$h_\alpha(\omega) = \frac{\lambda_\alpha}{\omega} + \frac{1}{\pi}\int d\omega_p p^3 v^2(p) \left\{ \frac{|h_\alpha(\omega_p)|^2}{\omega_p - \omega - i\epsilon} \right.$$
$$\left. + \sum_\beta A_{\alpha\beta}\frac{|h_\beta(\omega_p)|^2}{\omega_p + \omega} \right\}. \qquad (40)$$

Let us eliminate the pole at the origin by introducing a new (real) function

$$g_\alpha(z) = \frac{\lambda_\alpha}{z} - [h_\alpha(z)]^{-1}. \qquad (41)$$

The boundary condition at $z=0$ is now that $g_\alpha(0)=1$. Furthermore, g_α behaves like a constant at infinity. The unitarity condition (34) implies that

$$\lim_{z \to \omega_p + i\epsilon} g_\alpha(z) - \lim_{z \to \omega_p - i\epsilon} g_\alpha(z) = -2i\frac{\lambda_\alpha p^3}{\omega_p}v^2(p), \quad (42)$$

for $\omega_p \gtrless 1$. The crossing relation becomes

$$\sum_{\beta=1}^{3} B_{\alpha\beta}\frac{1}{g_\beta(z)} = \frac{1}{g_\alpha(-z)}, \quad (43)$$

where

$$B = \begin{bmatrix} -1/9 & 2/9 & 8/9 \\ 8/9 & -7/9 & 8/9 \\ 8/9 & 2/9 & -1/9 \end{bmatrix}. \quad (44)$$

Finally, consider the location of the singularities of $g_\alpha(z)$. Clearly, just as for $h_\alpha(z)$, there are branch points at $z=\pm 1$ with cuts along the real axis to $\pm\infty$. If $h_\alpha(z)$ has no zeros, these will be the only singularities of $g_\alpha(z)$.

With f^2 sufficiently small, there are certainly no zeros in $h_\alpha(z)$. For this case, the boundary conditions on $g_\alpha(z)$, together with the nature of its singularities, imply that this function may be written

$$g_\alpha(z) = 1 - \frac{z}{\pi}\int_1^\infty dx' \left[\frac{F_\alpha(x')}{x'-z} + \frac{G_\alpha(x')}{x'+z}\right], \quad (45)$$

where $F_\alpha(x)$ and $G_\alpha(x)$ are real weighting functions defined for $x \geqslant 1$. The functions $F_\alpha(x)$ and $G_\alpha(x)$ are power series in f^2, whose coefficients are continuous differentiable functions of the variable x. Since $F_\alpha(x)$ gives the jump in $g_\alpha(z)$ in going across the real axis for $z \geqslant 1$, the first of the two weighting functions is completely determined by the unitarity condition (42):

$$F_\alpha(\omega_p) = \lambda_\alpha p^3/\omega_p^2 v^2(p). \quad (46)$$

The final condition to be satisfied is the crossing relation (43), which is just sufficient to determine the second weighting function $G_\alpha(x)$. Thus, for values of f^2 which are such that the power series for $\int G_\alpha(x)dx/(x+z)$ converges, we have

$$h_\alpha(z)$$

$$= \frac{\lambda_\alpha/z}{1 - \frac{z}{\pi}\lambda_\alpha\int\frac{d\omega_p}{\omega_p^2}\frac{p^3 v^2(p)}{\omega_p - z} - \frac{z}{\pi}\int\frac{d\omega_p}{\omega_p^2}\frac{p^3 v^2(p)}{\omega_p + z}H_\alpha(\omega_p)}, \quad (47)$$

where

$$G_\alpha(\omega_p) = p^3\frac{v^2(p)}{\omega_p^2}H_\alpha(\omega_p).$$

For larger values of f^2, the function that replaces $\int H_\alpha(p)d\omega_p p^3 v^2/(\omega_p+z)\omega_p^2$ in (47) must be determined

by analytic continuation. We have been unable to do this for the symmetric pseudoscalar theory.[12]

It will be seen that in the region of convergence of $\int G_\alpha(x)dx/(x+z)$ there can be no zeros of $h_\alpha(z)$, so that the sign of the phase shifts must be the same as that of their Born approximations.

It has been pointed out by Castillejo, Dalitz, and Dyson[13] that (47) is probably not the only solution of (40). The extra solutions found by them, however, are not analytic continuations of the perturbation theory power series, as is our expression (47). If one assumes that the solution of the original field theoretic problem is unique and is the analytic continuation of the power series, then it is clear that our solution is the only one of physical interest.

G. Salzman has solved Eq. (40) by a numerical method which he will discuss in detail in a forthcoming paper. Here we concentrate on features of the solution which may be deduced from general considerations.

It should be noted that the real part of g_α is essentially the cotangent of the phase shift. More precisely,

$$\text{Re}\,g_\alpha(\omega_p) = \frac{\lambda_\alpha p^3 v^2(p)}{\omega_p}\cot\delta_\alpha(\omega_p) \quad (48)$$

[the imaginary part of $g_\alpha(\omega_p)$ is fixed by (42)]. It seems natural in relating theory to experiment to discuss as the primary experimental quantity the right-hand side of (48). One is then led, by analogy to the corresponding situation in nucleon-nucleon scattering theory, to what might be called an "effective-range" treatment of pion-nucleon scattering.

V. EFFECTIVE-RANGE APPROXIMATION

We consider

$$\text{Re}\,g_\alpha(\omega) = 1 - \omega\left\{\frac{\lambda_\alpha}{\pi}P\int\frac{d\omega_p p^3 v^2(p)}{\omega_p^2(\omega_p - \omega)}\right.$$
$$\left. + \frac{1}{\pi}\int\frac{d\omega_p}{\omega_p^2}p^3 v^2(p)\frac{H_\alpha(\omega_p)}{\omega_p + \omega}\right\}. \quad (49)$$

The effective range approximation is based on the weak dependence of $\text{Re}\,g_\alpha(\omega)$ on the ω occurring in the denominators of the integrands in (49). Neglect of this dependence seems reasonable a priori for values of ω which are small compared to ω_{max}, the maximum energy effectively allowed by the cutoff factor $v(p)$, provided only that $H_\alpha(\omega_p)$ maintains the type of smoothness indicated by the first few terms of the power series. One would suppose the error incurred to be of the order ω/ω_{max}, which may not be excessive for pion kinetic

[12] It has been kindly pointed out to us by Dr. T. D. Lee that an exact solution of the one-meson approximation can be obtained for the charged scalar theory, in which case one finds essentially $H_\alpha = -\lambda_\alpha$, and the problem of analytic continuation becomes trivial.

[13] Castillejo, Dalitz, and Dyson, Phys. Rev. 101, 453 (1956).

energies less than 200 Mev, since ω_{max} is in the neighborhood of 1 Bev. Explicit calculation of the first integral in (49) verifies this conjecture.

Referring back to (48), we see that the combination

$$\lambda_\alpha p^3 \cot\delta_\alpha / \omega_p \tag{50}$$

can be written in the form

$$1 - \omega r_\alpha(\omega), \tag{51}$$

where $r_\alpha(\omega)$ is almost a constant for small ω. The effective-range approximation corresponds to a complete neglect of the energy dependence of $r_\alpha(\omega)$. It can be tested experimentally by plotting $(p^3/\omega) \cot\delta_\alpha(\omega)$ against ω. According to (50) and (51), one should find a straight line with intercept at zero energy equal to λ_α^{-1}. Lindenbaum and Yuan[14] have made such a plot for δ_{33} (where, for reasons to be discussed in Sec. VII, ω_p has been replaced by $\omega_p{}^* = \omega_p + p^2/2M$, M being the nucleon mass). The expected linear dependence has been found and the intercept leads to a value for the renormalized (unrationalized) coupling constant of $f^2 = 0.08$.

The effective-range approximation that $r_\alpha(\omega) \approx r_\alpha(0)$ for $\omega \ll \omega_{max}$ may seem superficially equivalent to the statement that an expansion of $r_\alpha(\omega)$ in powers of ω has a radius of convergence $\gtrsim \omega_{max}$. Such is not the case, of course, because the branch points in $g_\alpha(z)$ at $z = \pm 1$ give a radius of convergence equal to 1. Serber and Lee[15] have pointed out that the part of $g_\alpha(z)$ which is not analytic at $z = +1$ can be isolated and evaluated, and one may extend their approach to separate also the part which is most singular at $z = -1$. The reason that the effective-range approximation works is that the remaining part of $g_\alpha(z)$, which has no singularity in the low-energy region, is larger than the nonanalytic parts by a factor of order ω_{max}.

If the experimental data were sufficiently accurate to warrant the effort, one could improve the coupling constant determination by correcting for the small terms which cannot be extrapolated from the physical region ($\omega > 1$) to the point $\omega = 0$. The recipe turns out to be the following: Introduce a quantity

$$\Gamma_\alpha(\omega_p) = \frac{p^3}{\omega_p} \cot\delta_\alpha(\omega_p) - \frac{1}{\omega_p} + \frac{3}{2}\omega_p + \begin{bmatrix} -\frac{1}{2} \\ -2 \\ +1 \end{bmatrix} \omega_p$$

$$\times \left[\frac{p^2}{\omega_p{}^2} \left(\frac{p}{\pi} \log \frac{1}{\omega_p - p} + \frac{1}{2} - \frac{\omega_p}{\pi} \right) + \frac{1}{4} - \frac{\omega_p}{3\pi} \right]. \tag{52}$$

One can show[16] that to an accuracy of order $1/\omega_{max}{}^2$

[14] S. J. Lindenbaum and L. C. L. Yuan, Phys. Rev. **100**, 306 (1955).

[15] R. Serber and T. D. Lee (private communication). See also Friedman, Lee, and Christian, Phys. Rev. **100**, 1494 (1955).

[16] A proof of this statement will be given in a forthcoming review article by G. F. Chew, *Encyclopedia of Physics* [Springer-Verlag, Berlin (to be published)], Vol. 43.

the new quantity $\Gamma_\alpha(\omega_p)$ will be of the following form:

$$\Gamma_\alpha(\omega_p) = \frac{1}{\lambda_\alpha}[1 - r_\alpha \omega_p + P_\alpha \omega_p{}^2], \tag{53}$$

where the coefficient P_α should be smaller than r_α by a factor of order $1/\omega_{max}$. That is, $\Gamma_\alpha(\omega_p)$ is an almost linear function in the low-energy region and extrapolates to the value λ_α^{-1} at $\omega_p = 0$.

Note that there is no point in making the above refinement unless the extrapolation is at least quadratic, because the term in (53) proportional to P_α is presumably of the same order as the modifications made by formula (52) to the original effective-range approximation. Existing experimental data probably do not warrant the refined extrapolation procedure.

Plots for δ_{11}, δ_{13}, and δ_{31} should of course lead to the same value of f^2, but unfortunately the experimental information on these phase shifts is only that they are small compared to δ_{33}. Our theory definitely predicts that they should all be negative for not too large values of f^2, and the first few terms of the power series for H_α suggests strongly that the phase shifts should be small.

The crossing relation (43) makes a prediction about the coefficients of the term proportional to ω in an expansion of g_α about $\omega = 0$. When translated into a statement about the effective range $r_\alpha(0)$, the prediction is that

$$r_\alpha = r_3 \begin{bmatrix} -1 + \frac{1}{4}x \\ -x \\ 1 \end{bmatrix}, \tag{54}$$

where x is an unknown parameter.

In the following section, it will be proved that for the (3,3) state the coefficient of the linear term is negative definite, that is to say, that r_3 is positive definite. One might say, then, that the theory "predicts" a resonance in the 33 state provided the coupling is sufficiently strong. Estimates based on the power series for H_α suggest that resonance will indeed occur at the right energy with the known value of f^2 if the cutoff energy ω_{max} is in the neighborhood of 6.

It actually can be shown from the form of the complete scattering equation, that is, the equation before the multi-meson terms are dropped, that the neglected terms will not interfere with the effective-range approach. If anything the energy dependence of integrals over the multi-meson states will be weaker than that of the one-meson terms considered in this section. The value of the effective ranges for a given cutoff is of course altered, but the relation (54), which depends only on the crossing symmetry, is preserved.

VI. TOTAL CROSS SECTION SUM RULES

Returning to the general problem, we now wish to point out some important relations involving total cross sections which may be derived in the static model without any approximation. The basis of these relations is formula (12) for the total cross section, or rather a generalization of (12) which we now write down.

The notation employed so far can accommodate arbitrary initial and final nucleon spin—isotopic-spin states. The meson states, however, must be such that the linear momentum and isotopic variable are well defined. If we wish to consider a more general set of one-meson states, say $\phi_a(q)$, then the quantity which describes the scattering from state a to state n is

$$T_a(n) = \sum_q T_q(n)\phi_a(q), \qquad (55)$$

while that to another state b, of the same set, is

$$\langle b|T|a\rangle = \sum_{q,p} \phi_b{}^*(p)T_q(p)\phi_a(q), \qquad (56)$$

if the standard normalization

$$\sum_q \phi_b{}^*(q)\phi_a(q) = \delta_{ba} \qquad (57)$$

is employed. The total cross section for the state a is

$$\sigma_a = \frac{2\pi}{v_a} \sum_n \delta(E_n - \omega_a)T_a{}^\dagger(n)T_a(n), \qquad (58)$$

where an expectation with respect to the initial nucleon state is understood. One may also write down the equation for $\langle b|T|a\rangle$ corresponding to (25):

$$\langle b|T|a\rangle = -\sum_n \left[\frac{T_b{}^\dagger(n)T_a(n)}{E_n - \omega_a - i\epsilon} + \frac{T_{a^*}{}^\dagger(n)T_{b^*}(n)}{E_n + \omega_a} \right], \qquad (59)$$

where a^* and b^* refer to states which are the complex conjugates of ϕ_a and ϕ_b, respectively.

For the special case $b=a$ and the same initial and final nucleon states, it is clear that (58) allows (59) to be rewritten in terms of total cross sections. One finds easily that

$$\langle a|T|a\rangle = \langle a|T^0|a\rangle - \frac{1}{2\pi} \frac{q_a{}^2 v^2(q_a)}{\omega_a} \int_1^\infty \frac{dE}{q_E v^2(q_E)}$$
$$\times \left[\frac{\sigma_a(E)}{E - \omega_a - i\epsilon} + \frac{\sigma_{a^*}(E)}{E + \omega_a} \right], \qquad (60)$$

where q_E is the momentum of a meson of energy E. The operator T^0 represents the zero-meson part of the sum over states in (59). It is also the zero-energy limit of T. The notation $\sigma_a(E)$ means the total cross section for an incident meson whose energy is E but whose other variables (isotopic spin and angular momentum) are those of the state a.

We concentrate here on two special cases of (60) although other applications may also be interesting. The first case is when ϕ_a and $\phi_a{}^*$ represent the same state of the incident meson. A neutral meson with well-defined linear momentum is the simplest example of this situation. [Note that for a charged meson,

$\phi_a \neq \phi_a{}^*$.] With no distinction between a and a^*, (60) reduces to

$$\langle a|T|a\rangle = \langle a|T^0|a\rangle$$
$$- \frac{q_a{}^2 v^2(q_a)}{\pi \omega_a} \int_0^\infty \frac{dq_E}{v^2(q_E)} \frac{\sigma_a(E)}{q_E{}^2 - q_a{}^2 - i\epsilon}, \qquad (61)$$

a result which is almost identical with the well-known dispersion relation.[17] Apart from the cut-off factor, it differs from the usual dispersion relation in that ϕ_a may be an arbitrary real (in this representation) state. If all important contributions to the integral come below $\omega_{\max}$, the cutoff factors are unimportant. It is interesting that a result so close to the dispersion relation is obtained, because the latter is a consequence of causality and with an extended source the static model is of course not causal.

A second special initial state of particular interest is one which is an eigenstate of total angular momentum and total isotopic spin. This can occur only in the (3,3) case for the kind of states we have considered, where for example ϕ_a may be chosen as a positive meson with its orbital angular momentum up. The nucleon must then be a proton with its spin up. $\phi_a{}^*$ then corresponds to a negative meson with orbital angular momentum down. For this special case, (60) shows that the quantity $h_3(\omega)$ defined by Eqs. (32) and (33) is given by

$$h_3(\omega) = \frac{\lambda_3}{\omega} + \frac{1}{12\pi^2} \int_1^\infty \frac{dE}{q_E v^2(q_E)} \left[\frac{\sigma_+(E)}{E - \omega - i\epsilon} + \frac{\sigma_-(E)}{E + \omega} \right], \qquad (62)$$

where σ_+ is the total cross section for the state ϕ_a and σ_- that for $\phi_a{}^*$. This result shows among other things that the quantity r_3, defined in connection with the effective-range discussion, is certainly positive. Comparison with (51) and (41) leads to

$$r_3 f^2 = \frac{1}{(4\pi)^2} \int_1^\infty \frac{dE}{E q_E v^2(q_E)} [\sigma_+(E) + \sigma_-(E)], \qquad (63)$$

a relation which may be used to determine r_3 experimentally if the appropriate parts of the total cross section can be isolated. It should be noted that according to (63) the contribution to r_3 from each type of state ($n=1$, 2, etc.) is individually positive. Thus the one-meson approximation probably requires an unnecessarily high cut-off energy to produce the required value of r_3.

We now derive a different type of sum rule to obtain a relation between the renormalized and unrenormalized coupling constants. Consider the expectation of the operator $\sigma_q \tau_q \sigma_q \tau_q = 1$ taken with respect to a single-nucleon state. If the latter is normalized, the expected value is of course unity. Starting with this apparently trivial fact and using the completeness of the set of

[17] R. Karplus and M. A. Ruderman, Phys. Rev. 98, 771 (1955); Goldberger, Miyazawa, and Oehme, Phys. Rev. 99, 979 (1955).

states $\Psi_n^{(-)}$, we obtain the following:

$$1 = (\Psi_0, \sigma_q \tau_q \cdot \sigma_q \tau_q \Psi_0) \qquad (64)$$

$$= \sum_n (\Psi_0, \sigma_q \tau_q \Psi_n^{(-)})(\Psi_n^{(-)}, \sigma_q \tau_q \Psi_0) \qquad (65)$$

$$= \sum_n \left[f_r^{(0)} \frac{iqv(q)}{(2\omega_q)^{\frac{1}{2}}} \right]^{-2} (\Psi_0, V_q^{(0)} \Psi_n^{(-)})(\Psi_n^{(-)}, V_q^{(0)} \Psi_0) \quad (66)$$

$$= [f_r^{(0)}]^{-2} \sum_n \frac{2\omega_q}{q^2 v^2(q)} T_q^\dagger(n) T_q(n), \qquad (67)$$

where an expectation with respect to the nucleon state is understood. Next multiply (67) by $(f_r^0)^2$ and separate the $n=0$ terms from the sum. The result is

$$(f_r^{(0)})^2 = f_r^2 + \sum_{n>0} \frac{2\omega_q}{q^2 v^2(q)} T_q^\dagger(n) T_q(n); \qquad (68)$$

finally, we employ (58) to introduce the total cross section and find

$$(f_r^{(0)})^2 = f_r^2 + \frac{1}{\pi} \int \frac{dE \sigma_q(E)}{q_E v^2(q_E)}. \qquad (69)$$

The result (69) demonstrates that $(f_r^0)^2 > f_r^2$, a fact already shown by Lee.[11] It also gives in principle a method for measuring experimentally the unrenormalized coupling constant. Note that the cross section involved could be that for a neutral pion or it could be the average for positive and negative pions. The target may be either a neutron or a proton.

VII. CONCLUSIONS AND POSSIBLE EXTENSION OF RESULTS

We summarize here the results of the preceding sections which we feel are the most significant:

(a) The function $p^3/\omega_p \cot\delta_\alpha$ should be approximately linear at low energies and should extrapolate to λ_α^{-1} at $\omega=0$. Here $\lambda_{11} = -(8/3)f^2$, $\lambda_{13} = \lambda_{31} = -(2/3)f^2$, $\lambda_{33} = (4/3)f^2$. This implies among other things that the p-wave phase shifts maintain the sign of the Born approximation, that is, δ_{33} is positive and all the others are negative.

(b) The effective ranges for the various p-states, defined by (50) and (51), are not completely independent but must obey the relation (54).

(c) The effective range in the $(3,3)$ state is certainly positive, so that a resonance will occur in this state if the coupling is sufficiently strong.

These results have been derived on the basis of a theory which completely neglects relativistic effects. The question naturally arises as to whether they will be maintained when nucleon recoil and nucleon pair formation, as well as the effects of other particles, are taken into account. It is not possible to give an unqualified answer to this question, but recent and independent investigations of the relativistic theory, based on such

general requirements as Lorenz invariance, give some indications as to the validity of the statements (a)–(c).

In the first place, it has been rigorously demonstrated for the relativistic Yukawa theory that the p-wave scattering amplitude approaches the renormalized Born approximation in the limit $\omega_q^* \to 0$, where

$$\omega_q^* = \omega_q + (q^2/2M) + O(1/M^2), \qquad (70)$$

if M is the nucleon mass. That the functional form of the individual phase shifts is as given in statement (a) has not really been proved but seems extremely likely to us in view of recent work by Thirring[18] and Oehme.[19] These authors have shown or have promised to show that the first derivatives of the scattering amplitude with respect to $\sin\theta$ and $\cos\theta$, evaluated at $\theta=0$, have an analytic form which corresponds to condition (E) of Sec. III above. Conditions (A) and (B), which correspond to unitarity and crossing, respectively, are certainly general. Condition (C) is the zero-energy limit theorem which, as stated above, has been proved to be general. At low energies it seems legitimate to neglect orbital angular momenta higher than one, in which case the Thirring-Oehme amplitude derivatives can be identified to order $1/M$ with the p-wave amplitude alone, and one seems almost to have reproduced the equations of the cut-off theory from a general point of view. The difficulty of course is that at high energies orbital angular momenta larger than one certainly contribute, and high energies are important under the integrals which occur in the scattering equations.

One cannot, then, make clear-cut statements about the behavior of the p-wave amplitude in the entire complex plane. It is hard to imagine, however, a form for the amplitude which differs at low energies from that in statement (a) above and still manages to be unitary, to satisfy the Thirring-Oehme equations and to approach the correct zero-energy limit.

Statement (b), which connects the various p-wave effective ranges, is not rigorously true in the relativistic case because s-wave and p-wave amplitudes occur together in the term linear in the energy when the appropriate covariant expansion about zero energy is made.[20] However, the s-wave amplitudes experimentally are sufficiently small so that Eq. (54) remains a good approximation.

We do not have confidence that statement (c) above, although known experimentally to be correct, is necessarily a consequence of the relativistic Yukawa theory because it depends sensitively on the behavior of the scattering at high energies.

In conclusion we should like to emphasize a distinction between two classes of meson phenomena: those that depend only on low-energy matrix elements (in the sense of this paper) and those whose calculation

[18] W. Thirring, private communication.
[19] R. Oehme, Phys. Rev. **100**, 1503 (1955).
[20] F. Low (to be published).

requires a knowledge of high-energy matrix elements. An example of the former is the theorem that the zero energy limit of the p-wave effective range extrapolation measures the same coupling constant as the zero-energy limit of the photomeson production amplitude (according to the Kroll-Ruderman theorem). On the other hand the problem of theoretically evaluating the effective range falls in the latter class. Formula (49), for example, shows clearly that the value of r_3 depends on high-energy phenomena.

We have made no serious attempt in this paper to calculate the effective ranges. Presumably the (3,3) effective range could be matched by an appropriate choice of the cut-off energy, whatever method of approximation were used, and the dominant role played by the (3,3) state at low energies guarantees the success of any approach which produces the correct value for r_3. The question naturally arises as to whether one should expect to be able to calculate r_3 and other quantities which involve integrals over high virtual energies with the conventional relativistic form of the Yukawa theory, which has no adjustable cut-off parameter. We think the answer is no, because this theory does not take account of the existence of hyperons and K-particles which interact strongly with the pion-nucleon system. Both the cutoff and the local forms of the Yukawa theory are incorrect (or at least incomplete) in the Bev energy region.

Our zero-energy results hold for both theories and we believe they will probably hold in future theories, although this last statement is of course little more than a guess. We also believe that the linear extrapolation of the cotangent of the phase shifts will be maintained because this is essentially a statement of ignorance: the more important are high-energy phenomena, the more nearly constant is the effective-range integral.

We hope to show in the paper on photomeson production, which follows, that many aspects of this latter process fall in the first (low virtual energy) class of phenomena. The same is true for Compton scattering by protons and probably for the nuclear force problem. Phenomena which belong to the second class presumably include s-wave scattering, π^0 decay, the charge and current density of nucleons, as well as the fundamental questions concerning the nature and interactions of curious particles.

PHYSICAL REVIEW VOLUME 106, NUMBER 6 JUNE 15, 1957

Application of Dispersion Relations to Low-Energy Meson-Nucleon Scattering*

G. F. Chew,† *University of Illinois, Urbana, Illinois and Institute for Advanced Study, Princeton, New Jersey*

M. L. Goldberger,‡ *Fermi Institute for Nuclear Studies, University of Chicago, Chicago, Illinois*

F. E. Low, *University of Illinois, Urbana, Illinois and Department of Physics and Laboratory for Nuclear Science, Massachusetts Institute of Technology, Cambridge, Massachusetts*

AND

Y. Nambu, *Fermi Institute for Nuclear Studies, University of Chicago, Chicago, Illinois*

(Received February 21, 1957)

Relativistic dispersion relations are used to derive equations for low-energy S-, P-, and D-wave meson-nucleon scattering under the assumption that the (3,3) resonance dominates the dispersion integrals. The P-wave equations so obtained differ only slightly from those of the static fixed-source theory. The conclusions of the static theory are re-examined in the light of their new derivation.

1. INTRODUCTION

DISPERSION relations for the scattering of π mesons by nucleons have been given by many authors.[1] Proofs[2] based on the field theory formalism have been given for the special case of forward scattering by Symanzik, Lehmann and Jost, and Bogoliubov, for angles infinitesimally near forward by Symanzik, and for arbitrary angles by Bogoliubov.

In the neighborhood of the forward direction, the dispersion relations express real parts of scattering amplitudes as integrals over sums of partial-wave cross sections. Since these quantities are at least in principle measurable, one has at hand a multiple infinity of sum rules which may be compared directly with experiment. This is the procedure which has been followed by Anderson, Davidon and Kruse,[3] by Haber-Schaim,[4] and by Davidon and Goldberger.[5]

Another use to which the dispersion relations may be put has been pointed out by Oehme[6] who showed that in the static limit, provided higher waves than $l=1$ are neglected, the equations of the static P-wave theory are obtained together with a similar set of static S-wave equations.

We will here consider the problem in an intermediate way, one which is motivated at the same time by the success of the static P-wave theory in correlating meson scattering and photoproduction experiments and by the observed dominance of the (3,3) resonance in dispersion integrals. In effect we shall assume that the (3,3) resonance not only dominates dispersion integrals but exhausts them. Once this assumption is made, the equations of the static theory follow naturally, since in the energy range of the resonance the nucleon recoil velocity is small, $v/c \lesssim \frac{1}{3}$. Including effects of order v/c gives us the contributions to the dispersion integrals of the resonance region accurate to about 10%. Our most important conclusion will be that these assumptions lead to effective range relations for the P-wave phase shifts; they do not make possible a determination of the actual location of the (3,3) resonance, which must be taken from experiment. Once the (3,3) phase shift is known the S-, D-, and small P-wave phase shifts may be directly calculated in our approximation. The validity of the approximation is hard to estimate without knowing the partial wave decomposition of cross sections in the Bev region; order of magnitude estimates based on known total cross sections indicate that the resonant state is correctly given to about 10%, but that the contribution of high-energy cross sections to a typical small amplitude, although less than 10% of the (3,3) amplitude, is still comparable to the small amplitude itself. More detailed studies of this question are now being made.

In Sec. II we introduce the necessary variables and describe the transformation from relativistically invariant scattering amplitudes to a partial wave decomposition in the center-of-mass system. This material is not new, nor is it related to meson theory; we include it only for convenient reference. In Sec. III we give dispersion relations and use them to derive equations for S-, P-, and D-wave scattering. The algebra in this section is quite complicated. We advise the interested reader (as opposed to the dedicated one) to read up to and including Eq. (3.18), by which time the method of calculation should be clear. The essential results of the remaining algebra are contained in Eq. (3.20), (3.30), (3.33), (3.34), and (3.35). In Sec. IV we discuss the conditions imposed on the P-wave scattering amplitude by the P-wave equations, and re-examine the solutions of the static theory in the light of their new derivation.

* This work was supported in part by the joint program of the Office of Naval Research and the U. S. Atomic Energy Commission.

† Present address: Department of Physics, University of California, Berkeley, California.

‡ Present address: Department of Physics, Princeton University, Princeton, New Jersey.

[1] For a complete list of references see R. H. Capps and G. Takeda, Phys. Rev. **103**, 1877 (1956).

[2] Reported at the International Conference on Theoretical Physics, Seattle, Washington, September, 1956.

[3] Anderson, Davidon, and Kruse, Phys. Rev. **100**, 339 (1955).

[4] U. Haber-Schaim, Phys. Rev. **104**, 1113 (1956).

[5] W. Davidon and M. L. Goldberger, Phys. Rev. **104**, 1119 (1956).

[6] R. Oehme, Phys. Rev. **100**, 1503 (1955); **102**, 1174 (1956).

1337

2. KINEMATICAL CONSIDERATIONS

Let the four-vector momenta of the incident and outgoing pion be q_1 and q_2, respectively, while those of the initial and final nucleon are p_1 and p_2. Momentum-energy conservation,

$$p_1 + q_1 = p_2 + q_2, \tag{2.1}$$

means that only three of the four vectors are independent.[7] We choose to consider the combinations

$$P = \tfrac{1}{2}(p_1 + p_2), \quad Q = \tfrac{1}{2}(q_1 + q_2), \quad \kappa = \tfrac{1}{2}(q_1 - q_2), \tag{2.2}$$

as the three independent four-vectors.

The mass shell restrictions, $p_1{}^2 = p_2{}^2 = -M^2$, $q_1{}^2 = q_2{}^2 = -1$, mean that

$$P \cdot \kappa = Q \cdot \kappa = 0,$$
$$P^2 + \kappa^2 = -M^2, \quad Q^2 + \kappa^2 = -1, \tag{2.3}$$

so that there are only two independent scalars, which we may take as

$$\nu = -P \cdot Q/M \quad \text{and} \quad \kappa^2. \tag{2.4}$$

The second of these variables is one-quarter of the invariant momentum-transfer squared:

$$\kappa^2 = \tfrac{1}{4}(q_1 - q_2)^2 = \tfrac{1}{2}\mathbf{q}^2(1 - \cos\theta), \tag{2.5}$$

where $\mathbf{q}$ is the three-vector momentum and θ the scattering angle in the center-of-mass system. Also

$$\nu = \nu_L - (\kappa^2/M), \tag{2.6}$$

where ν_L is the incident meson energy in the lab system. Thus ν is almost equal to the lab energy for moderate momentum transfer.

To form further invariants, we must use the Dirac operators. By virtue of the Dirac equation,

$$
\begin{aligned}
(i\gamma \cdot p_1 + \mathrm{M})u_1 &= 0, \\
(i\gamma \cdot p_2 + \mathrm{M})u_2 &= 0.
\end{aligned}
\tag{2.7}
$$

Thus the invariants $i\gamma \cdot p_1$ and $i\gamma \cdot p_2$ may be anticommuted through the matrix element until they act on the initial or final spinor, respectively, where by (2.7) they give a constant. The same is true for $i\gamma \cdot \kappa = i\gamma \cdot (p_2 - p_1)/2$, so that only $i\gamma \cdot Q$ remains as an independent scalar.

The S matrix can be written

$$S = \delta_{fi} - (2\pi)^4 i \delta^4(p_2 + q_2 - p_1 - q_1)$$
$$\times \left(\frac{M^2}{4E_1 E_2 \omega_1 \omega_2} \right)^{\frac{1}{2}} \bar{u}_2 T u_1, \tag{2.8}$$

where E_1 and E_2 are initial and final nucleon energies and ω_1 and ω_2 the initial and final meson energies. The

spinor normalization is

$$\bar{u}_2 u_2 = \bar{u}_1 u_1 = 1. \tag{2.9}$$

Our considerations have shown that T may be written

$$T = -A + i\gamma \cdot Q B, \tag{2.10}$$

where A and B are functions of ν and κ^2, and are also matrices in charge space. Charge independence limits this last complication to a doubling of the number of functions. Let β be the state of the final meson ($\beta = 1, 2, 3$) and α that of the initial. Then

$$A_{\beta\alpha} = \delta_{\beta\alpha} A^{(+)} + \tfrac{1}{2}[\tau_\beta, \tau_\alpha] A^{(-)}, \tag{2.11}$$

and

$$B_{\beta\alpha} = \delta_{\beta\alpha} B^{(+)} + \tfrac{1}{2}[\tau_\beta, \tau_\alpha] B^{(-)}, \tag{2.12}$$

where $A^{(\pm)}$ and $B^{(\pm)}$ are simply functions of ν and κ^2.

It is frequently useful to express the $(\pm)$ amplitudes in terms of the total isotopic spin. One finds easily

$$A^{(+)} = \tfrac{1}{3}(A^{(\frac{1}{2})} + 2A^{(\frac{3}{2})}),$$
$$A^{(-)} = \tfrac{1}{3}(A^{(\frac{1}{2})} - A^{(\frac{3}{2})}), \text{ etc.} \tag{2.13}$$

Finally we state without proof the relation between the A's and B's and the conventional scattering amplitudes in states of definite parity and angular momentum. For this relation it is convenient to introduce the center-of-mass variables

$$
\begin{aligned}
W &= \text{total energy}, \\
E &= \text{total nucleon energy}, \\
x &= \cos\theta.
\end{aligned}
\tag{2.14}
$$

In terms of these variables,

$$\frac{1}{4\pi} A^{(\pm)} = \frac{W+M}{E+M} f_1{}^{(\pm)} - \frac{W-M}{E-M} f_2{}^{(\pm)}, \tag{2.15}$$

$$\frac{1}{4\pi} B^{(\pm)}{}^{\frac{1}{2}} = \frac{1}{E+M} f_1{}^{(\pm)} + \frac{1}{E-M} f_2{}^{(\pm)}. \tag{2.16}$$

Here f_1 and f_2 are simply related to the scattering cross sections in the center-of-mass system:

$$\frac{d\sigma}{d\Omega} = \sum \left| \left\langle f \left| f_1 + \frac{\boldsymbol{\sigma} \cdot \mathbf{q}_2 \boldsymbol{\sigma} \cdot \mathbf{q}_1}{q_2 q_1} f_2 \right| i \right\rangle \right|^2 \tag{2.17}$$

where the symbol $\sum$ represents a sum over final and average over initial spin states. In (2.17) we have suppressed the superscripts referring to charge states.

Finally

$$f_1 = \sum_{l=0}^{\infty} f_{l+} P_{l+1}'(x) - \sum_{l=2}^{\infty} f_{l-} P_{l-1}'(x), \tag{2.18}$$

$$f_2 = \sum_{l=1}^{\infty} (f_{l-} - f_{l+}) P_l'(x), \tag{2.19}$$

where $f_{l\pm}$ is the scattering amplitude in the state of

---

[7] In this paper four-vectors are represented by italicized symbols, thus: p. Three-dimensional vectors are represented by bold-face symbols, thus: $\mathbf{p}$. The four-dimensional inner product is $p \cdot q = \sum p_\lambda q_\lambda = \mathbf{p} \cdot \mathbf{q} - p_0 q_0$. We also set $\hbar = c = \mu = 1$, where μ is the pi-meson mass. The nucleon mass is M.

parity $-(-1)^l$ and total angular momentum $j=l\pm\frac{1}{2}$. $P_l'(x)$ is the first derivative of the conventionally normalized Legendre polynomial.

The $f_\pm$ are normalized so that

$$(j+\tfrac{1}{2})\,\mathrm{Im}f_{l\pm}=\frac{q}{4\pi}\sigma_{l\pm}, \qquad (2.20)$$

where $\sigma_{l\pm}$ is the total cross section of the partial wave involved. Thus, for energies below the two-meson threshold,

$$f_{l\pm}=e^{i\delta_{l\pm}}\frac{\sin\delta_{l\pm}}{q}, \qquad (2.21)$$

where $\delta_{l\pm}$ is the real phase shift in the state $l_\pm$; above this threshold a representation of the form (2.21) still holds, but with complex $\delta_{l\pm}$.

3. DISPERSION RELATIONS

The form that the dispersion relations take depends on the behavior of $A(\nu,\kappa^2)$ and $B(\nu,\kappa^2)$ for very large values of ν. In what follows we shall make a drastic (and probably incorrect) assumption about this high-frequency behavior: we shall assume that A and B approach zero sufficiently rapidly so that all the integrals we introduce converge uniformly (as functions of κ^2) in the neighborhood of $\kappa^2=0$. With this assumption, we have

$$\mathrm{Re}A^{(\pm)}(\nu,\kappa^2)$$
$$=\frac{P}{\pi}\int_{1-\kappa^2/M}^{\infty}d\nu'\,\mathrm{Im}A^{(\pm)}(\nu',\kappa^2)\left(\frac{1}{\nu'-\nu}\pm\frac{1}{\nu'+\nu}\right), \quad (3.1)$$

and

$$\mathrm{Re}B^{(\pm)}(\nu,\kappa^2)=\frac{g_r^2}{2M}\left(\frac{1}{\nu_B-\nu}\mp\frac{1}{\nu_B+\nu}\right)$$
$$+\frac{P}{\pi}\int_{1-\kappa^2/M}^{\infty}d\nu'\,\mathrm{Im}B^{(\pm)}(\nu',\kappa^2)\left(\frac{1}{\nu'-\nu}\mp\frac{1}{\nu'+\nu}\right). \quad (3.2)$$

Here $\nu_B=-(1/2M)-(\kappa^2/M)$ and g_r^2 is the rationalized, renormalized (according to the Lepore-Watson renormalization convention[8]) pseudoscalar coupling constant. Experimentally, $g_r^2/4\pi\approx14$. As usual the symbol P stands for principal value.

In this paper we wish to exhibit the consequences of augmenting Eq. (3.1) and Eq. (3.2) with the assumption of the dominance of the (3,3) state. Put differently, we shall investigate the effects of (3,3) contributions to integrals in Eq. (3.1) and Eq. (3.2). Now if the integrals in question do not converge sufficiently rapidly to make (3.1) and (3.2) meaningful equations, it is still possible to obtain valid equations by subtraction at some fixed value of ν. The new equations so obtained

[8] K. M. Watson and J. V. Lepore, Phys. Rev. **76**, 1157 (1949).

will contain arbitrary functions of κ^2 which do not arise from resonant integrals, and which at present cannot be predicted by dispersion theory. We shall therefore take the point of view that the existence of these arbitrary functions is one among many high-energy effects which we shall not attempt to evaluate; we shall therefore use Eq. (3.1) and Eq. (3.2). Finally, in order to simplify the results, we shall make the valid approximation that the nucleon velocity is small in the resonance region; we shall, however, carry our results to include first order terms in this velocity (as opposed to the static limit, which keeps only zero-order terms). It should be understood, then, that in the following that although we shall continue to write integrals with an infinite upper limit we really have in mind as an upper limit the energy at which the (3,3) state fails to dominate, say three or four hundred Mev lab energy.

We first change the variables in (3.1) and (3.2) to ν_L and κ^2 by means of (2.6). We find

$$\mathrm{Re}A^{(\pm)}(\nu_L,\kappa^2)=-\frac{P}{\pi}\int_1^{\infty}d\nu_L'\,\mathrm{Im}A^{(\pm)}(\nu_L',\kappa^2)$$
$$\times\left(\frac{1}{\nu_L'-\nu_L}\pm\frac{1}{\nu_L'+\nu_L-2\kappa^2/M}\right), \quad (3.3)$$

and

$$\mathrm{Re}B^{(\pm)}(\nu_L,\kappa^2)=\frac{g_r^2}{2M}\left(\frac{1}{\nu_0-\nu_L}\mp\frac{1}{\nu_0+\nu_L-2\kappa^2/M}\right)$$
$$+\frac{P}{\pi}\int_1^{\infty}d\nu_L'\,\mathrm{Im}B^{(\pm)}(\nu_L',\kappa^2)$$
$$\times\left(\frac{1}{\nu_L'-\nu_L}\mp\frac{1}{\nu_L'+\nu_L-2\kappa^2/M}\right), \quad (3.4)$$

where $\nu_0=-1/2M$.

Next we solve (2.15) and (2.16) for f_1 and f_2. We find easily that

$$f_1=\left(\frac{E+M}{2W}\right)\left(\frac{A+(W-M)B}{4\pi}\right), \qquad (3.5)$$

and

$$f_2=\left(\frac{E-M}{2W}\right)\left(\frac{-A+(W+M)B}{4\pi}\right). \qquad (3.6)$$

As long as we intend to carry our calculation only to a finite order in v/c (or equivalently $1/M$), we save ourselves a great deal of trouble by expanding in powers of κ^2. It will become obvious that S-wave amplitudes may be expected to be of order $g_r^2/M\approx Mf_r^2$ (although in fact they are greatly reduced from this value for reasons that are still obscure), P waves of order f_r^2 (although the enhanced state here is of order unity, of course) and D waves of order f_r^2/M, as long as high-energy contributions to (3.3) and (3.4) are negligible. In any case, these orders of magnitude should be kept in mind as a basis for comparison of different terms. Here $f_r{}^2$ is, of course, the rationalized pseudovector coupling constant, $f_r{}^2 = g_r{}^2/4M^2$.

Let us write (2.18) and (2.19) in terms of κ^2 rather than the c.m. angle θ. Including D waves (but no higher), we have

$$f_1 = f_S - f_{D\frac{3}{2}} + 3f_{P\frac{3}{2}}\left(1 - \frac{2\kappa^2}{q^2}\right)$$

$$+\tfrac{1}{2}f_{D\frac{5}{2}}\left[15\left(1 - \frac{2\kappa^2}{q^2}\right)^2 - 3\right] + \cdots, \quad (3.7)$$

and

$$f_2 = f_{P\frac{3}{2}} - f_{P\frac{1}{2}} + 3\left(1 - \frac{2\kappa^2}{q^2}\right)(f_{D\frac{5}{2}} - f_{D\frac{3}{2}}) + \cdots, \quad (3.8)$$

where q^2 is the c.m. meson momentum.

If we set $\kappa^2 = 0$ in (3.7), we obtain

$$f_1(0) \cong f_S + 3f_{P\frac{3}{2}}, \quad (3.9)$$

since $f_D \ll f_S$ in the region of interest. Differentiating (3.7) with respect to κ^2 and then setting $\kappa^2 = 0$, we find

$$f_1'(0) \cong -6f_{P\frac{3}{2}}/q^2, \quad (3.10)$$

where again D waves have been neglected. Combining (3.9) and (3.10), we find

$$f_S = f_1(0) + \tfrac{1}{2}q^2 f_1'(0) + \sim D \text{ waves.} \quad (3.11)$$

The prime here stands for differentiation with respect to κ^2. The argument zero also stands for κ^2. The amplitudes $f(0)$, $f'(0)$, etc., are of course still functions of energy. Similarly,

$$-\frac{6}{q^2}f_{P\frac{1}{2}} = f_1'(0) + \tfrac{1}{2}q^2 f_1''(0) + \sim F \text{ waves}, \quad (3.12)$$

$$f_{P\frac{3}{2}} - f_{P\frac{1}{2}} = f_2(0) + \tfrac{1}{2}q^2 f_2'(0) + \sim F \text{ waves}, \quad (3.13)$$

$$\frac{60}{q^4}f_{D\frac{5}{2}} = f_1''(0) + \sim F \text{ waves}, \quad (3.14)$$

$$-\frac{6}{q^2}(f_{D\frac{5}{2}} - f_{D\frac{3}{2}}) = f_2'(0) + \sim F \text{ waves}. \quad (3.15)$$

Finally, on the right-hand side of Eq. (3,3) we shall, at least for the moment, keep only the (3.3) state—that is, we write

$$\frac{1}{4\pi}\operatorname{Im}A^{(\pm)}(\nu_L', \kappa^2)$$

$$= \left[\frac{3(W'+M)(1-2\kappa^2/q^2)}{E'+M} + \frac{W'-M}{E'-M}\right]\operatorname{Im}f_3^{(\pm)}, \quad (3.16)$$

$$\frac{1}{4\pi}\operatorname{Im}B^{(\pm)}(\nu_L', \kappa^2)$$

$$= \left[\frac{3(1-2\kappa^2/q^2)}{E'+M} - \frac{1}{E'-M}\right]\operatorname{Im}f_3^{(\pm)}, \quad (3.17)$$

where

$$f_3^{(+)} = \tfrac{2}{3}f_{33}, \quad f_3^{(-)} = -\tfrac{1}{3}f_{33}. \quad (3.18)$$

We first calculate the resonance contribution to the S-wave amplitudes:

$$\operatorname{Re}f_S^{(\pm)} \cong f_1^{(\pm)}(0) + \tfrac{1}{2}q^2 f_1^{(\pm)\prime}(0) = \frac{1}{4\pi}\cdot\frac{E+M}{2W}\left[A^{(\pm)}(0) + \tfrac{1}{2}q^2 A^{(\pm)\prime}(0) + (W-M)(B^{(\pm)}(0) + \tfrac{1}{2}q^2 B^{(\pm)\prime}(0))\right]$$

$$= \frac{E+M}{2W}\left\{\frac{(W-M)g^2}{2M}\left(\frac{1}{\nu_0-\nu_L} \mp \frac{1}{\nu_0+\nu_L} \mp \frac{q^2/M}{(\nu_0+\nu_L)^2}\right) + \frac{P}{\pi}\int_1^\infty d\nu_L'\,\operatorname{Im}f_3^{(\pm)}(\nu_L')\left[\frac{1}{\nu_L'-\nu_L}\left(\frac{3(W'+M)(1-q^2/q'^2)}{E'+M}\right.\right.\right.$$

$$\left. + \frac{W'-M}{E'-M} + (W-M)\left(\frac{3(1-q^2/q'^2)}{E'+M} - \frac{1}{E'-M}\right)\right) \pm \frac{1}{\nu_L'+\nu_L}\left(\frac{3(W'+M)(1-q^2/q'^2)}{E'+M} + \frac{W'-M}{E'-M}\right.$$

$$\left. - (W-M)\left(\frac{3(1-q^2/q'^2)}{E'+M} - \frac{1}{E'-M}\right)\right) \pm \frac{q^2/M}{(\nu_L'+\nu_L)^2}\left(\frac{3(W'+M)}{E'+M} + \frac{W'-M}{E'-M}\right.$$

$$\left.\left.\left. - (W-M)\left(\frac{3}{E'+M} - \frac{1}{E'-M}\right)\right)\right]\right\}, \quad (3.19)$$

where g^2 is the nonrationalized coupling constant,

$$g^2 = g_r{}^2/4\pi.$$

We next express everything in terms of center-of-mass energies, using the relation $\nu_L = (W^2 - M^2 - 1)/2M$. There results, to the desired order in $1/M$,

$$f_S^{(\pm)} \cong \frac{M}{W}\left\{-\frac{g^2}{2M}\left[\left(1 - \frac{\omega}{2M}\right) \pm \left(1 + \frac{\omega}{2M}\right)\right] + \frac{2M}{\pi}\int_1^\infty \frac{d\omega'}{q'^2}\left[\left(1 + \frac{2\omega'}{M} + \frac{\omega}{M}\right) \pm \left(1 + \frac{2\omega'}{M} - \frac{\omega}{M}\right)\right]\operatorname{Im}f_3^{(\pm)}(\omega')\right\}, \quad (3.20)$$

where $\omega = W - M$, $\omega' = W' - M$, and terms of order $1/M$ relative to those kept are left out.

We see that the strong energy dependence of f_3 produces no reflection on the S-wave energy dependence, which to a high degree of approximation is given (except for the trivial phase space factor M/W) by a constant for $f^{(+)}$ and a constant times ω for $f^{(-)}$.

It is a simple matter to add the low-energy contribution of the S-wave amplitude under the integrals. Here we may be very careless in taking nonrelativistic limits, since the term $f_2/(E-M)$ in Eq. (2.15) and (2.16) receives no contribution from S waves. Thus, to lowest order in $1/M$,

$$f_S \cong A(\kappa^2 = 0)/4\pi, \qquad (3.21)$$

so that Eq. (3.20) becomes

$$\mathrm{Re} f_S{}^{(\pm)} \cong -\left[\left(\lambda^{(+)} - \frac{\omega}{2M}\lambda^{(-)}\right) \pm \left(\lambda^{(+)} + \frac{\omega}{2M}\lambda^{(-)}\right)\right]$$
$$+ \frac{P}{\pi}\int d\omega' \, \mathrm{Im} f_S{}^{(\pm)}(\omega')\left[\frac{1}{\omega' - \omega} \pm \frac{1}{\omega' + \omega}\right], \quad (3.22)$$

where explicit expressions for $\lambda^{(\pm)}$ are provided by Eq. (3.20):

$$\lambda^{(+)} = \frac{g^2}{2M} - \frac{2M}{\pi}\int \frac{d\omega'}{q'^2}\left(1 + \frac{2\omega'}{M}\right)\mathrm{Im} f_3{}^{(+)}(\omega'), \quad (3.23)$$

$$\lambda^{(-)} = \frac{g^2}{2M} + \frac{4M}{\pi}\int \frac{d\omega'}{q'^2}\mathrm{Im} f_3{}^{(-)}(\omega'). \quad (3.24)$$

These are the Oehme[6] equations. Now the contribution of the S-wave integrals in (3.22) at threshold is small, so that $\lambda^{(\pm)}$ may be determined by experiment.

According to Orear,[9]

$$f^{(\frac{1}{2})}(q=0) = -0.11,$$
$$\qquad\qquad\qquad\qquad (3.25)$$
$$f^{(\frac{3}{2})}(q=0) = +0.16.$$

thus

$$f^{(+)}(0) = -0.02, \quad f^{(-)}(0) = +0.09,$$

and

$$\lambda^{(+)} = 0.01, \quad \lambda^{(-)} = 0.6. \qquad (3.26)$$

These numbers for $\lambda^{(\pm)}$ are in fact not inconsistent with (3.23) and (3.24) integrated over the (3.3) reso-

nance. We prefer not to take such an agreement seriously since the approximate constancy of high-energy cross sections (as observed in cosmic rays) argues strongly against the validity of Eq. (3.1) as it stands, so that at least one subtraction must presumably be made in Eq. (3.1).

Of course, if one is primarily interested in calculating threshold scattering the partial wave reduction we have performed is unnecessary since only S waves contribute at zero kinetic energy. The constants $\lambda^{(\pm)}$ should therefore be calculated directly from (3.1) and (3.2). The result of such a calculation for $\lambda^{(+)}$ is inconclusive, whereas for $\lambda^{(-)}$, as carried out by Haber-Schaim,[4] it yields a number in surprisingly close agreement with (3.25). This presumably means that in practice no subtraction need be carried out to make (3.1)$^{(-)}$ a correct equation.

We turn next to the derivation of P-wave equations. We calculate first

$$f_{P_{\frac{1}{2}}}{}^{(\pm)} - f_{P_{\frac{3}{2}}}{}^{(\pm)} \cong f_2{}^{(\pm)}(0) + \frac{1}{2}q^2 f_2{}^{(\pm)\prime}(0)$$
$$= \frac{1}{4\pi}\left(\frac{E-M}{2W}\right)\{-A^{(\pm)}(0) - \frac{1}{2}q^2 A^{(\pm)\prime}(0)$$
$$+ (W+M)[B^{(\pm)}(0) + \frac{1}{2}q^2 B^{(\pm)\prime}(0)]\}. \quad (3.27)$$

The Born approximation to (3.27) is easily found, since (3.19) and (3.20) inform us that

$$\frac{1}{4\pi}\left(\frac{E+M}{2W}\right)(W-M)[B^{(\pm)}(0) + \frac{1}{2}q^2 B^{(\pm)\prime}(0)]|_{\mathrm{Born}}$$
$$\cong \frac{M}{W}\left(-\frac{g^2}{2M}\right)\left[\left(1 - \frac{\omega}{2M}\right) \pm \left(1 + \frac{\omega}{2M}\right)\right].$$

Thus

$$\left(\frac{E-M}{2W}\right)(W+M)[B^{(\pm)}(0) + \frac{1}{2}q^2 B^{(\pm)\prime}(0)]|_{\mathrm{Born}}$$
$$\qquad\qquad\qquad\qquad\qquad (3.28)$$
$$\cong -\frac{q^2}{\omega}f^2\left(1 - \frac{\omega}{2M}\right)\left[\left(1 - \frac{\omega}{2M}\right) \pm \left(1 + \frac{\omega}{2M}\right)\right],$$

where $f^2 = g^2/4M^2 \cong 0.08$.

The contribution of the (3.3) integrals is, however, different from the S-wave case:

$$\mathrm{Re}(f_{P_{\frac{1}{2}}}{}^{(\pm)} - f_{P_{\frac{3}{2}}}{}^{(\pm)}) \cong -\frac{q^2 f^2}{\omega}\left(1 - \frac{\omega}{2M}\right)\left[\left(1 - \frac{\omega}{2M}\right) \pm \left(1 + \frac{\omega}{2M}\right)\right] + \frac{q^2}{4M} \cdot \frac{P}{\pi}\int_1^\infty \frac{d\nu_L'}{W}$$

$$\times \mathrm{Im} f_3{}^{(\pm)}(\nu_L')\left\{\frac{1}{\nu_L' - \nu_L}\left[-\frac{3(W'+M)(1-q^2/q'^2)}{M+E'} - \frac{W'-M}{E'-M} + (W+M)\left(\frac{3(1-q^2/q'^2)}{M+E'} - \frac{1}{E'-M}\right)\right]\right.$$

$$\pm \frac{1}{\nu_L' + \nu_L}\left[-\frac{3(W'+M)}{E'+M}(1-q^2/q'^2) - \frac{W'-M}{E'-M} - (W+M)\left(\frac{3(1-q^2/q'^2)}{E'+M} - \frac{1}{E'-M}\right)\right]$$

$$\left. \pm \frac{q^2/M}{(\nu_L' + \nu_L)^2}\left[-\frac{3(W'+M)}{E'+M} - \frac{W'-M}{E'-M} - (W+M)\left(\frac{3}{E'+M} - \frac{1}{E'-M}\right)\right]\right\}. \quad (3.29)$$

---

[9] J. Orear, Phys. Rev. **100**, 288 (1955).

Again we express ν_L and ν_L' as functions of ω and ω' and expand in powers of ω/M and ω'/M. The result is

$$\mathrm{Re}(f_{P\frac{3}{2}}^{(\pm)}-f_{P\frac{1}{2}}^{(\pm)})=-\frac{q^2f^2}{\omega}\left[\left(1-\frac{\omega}{2M}\right)\pm\left(1+\frac{\omega}{2M}\right)\right]\left(1-\frac{\omega}{2M}\right)$$

$$+q^2\frac{P}{\pi}\int_1^\infty\frac{d\omega'}{q'^2}(-\mathrm{Im}f_3^{(\pm)}(\omega'))\left(\frac{1}{\omega'-\omega}+\frac{1}{M}\mp\frac{1}{\omega'+\omega}\right). \quad (3.30)$$

Notice that except for the added constant $1/M$, all $1/M$ corrections under the integral have gone into the energy variable $\omega=W-M$. That is, the form of (3.30), except for the added constant, is the same as in the static theory[10]; the difference is in the meaning of the variable ω, which is now obviously the appropriate one for describing low-energy P-wave scattering.

Next we calculate $f_{P\frac{3}{2}}$:

$$-\frac{6}{q^2}\mathrm{Re}f_{P\frac{3}{2}}\cong f_1^{(\pm)\prime}(0)+\tfrac{1}{2}q^2f_1^{(\pm)\prime\prime}(0)$$

$$=\left(\frac{E+M}{2W}\right)[A^{(\pm)\prime}(0)+\tfrac{1}{2}q^2A^{(\pm)\prime\prime}(0)+(W-M)(B^{(\pm)\prime}(0)+\tfrac{1}{2}q^2B^{(\pm)\prime\prime}(0))], \quad (3.31)$$

or

$$-\frac{6}{q^2}\mathrm{Re}f_{P\frac{3}{2}}^{(\pm)}=\left(\frac{E+M}{2W}\right)\left\{\mp\frac{g^2}{2M}(W-M)\left(\frac{2/M}{(\nu_0+\nu_L)^2}+\frac{4q^2/M^2}{(\nu_0+\nu_L)^3}\right)+\frac{P}{\pi}\int_1^\infty d\nu_L'\,\mathrm{Im}f_3^{(\pm)}(\nu_L')\right.$$

$$\times\left[\frac{1}{\nu_L'-\nu_L}\left(-\frac{6(W'+M)}{E'+M}\cdot\frac{1}{q'^2}-\frac{6(W-M)}{E'+M}\cdot\frac{1}{q'^2}\right)\pm\frac{1}{\nu_L'+\nu_L}\left(-\frac{6(W'+M)}{E'+M}\cdot\frac{1}{q'^2}+\frac{6(W-M)}{E'+M}\cdot\frac{1}{q'^2}\right)\right.$$

$$\pm\frac{2/M}{(\nu_L'+\nu_L)^2}\left(\frac{3(W'+M)}{E'+M}+\frac{W'-M}{E'-M}-(W-M)\left(\frac{3}{E'+M}-\frac{1}{E'-M}\right)\right)\pm\frac{\tfrac{1}{2}q^2}{(\nu_L'+\nu_L)^2}\left(-\frac{24}{Mq'^2}\right)$$

$$\left.\left.\times\left(\frac{W'+M}{E'+M}-\frac{W-M}{E'+M}\right)\pm\frac{4q^2/M^2}{(\nu_L'+\nu_L)^3}\left(\frac{3(W'+M)}{E'+M}+\frac{W'-M}{E'-M}-(W-M)\left(\frac{3}{E'+M}-\frac{1}{E'-M}\right)\right)\right]\right\}. \quad (3.32)$$

An entirely analogous reduction of this equation gives

$$-6\,\mathrm{Re}\frac{f_{P\frac{3}{2}}^{(\pm)}}{q^2}=\mp\frac{4f^2}{\omega}+\frac{P}{\pi}\int_1^\infty\frac{d\omega'}{q'^2}$$

$$\times\mathrm{Im}f_{P\frac{3}{2}}^{(\pm)}(\omega')\left[-\frac{6}{\omega'-\omega}-\frac{6}{M}\mp\frac{2}{\omega'+\omega}\right]. \quad (3.33)$$

Before going on to a discussion of these results, let us find expressions for the D waves:

$$\frac{60}{q^4}f_{D\frac{3}{2}}^{(\pm)}\cong f_1^{(\pm)\prime}(0),$$

or

$$f_{D\frac{3}{2}}^{(\pm)}\cong-\frac{q^4}{15M}\left[\frac{4f^2}{\omega^2}+\frac{2}{\pi}\int\frac{d\omega'}{q'^2}\frac{\mathrm{Im}f_3^{(\pm)}(\omega')}{(\omega'+\omega)^2}\right], \quad (3.34)$$

and

$$-\frac{6}{q^2}(f_{D\frac{3}{2}}^{(\pm)}-f_{D\frac{1}{2}}^{(\pm)})\cong f_2^{(\pm)\prime}(0),$$

[10] G. F. Chew and F. E. Low, Phys. Rev. 101, 1570 (1956).

or

$$-(f_{D\frac{3}{2}}^{(\pm)}-f_{D\frac{1}{2}}^{(\pm)})$$

$$=\pm\frac{1}{3}\frac{q^4}{m}\left[\frac{f^2}{\omega^2}-\frac{1}{\pi}\int\frac{d\omega'}{q'^2}\frac{\mathrm{Im}f_3^{(\pm)}(\omega')}{(\omega'+\omega)^2}\right]. \quad (3.35)$$

4. DISCUSSION OF RESULTS

The P-wave equations may be rewritten (using (2.13))

$$\mathrm{Re}f_{11}=-\frac{8}{3}\frac{f^2q^2}{\omega}+\frac{3}{M}f^2q^2+\frac{16}{9}\frac{q^2}{\pi}\int\frac{d\omega'\,\mathrm{Im}f_{33}(\omega')}{q'^2}\frac{1}{\omega'+\omega},$$

$$\mathrm{Re}f_{13}=\mathrm{Re}f_{31}=\tfrac{1}{4}\,\mathrm{Re}f_{11}-\frac{3}{4M}f^2q^2,$$

$$\mathrm{Re}f_{33}=\frac{4}{3}\frac{f^2q^2}{\omega}+\frac{q^2}{\pi}P\int\frac{d\omega'}{q'^2}$$

$$\times\mathrm{Im}f_{33}(\omega')\left[\frac{1}{\omega'-\omega}+\frac{1}{M}+\frac{1}{9}\left(\frac{1}{\omega'+\omega}\right)\right]. \quad (4.1)$$

These equations are almost the same as those of the static theory[10] (with the same assumption, of course, that the resonance region dominates so that the contributions of the "small phase shift" states may be neglected). To obtain the static equations from (4.1), one simply replaces q^2 by $v^2(q)q^2$ where $v^2(q)$ is the cutoff function, ω by $(1+q^2)^{\frac{1}{2}}$ [the reader will recall that here ω is the total center-of-mass energy: $\omega = q^2/2M + (1+q^2)^{\frac{1}{2}}$], and finally one drops the explicit $1/M$ terms in (4.1). Since all of these changes are small ones in the resonance region, we have in (4.1) a "derivation" of the static theory and a method of partially understanding its agreement with experiment.

Let us repeat this last point, since it is an important one. The reason the static theory agrees with experiment is that in the integrals on the right-hand side of the dispersion relations the resonance region dominates, so that Eq. (4.1) holds; Eq. (4.1) is in turn a consequence of static meson theory, provided that there also one assumes the dominance of the resonance integral for low-energy phenomena. What we have achieved, therefore, is not really a complete derivation of the static meson theory, but a set of instructions on how that theory must be used and which of its results are believable.

Let us replace Eq. (4.1) by the static equations [see reference 10, Eq. (40)]:

$$\mathrm{Re}f_\alpha = \lambda_\alpha \frac{q^2 v^2(q)}{\omega} + \frac{q^2 v^2(q)}{\pi} P \int \frac{d\omega'}{q'^2 v^2(q')}$$
$$\times \left[\frac{\mathrm{Im}f_\alpha(\omega')}{\omega'-\omega} + \sum_\beta \frac{A_{\alpha\beta}\,\mathrm{Im}f_\beta(\omega')}{\omega'+\omega} \right], \quad (4.2)$$

and look for solutions subject to the one-meson approximation and to the condition that the (3.3) resonance be properly located. These may or may not be such that the (3.3) resonance integral dominates for low values of ω, depending on $v^2(q)$ and, in the event of the existence of several solutions,[11] on which one is chosen. All those solutions which have substantial non-(3-3) resonance contributions to the right-hand side of (4.2) for small ω we throw out. This will include all of those solutions having zeros in the low-energy region since these will necessarily also have extra resonances which will contribute to the integrals.

As shown in reference 10, the solutions of (4.2) for ω not too large are roughly of the form

$$f_\alpha \cong \frac{\lambda_\alpha q^2/\omega}{1 - r_\alpha \omega - i\lambda_\alpha q^3/\omega}, \quad (4.3)$$

independent of the shape of the cutoff function (provided it is singular enough to produce the observed resonance), and provided there are no zeros of f_α. As shown in reference (11), however, any zeros whose

[11] Castillejo, Dalitz, and Dyson, Phys. Rev. **101**, 453 (1956).

corresponding resonances do not contribute to the right-hand side of (4.2) must be on the real axis, but quite far from $\omega=0$, so that they result only in new effective ranges for the three states.

If one actually tries to solve Eq. (4.2) for the location of the resonance, one finds

$$1/r_{33} = \omega_r \approx 1/f^2 \omega_{\mathrm{max}}, \quad (4.4)$$

where ω_{max} is the cutoff energy. The location of the resonance at a moderate energy with a small coupling constant $f^2 = 0.08$, therefore, necessitates a high cutoff, $\omega_{\mathrm{max}} \gg 1$. This circumstance in turn insures the approximate constancy of the effective ranges r_α.

This set of statements may be approximately deduced directly from Eq. (4.1). In particular, if we consider $1/9 \ll 1$ and $1/M \gg 1$, then the equation for f_{33} becomes

$$f_{33} = \frac{\lambda q^2}{\omega} + \frac{q^2}{\pi} \int \frac{d\omega'}{q'^2} \frac{\mathrm{Im}f_{33}(\omega')}{\omega'-\omega-i\epsilon}, \quad (4.5)$$

with $\lambda = (4/3)f^2$ or, for a narrow resonance,

$$f_{33} \cong \frac{\lambda q^2}{\omega} + \frac{q^2}{\omega_r - \omega} \cdot \frac{1}{\pi} \int \frac{d\omega'}{q'^2} \mathrm{Im}f_{33}(\omega') \quad (4.6)$$

$$\cong \frac{\lambda q^2}{\omega} + \frac{\lambda q^2}{\omega_r - \omega} = \frac{\lambda q^2/\omega}{1 - \omega/\omega_r}, \quad (4.7)$$

where (4.6) follows from (4.5) provided $\omega_r - \omega \gg \Gamma$ where Γ is the width of the resonance; in going from (4.6) to (4.7) we have actually integrated (4.3) over the resonance, again assuming the width to be small. One may also derive (4.7) directly from (4.5) by noting that for $\omega \gg \omega_r$, $f_{33} \approx 0$; thus the integral in (4.6) must equal λ.

Thus the effective-range formula is approximately consistent with (4.1) for any resonance energy; the equation does not, therefore, determine the resonance, since the left and right sides are approximately equal for any ω_r. Thus small terms in the equation, such as $1/9$, or $1/M$, or f^2, or the high-energy contribution, will actually determine the precise location of the resonance. In this way we can reconcile the cutoff dependence of the resonance energy predicted by Eq. (4.4) and the dominance of the resonance integrals in Eq. (4.2).

We conclude that the shape of the resonance curve is determined by our considerations once the position of the resonance has a given value; this position however, we have not been able to determine from first principles.

For the small phase shifts, we have only to do the integrals over $d\omega'$. We find easily

$$f_{11} \cong -\frac{8}{3} \frac{f^2 q^2/\omega}{1 + \omega/\omega_r}, \quad (4.8)$$

$$f_{13} \cong_{31} \cong \frac{1}{4} f_{11},$$

where we have again set Γ, $1/M$, and $1/9$ approximately equal to zero.

Let us summarize: the assumption of the dominance of the (3.3) contribution to low-energy dispersion integrals, together with the experimental location of the (3.3) resonance, leads to the following results:

(1) The P-wave phase-shifts approximately satisfy effective-range formulas

$$\frac{\lambda_\alpha q^3}{\omega}\cot\delta_\alpha\cong 1-\omega r_\alpha, \qquad (4.9)$$

with

$$\lambda_{11}=-(8/3)f^2, \quad \lambda_{33}=\tfrac{4}{3}f^2, \quad \lambda_{13}=\lambda_{31}=-\tfrac{2}{3}f^2,$$

$$r_{33}=1/\omega_r, \quad r_{11}\cong r_{13}=r_{31}\cong-r_{33}.$$

Also, to order $1/M$, $f_{13}=f_{31}$, as in the static theory.

(2) The S-wave amplitudes should be approximately given by the two zero energy scattering lengths, $\lambda^{(\pm)}$; since we have not considered the addition of any arbitrary constants, the only interesting fact that emerges is that even to order $1/M$ the strong energy dependence of the (3,3) state has no reflection in the S-wave energy dependence.

(3) The D-wave phase shifts are approximately given by[12]

$$\delta_{D\tfrac{3}{2}}=-\lambda_D\left[1+\frac{112}{9}\left(\frac{\omega}{\omega+\omega_r}\right)^2\right], \qquad (4.10)$$

---

[12] The D-wave phase shifts given here have been previously obtained by V. Wataghin (unpublished). We would like to thank Dr. Wataghin for informing us of his results.

$$\delta_{D\tfrac{3}{2}}=\lambda_D\left[2-\frac{28}{9}\left(\frac{\omega}{\omega+\omega_r}\right)^2\right], \qquad (4.11)$$

$$\delta_{D\tfrac{5}{2}}=\lambda_D\left[4-\frac{32}{9}\left(\frac{\omega}{\omega+\omega_r}\right)^2\right], \qquad (4.12)$$

$$\delta_{D\tfrac{5}{2}}=-\lambda_D\left[8+\frac{8}{9}\left(\frac{\omega}{\omega+\omega_r}\right)^2\right], \qquad (4.13)$$

[with $\lambda_D=(1/15)(f^2/M)q^5/\omega^2$], as one finds simply by carrying out the integrals in (3.34) and (3.35) in the zero-width approximation. Now $\lambda_D\approx 0.21°$ at $\omega=\omega_r$, so that the D wave phase shifts are all very small. Since, however, $\delta_{D\tfrac{3}{2}}$ is of the order of magnitude two degrees at $\omega\cong\omega_r$, and since the weighting factor for a $j=5/2$ state is 3, the present analysis in terms of P and S states is unreliable as far as the small P phase shifts and S phase shifts are concerned in the resonance region.

Finally, which of these results will survive the addition of contributions from high-energy cross sections? It is our tentative opinion that only the (3.3) amplitude will stand this test, since the others, with the possible exception of f_{11}, are so small that very small corrections can change them by their own order of magnitude: the present theory predicts $\delta_{13}\approx\delta_{31}-4°$ at the resonance, which is just the order of magnitude of the high-energy contributions we have estimated from the known total cross sections at 1 Bev. The chances of the present theory adequately describing δ_{13}, δ_{31}, and the D waves are thus very small. A slight consolation is perhaps that the argument can be turned around, and eventual measurement of δ_{13} and δ_{31} used to provide information on the high-energy cross sections.

REGGE TRAJECTORIES AND THE PRINCIPLE OF MAXIMUM STRENGTH FOR STRONG INTERACTIONS*

Geoffrey F. Chew

Lawrence Radiation Laboratory and Department of Physics, University of California, Berkeley, California

and

Steven C. Frautschi

Department of Physics, Cornell University, Ithaca, New York

(Received December 4, 1961)

In previous publications, the authors have discussed the possibility that strong interactions "saturate" the unitarity condition; i.e., that they have the maximum possible strength consistent with the unitarity and analyticity of the S matrix.[1] Our earlier discussion was confined to elastic scattering, however, and although the conjectured existence of Regge poles underlay our arguments, we did not at the time of the earlier work appreciate certain essential properties of these poles. We wish here, therefore, to give a general statement of the principle of maximum strength in terms of Regge poles and to explain certain qualitative and quantitative experimental predictions that follow.

In a recent Letter it was proposed that all baryons and mesons (stable or unstable) are associated with Regge poles that move in the complex angular-momentum plane as a function of energy.[2] The trajectory of a particular pole is characterized by a set of internal quantum numbers and by the even-ness or oddness of physical J for mesons or $J - \frac{1}{2}$ for baryons; but all S-matrix elements, regardless of multiplicity, are supposed to contain any pole whose quantum numbers are appropriate. (The residues of corresponding poles in different S-matrix elements will of course differ.) The position α_i of each Regge pole in the J plane is conjectured to be an analytic function of $s = E^2$, and Re$\alpha(s)$ is supposed to be monotonically increasing for $s < 0$ as well as throughout the (real) positive region of s in which stable and metastable particles occur. The imaginary part of α_i vanishes below the threshold for the lowest energy channel with the quantum numbers in question and is positive definite above this threshold. (Throughout the region of reasonably sharp resonances, we have Im$\alpha_i \ll 1$.) Stable or metastable particles occur at energies where Reα_i is equal to a possible physical value of J, the half-width of a resonance (metastable particle) being given by

$$\tfrac{1}{2}\Gamma_i = \frac{\mathrm{Im}\,\alpha_i}{(d\mathrm{Re}\,\alpha_i/dE)}, \tag{1}$$

where the right-hand side is evaluated at the resonance energy. All the above conjectures are motivated by the properties of poles in potential-scattering amplitudes—as deduced by Regge.[3] [See note at end of Letter.]

Figure 1 is a plot of the angular momentum of all strongly interacting particles for which spin evidence exists (and which have a baryon number less than two) as a function of mass squared. Each point is supposed to lie on a Regge trajectory, but if the above rules are followed with respect to quantum numbers and slope of trajectory, one concludes that only two particles—the nucleon and the N_3^*—could belong to the same trajectory.[4] This circumstance is not surprising if the low-energy slopes of all trajectories are similar in magnitude. The average displacement in m^2 between two members of the same family ($\Delta J = 2$) would then be of the order of $100\,m_\pi{}^2$; so the second member of any family—if it exists—will always lie well inside the continuum and be correspondingly difficult to find experimentally. Below, we discuss tentative evidence that $(d\alpha/ds)_{s=0}$ is of the order of $1/(50\,m_\pi{}^2)$ for trajectories other than that to which the nucleon belongs, and a theoretical motivation for such uniformity of slope is provided by Regge's potential-scattering formula,[3]

$$d(\alpha + \tfrac{1}{2})^2/dp^2 = R^2, \tag{2}$$

where p is the momentum and R an average radius of the bound state. All the baryons and mesons in Fig. 1 are expected to have similar "sizes," and the slope in question corresponds through formula (2) to $R \approx 1/(2\,m_\pi)$, a plausible order of magnitude.[5]

The principle of maximum strength for strong interactions depends on the assumption that Regge trajectories can be continued to the region $s \leq 0$ and on the result of Froissart that in this region $\alpha_i(s) \leq 1$ for all trajectories.[6] The point is that a given Regge pole gives rise to high-energy amplitudes in "crossed reactions" which are proportional to $E_{\mathrm{lab}}{}^{\alpha_i(s)}$, where now $s = -\Delta^2$ (the negative square of momentum transfer); and amplitudes that asymptotically increase as a power of energy

41

Reprinted with permission from: G. F. Chew and S. C. Frautschi, *Phys. Rev. Lett.* **8**, 41–44 (1962).
https://doi.org/10.1103/PhysRevLett.8.41. ©1962 American Physical Society.

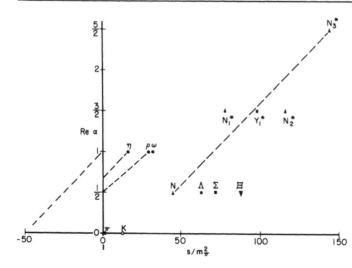

FIG. 1. The spin of particles of baryon number less than two, plotted against the square of their mass in units of m_π^2. In order to give a rough indication of slopes, the dashed lines connect pairs of points supposedly on the same trajectories, as explained in the text, but a strict linear behavior of the trajectories is not to be inferred.

greater than 1 violate the combined requirements of unitarity and analyticity.[6] From a glance at Fig. 1 it is evident that none of the trajectories associated with known particles is likely to reach the Froissart limit if all slopes are of the order of magnitude $1/(50\,m_\pi^2)$. Where then is there evidence for saturation of the unitarity condition?

The evidence, of course, lies in the fact that total cross sections actually appear to approach constants at high energy, implying an imaginary part of forward amplitudes $\propto E_{lab}$; so we have conjectured that a Regge pole with the quantum numbers of the vacuum is responsible—with a trajectory such that $\alpha_{vac}(s=0)=1$.[2] The slope of this vacuum trajectory is expected to be positive at low s (and similar in order of magnitude to the slopes of other trajectories), and it was explained in the previous Letter and is amplified below why it is plausible to have the vacuum trajectory lie above all others.[2] Thus the condition $\alpha_{vac}(s=0)=1$ represents a saturation of Froissart's limit.

Another way of looking at the situation is in terms of the binding forces responsible for the existence of baryons and mesons, all of which are composite in our picture. For quantum numbers where the net forces are weak or repulsive, the Regge trajectories never cross any physical values of J, and no particles appear. The stronger the attractive force, the higher in Fig. 1 the corresponding trajectory occurs, and in ordinary potential scattering there is no limit to the "level"

of a trajectory if the force strength is unbounded. In the relativistic case, however, unitarity in crossed channels leads to the Froissart limit, which constitutes an upper bound on the level of any trajectory. We believe that further study of crossing conditions will confirm that the strength of forces is in general correlated with simplicity of quantum numbers.[2] If so, the greatest attraction occurs for the quantum numbers of the vacuum, and here the Froissart limit is reached. In this sense, the forces are "as strong as possible."

The empirical association of quantum numbers with the ordering of trajectory levels in Fig. 1 is impressive. Following the "queen" of all trajectories, the vacuum, come four trajectories $(\eta, \rho, \omega, \pi)$ with zero strangeness (S) and zero baryon number (B). The isotopic spin (I) for the "prince consort" trajectory (η) is not yet definitely known,[7] but $I_\eta=0$ would fit naturally with the circumstance that exchange of these quantum numbers leads to a maximum coherence in high-energy scattering [i.e., the maximum value for $\alpha(s=0)$], next to the quantum numbers of the vacuum. Because $I_\rho=1$, the ρ trajectory should give less coherence, as also should the π. If the[8] ω has the same quantum numbers as the η, it must belong to a "second-rank" trajectory.[9] The next trajectories, K and K^* (if the K^* spin is 1), have $B=0$ and the lowest possible isotopic spin ($I=\frac{1}{2}$) consistent with one unit of strangeness. For the trajectories with $B=1$, there is a definite correlation of level order with strangeness,[10] and although

VOLUME 8, NUMBER 1   PHYSICAL REVIEW LETTERS   JANUARY 1, 1962

the correlation with isotopic spin is not clean, a preference for low I is manifest.

In our first published discussion of the principle of maximum strength, we failed to realize the crucial circumstance that Regge poles move with energy. Mandelstam reminded us of this feature,[11] which invalidates our original conclusion that particles should not occur with angular momentum greater than unity. A modified statement can still be made, however, to the effect that high spin should not occur in conjunction with low mass. From Fig. 1, for example, and our assumption about similarity of slope for all trajectories at low energy, it appears that the best chance for a spin-2 meson lies with the quantum numbers of the vacuum at a mass of the order of $7 m_\pi$. If the vacuum trajectory reaches a maximum below $\text{Re}\,\alpha = 2$, no such particle exists, but an experimental search seems worthwhile.

The first-order deviation from the Pomeranchuk high-energy limits[12] for certain particle combinations should be associated with the η trajectory at $s = 0$.[13] If the G parity turns out to be (-1), this pole may account for the substantial and slowly decreasing difference between $K^- p$ and $K^+ p$ as well as pp and $p\bar{p}$ total cross sections at high energy.[14] This difference will be proportional to $E_{\text{lab}}^{-[1-\alpha_\eta(0)]}$, and since from Fig. 1 we see that $1 - \alpha_\eta(0)$ is likely to be a good deal smaller than 1, a slow approach to equality of particle and antiparticle cross sections is easily understandable for the KN and NN combinations. In contrast, as pointed out to us by Udgaonkar,[15] the first-order difference between $\pi^- p$ and $\pi^+ p$ total cross sections at high energy (also the np-pp difference if $I_\eta = 0$) will be due to the trajectory associated with the ρ meson, where $I = 1$ and $G = +1$. Here the difference will be proportional to $E_{\text{lab}}^{-[1-\alpha_\rho(0)]}$ and should die out slightly more rapidly. Detailed and quantitative calculations are now being carried out to see if such a simple mechanism is capable of explaining the facts. A crude fitting of the observed deviations[14,16] from the Pomeranchuk limits suggests that $\alpha_\eta(0) \approx \frac{2}{3}$, while $\alpha_\rho(0) \approx \frac{1}{2}$—numbers that have been used in constructing Fig. 1.

No systematic effort has been made to milk all possible experimental suggestions out of Fig. 1, and readers may well observe significant features that we have overlooked—particularly in connection with strange particles, where our arguments no doubt lead to a prejudice about spin assignments for resonances. For example, $J = \frac{3}{2}$ hyperon resonances with both $I = 0$ and $I = 1$ are suggested for

masses near 13 to $14 m_\pi$.

A final remark concerns the slope of the vacuum trajectory. Frautschi, Gell-Mann, and Zachariasen[17] have analyzed recent data of Cocconi et al.[18] on high-energy pp elastic scattering in terms of formula (1) of reference 2 and have deduced therefrom $\alpha_{\text{vac}}(s = -45 m_\pi{}^2) \approx 0.1$. This result is used in Fig. 1, and further supports the notion that where Regge trajectories are vigorously rising, their slopes tend to be of the order $1/(50 m_\pi{}^2)$.

In conclusion we wish to state our belief that all of strong-interaction physics will flow from: (a) the principle of maximal analyticity of the S matrix, in angular as well as linear momenta,[2] (b) the principle of maximum strength, and (c) the conservation of B, S, and I. There should be no arbitrary dimensionless parameters and only one constant with the dimensions of length (or mass) to be added to $\hbar$ and c; there are no elementary particles. It seems conceivable to us that principles (b) and (c) above will eventually be shown to have a close relationship to (a), but at present we have no proposals in this direction. That (a), (b), and (c) form the basis for a complete theory of strong interactions has, of course, not been established by this Letter; but we feel greatly encouraged by the above-discussed internal consistency of the experimental facts when viewed in terms of principle (b) together with the notion of Regge-pole trajectories—which we are confident will soon be derived from principle (a).

Note added. We wish to apologize to R. Blankenbecler and M. L. Goldberger for insufficient reference in our previous Letter[2] to their work on Regge poles. In a report of their joint work delivered at the La Jolla Conference on the Theory of Weak and Strong Interactions in June, 1961, Goldberger stressed the existence of a family of particles associated with a given trajectory and the importance of J parity. He also mentioned the occurrence of a given pole in all amplitudes with the same quantum numbers and the possibility of deciding whether particles are elementary by continuation to a crossed channel. We are sincerely sorry for having omitted reference to this talk, which one of us (G.F.C.) heard but did not fully appreciate. We also would like to express belated appreciation for a remark from K. Wilson, Department of Physics, Harvard University, pointing out that the width shrinkage with increasing energy of forward and backward peaks due to Regge poles is only logarithmic.

*This work was done under the auspices of the U. S. Atomic Energy Commission.

[1]G. F. Chew and S. C. Frautschi, Phys. Rev. Letters $\underline{5}$, 580 (1960); Phys. Rev. $\underline{123}$, 1478 (1961).

[2]G. F. Chew and S. C. Frautschi, Phys. Rev. Letters $\underline{7}$, 394 (1961).

[3]T. Regge, Nuovo cimento $\underline{14}$, 951 (1959); $\underline{18}$, 947 (1960).

[4]We assume that $N_3{}^*$ is associated with an $F_{5/2}$, $I=\frac{1}{2}$ resonance in πN scattering at 900-Mev lab kinetic energy, and that $N_2{}^*$ is the 600-Mev $D_{3/2}$, $I=\frac{1}{2}$ resonance. For a recent survey of evidence concerning these two resonances, see R. Omnés and G. Valladas, Centre d'Etudes Nucléaires de Saclay Report 61-6 Rev., 1961 (unpublished). The particle we have designated $N_1{}^*$ is the 33 resonance.

[5]A singularity in the Regge trajectory occurs at each physical threshold, but we do not anticipate an important change in the average slope when crossing a threshold. We expect all trajectories to turn over at sufficiently high energy, and those with weak or repulsive forces rise little, if at all, so the concept of homogeneity of slope must be used with caution. Certainly it can only be used where the binding energies of particles are comparable to m_π, so that the particle radii are not much larger than $\sim 1/2\,m_\pi$.

[6]M. Froissart, Phys. Rev. $\underline{123}$, 1053 (1961).

[7]A. Pevsner, R. Kraemer, M. Nussbaum, C. Richardson, P. Schlein, R. Strand, T. Toohig, M. Block, A. Engler, R. Gessaroli, and C. Meltzer, Phys. Rev. Letters $\underline{7}$, 421 (1961); P. Bastien, J. P. Berge, O. Dahl, M. Ferro-Luzzi, D. H. Miller, J. J. Murray, A. H. Rosenfeld, R. D. Tripp, and M. B. Watson, Lawrence Radiation Laboratory (private communication).

[8]B. Maglić, L. W. Alvarez, A. H. Rosenfeld, and M. L. Stevenson, Phys. Rev. Letters $\underline{7}$, 178 (1961); and to be published.

[9]When the attractive forces for a given set of quantum numbers are sufficiently strong, there may be more than one Regge trajectory that reaches physical J values. A "second-rank" trajectory with the quantum numbers of the vacuum is presumably responsible for the $I=0$ two-pion anomaly observed near $s=4m_\pi{}^2$ by A. Aba-

shian, N. E. Booth, and K. M. Crowe, Phys. Rev. Letters $\underline{7}$, 35 (1961). Here the trajectory probably reaches its maximum near this value of s and turns over just before reaching $\mathrm{Re}\alpha=0$, producing what is usually called a "virtual state" rather than a resonance. The origin of the phenomenon is still a Regge pole, however. The fact that the highest level trajectories of the second rank are siblings of the "queen" and "prince consort," respectively, gives additional support to the notion that force strength is correlated with simplicity of configuration.

[10]We have assigned $Y_1{}^*$ a spin of $\frac{3}{2}$ on the basis of tentative evidence recently obtained by R. P. Ely, S. Y. Fung, G. Gidal, Y. L. Pan, W. M. Powell, and H. S. White, Phys. Rev. Letters $\underline{7}$, 461 (1961).

[11]S. Mandelstam, Department of Physics, University of Birmingham (private communication).

[12]S. Pomeranchuk, J. Exptl. Theoret. Phys. (U.S.S.R.) $\underline{34}$, 725 (1958) [translation: Soviet Phys. – JETP $\underline{34(7)}$, 499 (1958)].

[13]This observation has been made independently by M. Gell-Mann and F. Zachariasen, California Institute of Technology (private communication).

[14]G. Von Dardel, D. H. Frisch, R. Mermod, R. H. Milburn, P. A. Piroué, M. Vivargent, G. Weber, and K. Winter, Phys. Rev. Letters $\underline{5}$, 333 (1960); S. J. Lindenbaum, W. A. Love, J. A. Niederer, S. Ozaki, J. J. Russell, and L. C. L. Yuan, Phys. Rev. Letters $\underline{7}$, 184 (1961).

[15]B. M. Udgaonkar, Lawrence Radiation Laboratory, Berkeley, California (private communication).

[16]G. Von Dardel, R. Mermod, P. A. Piroué, M. Vivargent, G. Weber, and K. Winter, Phys. Rev. Letters $\underline{7}$, 127 (1961); S. J. Lindenbaum, W. A. Love, J. A. Niederer, S. Ozaki, J. J. Russell, and L. C. L. Yuan, Phys. Rev. Letters $\underline{7}$, 352 (1961).

[17]S. Frautschi (Cornell University) and M. Gell-Mann and F. Zachariasen (California Institute of Technology) (private communication).

[18]G. Cocconi, Proceedings of the International Conference on Theoretical Aspects of Very High Energy Phenomena, CERN Report 61-22, 1961 (unpublished); and private communication.

PHYSICAL REVIEW D   VOLUME 4, NUMBER 8   15 OCTOBER 1971

Hadron Bootstrap Hypothesis*

Geoffrey F. Chew

Department of Physics and Lawrence Berkeley Laboratory, University of California, Berkeley, California 94720

(Received 15 January 1971)

A discussion is given of the conjecture that classical space-time properties prescribe a unique S matrix which approximates strong-interaction phenomena.

I. INTRODUCTION

An esthetically compelling speculation is that the laws of nature might be determined uniquely by requirements of self-consistency or, phrased more picturesquely, by a "bootstrap." This paper puts foward and analyzes a "partial bootstrap" conjecture that has for some time been the subject of informal discussion, but that heretofore has not found its way into research publication. The conjecture is the following: *Quantum superposition, when expressed through a nontrivial S matrix, can achieve compatibility with the real (classical) world in only one possible way — close to the way exhibited by nature for hadrons.* Recent progress by Stapp and collaborators[1-3] in clarifying the relation between the S matrix and classical space-time suggests that the moment may be ripe for systematic analysis of this uniqueness conjecture.

From the standpoint of hard science, the complete bootstrap idea is inadmissible because science requires the *a priori* acceptance of certain language-defining concepts, so that "questions" can be formulated and experiments performed to give "answers." The role of theory is to provide a set of rules for predicting the results of experiment, but rules necessarily are formulated in a language of accepted ideas. Among currently unquestioned notions prerequisite to the conduct of science are:

(1) Three-dimensional space and a unidirectional time, with an associated cause-effect event structure; the existence of suitable measuring rods and clocks is corollary.

(2) The arrangement of macroscopic matter into blobs of sufficiently well-defined shape and permanency that the isolated system or "object" concept becomes meaningful.

(3) The existence of weak long-range interactions like electromagnetism and gravity that allow "measurements" to be made upon "objects" without the objects losing identity; the observer's integrity must also be preserved.

The foregoing detailed prerequisites may deceptively be summarized by the single term, "measurement," but the concept of measurement, on which hard science is based, is admissible only because of certain special attributes of nature, attributes that a complete bootstrap theory would have to explain as necessary components of self-consistency. It is in this sense that the idea of a complete bootstrap, while not obviously foolish, is intrinsically unscientific.

Although natural philosophy eventually will no doubt identify a framework more general than that of observational science, such a development cannot be expected soon. In the meantime it may be possible to find an area of relevance for a "partial bootstrap" that is explorable within the framework of conventional science, accepting without question the measurement process and the surrounding space-time attributes, but avoiding further specific and arbitrary ingredient concepts — such as elementary constituents of matter or a fundamental equation of motion.

The world of quantum phenomena constitutes a natural possibility for such a partial bootstrap if one regards as philosophically uncrossable the gap between quantum and "real" (classical) worlds. Measurement, the concept that science requires us to accept without question, belongs to the real world. Insofar as the quantum world can be described by a collection of amplitudes[4] — the scattering matrix — one may pose questions of self-consistency within the mathematical structure of the S matrix and temporarily ignore the puzzle of why nature arranges itself so as to permit those measurements that give physical meaning to scattering amplitudes. A further sense in which an S-matrix bootstrap would be only partial is, of course, that the superposition principle is accepted on an *a priori* basis and not explained. In other words, we take for granted the existence of a quantum world.

In the achievement of a separate meaning for real and quantum worlds, the role of electromagnetism — as reviewed in the following section — is mysterious but essential. We shall reason, correspondingly, that a scientific (partial) bootstrap is unlikely to shed light on the origin of electromagnetism. Our argument will suggest that the most promising possibility for a partial bootstrap is an idealized hadronic domain of purely strong interactions, confined entirely within the quantum world.

II. ELECTROMAGNETISM AS THE BRIDGE BETWEEN REAL AND QUANTUM WORLDS

Attempts to understand the relation between scattering amplitudes and the real world usually ignore the detailed mechanisms of interaction between matter. It appears, nonetheless, that special attributes of electromagnetism are vital both to the conceptual separation between quantum world and real world and to the practical linkage of the two.

If electromagnetic forces were of short range, it is hard to imagine how matter and the interaction between pieces of matter could assume a form consistent with the (real world) concept of measurement. A piece of measuring apparatus

based entirely on nuclear forces, that is to say, is extremely difficult to conceive. The long-range electromagnetic interaction, associated with the zero[5] photon mass, appears essential. A sharper formulation of the question is to inquire under what conditions a quantum picture of interactions in terms of scattering amplitudes may become compatible with a classical description. In appropriate circumstances it is known that the classical electromagnetic-field concept is consistent with the quantum picture of photons,[6] but the connection depends on special properties of the photon, especially the zero mass. It seems most unlikely that classical interaction-transmitting observables analogous to the electromagnetic field can be associated with particles other than photons.[7] Our argument here, in summary, is that measurement is a classical concept and that electromagnetism is unique among particle interactions in possessing a classical manifestation.

An important corollary is that particles are observed in the real world only through their electromagnetic interactions. Without electromagnetism there would exist no mechanism for contact between the quantum world and real world; there would be, in other words, no way to attach physical significance to the scattering matrix.

Attributing to electromagnetism an essential role in measurement suggests that a bootstrap effort to explain the zero photon mass would involve the nonscientific task of explaining the measurement concept itself. *Given* the zero photon mass, furthermore, the small value of the fine-structure constant appears essential to our picture of the quantum world based on scattering amplitudes. This latter point requires elaboration.

It is familiar that for compatibility with the event relationships of the real world, as well as with quantum superposition, the scattering matrix should be both Poincaré invariant and unitary.[8] But the existence of zero-mass particles gives trouble with unitarity because there then exist, at all energies, infinite numbers of accessible asymptotic states (open channels). The very concept of "asymptotic state," in fact, becomes imprecise. This difficulty is obscured by quantum electrodynamics because of the power-series expansion in the fine-structure constant α. A finite order in the α expansion corresponds to a finite number of photons and restores meaning to asymptotic states. Although the smallness of α allows superficial use of S-matrix machinery, the scattering matrix has been defined only in the limit $\alpha \to 0$, where photons can be ignored. In the absence of reliance on a truncated power series in the fine-structure constant, the zero-mass difficulty becomes even more severe with respect to a third major category of S-matrix properties, loosely described as "analyticity," that will be discussed in Sec. III.

There should be no surprise at the existence of a dilemma here if one accepts that the scattering matrix, designed to describe the quantum world in terms of measurements carried out in the real world, is incapable of describing the real world itself. To the extent that certain aspects of electromagnetism constitute defining character- istics of the real world, the S matrix should not be expected to encompass electromagnetism in totality. Of course, as already remarked, certain quantum-world aspects of electromagnetism, as embodied in quantum electrodynamics, can be given a superficial S-matrix description to the extent that the smallness of α permits neglect of all but a finite number of terms in the Feynman expansion. The amazing accuracy of this descrip- tion is an unending source of confusion for the subject under discussion.

If the smallness of the fine-structure constant is somehow necessary for our picture of the quantum world, one anticipates that a bootstrap effort to explain the value of α will become en- tangled with an explanation of the origin of quantum superposition.[9] Here is a further basis to believe that it would be futile to seek a scien- tific bootstrap theory of electromagnetism.

Should the logical interrelation of the points made in this section be obscure to the reader, let him be aware that the author fares no better. The intended message is that electromagnetism is deeply mysterious and its origin unlikely to be explained within our current scientific framework because the unique attributes of this interaction are inextricably enmeshed with the framework itself.

III. THE HADRON S-MATRIX BOOTSTRAP

Abandoning hope of explaining electromagnetism through a scientific bootstrap, we are led to con- sider an idealized quantum world in which the fine-structure constant becomes vanishingly small. Sending α to zero would change the real world beyond recognition, but it is plausible to postulate that the collection of hadron amplitudes would approach a meaningful limit, the "hadron S matrix," as $\alpha \to 0$ and that this limit would bear a recognizable relation to actually observed hadronic phenomena.

The experimental motivation for such a postulate is the observation of approximate iso- spin symmetry for hadrons. Since the symmetry breaking appears to be of electromagnetic origin,[10]

the difference between the idealized S-matrix limit and actual hadron phenomena may plausibly be presumed to be of the same order of magnitude as the observed differences within an isospin multiplet and thereby tolerably small. With respect to leptons there is no experimental basis for postulating a significant limit as electro- magnetism is "turned off." Our partial bootstrap is, therefore, not expected to encompass leptonic phenomena.[11]

The reader may be concerned that in turning off electromagnetism we have completely de- coupled the real and quantum worlds and thereby undercut the physical significance of the S matrix. Hadrons are observed, that is to say, only through their electromagnetic interactions. We need not require, however, that our idealized hadron S matrix make direct contact with the real world. We may compare its elements to experimentally measured "amplitudes" whose precise significance is, in principle, blurred by electromagnetic complications, but whose numerical value is supposed to be meaningfully "close" to the value of the ideal matrix elements.

Beyond Poincaré invariance and unitarity, if one considers in detail the cause-effect relation- ship of space-time events when massive particles are multiply scattered, it has been argued by Iagolnitzer and Stapp[1] that in momentum space the S matrix needs to be an analytic function, with only those physical-region singularities that correspond to macroscopically separated space- time events. The locations of these singularities and the associated discontinuities satisfy require- ments bearing the names of Landau and Cutkosky. Now, once one accepts that S-matrix analyticity, as well as Poincaré invariance and unitarity, is implied by the observed characteristics of the real world, it becomes conceivable that there may exist only one possible S matrix compatible with the classical (flat) space-time structure of the real world. This is the hypothesis stated in our Introduction.

We are dealing here with an extreme version of hadron bootstrap hypothesis. During the past decade many forms of bootstrap hypothesis have been advanced that involve more elaborate re- quirements. The hypothesis under consideration in this paper will be regarded by many readers as implausibly simple, or even as untenable on the grounds that indefinitely many hypothetical S matrices can surely be constructed. What is the basis for this latter opinion?

Awareness of analyticity arose historically from Lagrangian models; only recently have attempts been made to connect this S-matrix attribute directly with the real world. Such a history makes hard to swallow the idea that analyticity, together with Poincaré invariance and unitarity, might determine a unique S matrix, because there is nothing unique about a Lagrangian. It is well-known, at the same time, that no Lagrangian has ever been shown to lead to an S matrix which is satisfactory on all three counts. Should such a Lagrangian ever *be* found, containing any degree of arbitrariness, the conjecture in question would collapse. In recent years a variety of relativistic non-Lagrangian models, containing arbitrary aspects, have been formulated. Were any of these to lead to an acceptable S matrix, the conjecture similarly would become untenable. So far, none has approached success as closely as have conventional local-Lagrangian models.

It is a remarkable fact that, more than forty years after the discovery of the quantum super-position principle, no theoretical model has been constructed that is demonstrably compatible both with superposition and with physical (relativistic) space-time. Perhaps the reason is that all here-tofore-created models contain arbitrary aspects. It is correspondingly conceivable that quantum superposition, *as embodied in an S matrix*, can be made compatible with relativistic space-time principles in only one possible way – close to the way exhibited by nature for hadrons.

It must, of course, not be forgotten that even should a unique S matrix exist, it can be no more than an imperfect model of strong interactions, depending for its potential physical relevance, in parallel with quantum electrodynamics, on the smallness of the fine-structure constant. One can imagine that a framework broader than the S matrix (perhaps broader than conventional science) and capable of including zero-mass phenomena, will ultimately be developed. Within such a framework a self-consistency hypothesis might be feasible and might lead to an understanding of the hereto-fore arbitrary aspects of electromagnetism (and weak interactions). At present, however, physicists are not in possession of any such framework. The concept of an analytic S matrix, though imperfect, appears to be a natural model to describe a subset of physical phenomena where-in the absence of zero-mass particles is striking.

IV. SUPPLEMENTARY S-MATRIX PRINCIPLES

A variety of "global" hadron S-matrix attributes more detailed than "analyticity," unitarity, and Poincaré invariance has over the years been identified.[12] Examples are cluster decomposition, crossing, Hermitian analyticity, the Landau-Čut-kosky rules generalized to unphysical regions, the connection between spin and statistics, conserva-

tion of baryon number and hypercharge, the con-nection between baryon number and spin, time re-versal and parity invariance, isospin symmetry, and the principle of "second-degree analyticity" or "Regge asymptotic behavior" by which the S matrix may be constructed from a knowledge of its discontinuities. All of these attributes have substantial experimental support; some have been connected, to a greater or lesser extent, with classical space-time; all have been subjected to nontrivial tests of mutual consistency. A reason-able guess is that all are true – to the extent that the analytic S matrix constitutes a viable descrip-tion of hadrons. Additional global S-matrix princi-ples may be discovered in the future, either by logical deduction, by guesswork based on models, or as a result of experiment. What relation do such "supplementary principles" bear to the boot-strap hypothesis under consideration?

The hypothesis implies that all such supplemen-tary principles should be derivable from the re-quirement of compatibility with the cause-effect event structure of the real world, in the same sense that the Landau-Cutkosky rules for physical-region singularities have been derived. A demon-stration that any principle cannot be so derived would imply either that the extreme version of the hadron bootstrap hypothesis is inadequate or that the presumed supplementary principle does not in fact apply to nature.

Historically one may divide supplementary principles into two categories, those suggested by Lagrangian models and those discovered by other routes. It is so far only within the former category that substantial progress toward "derivation" has been achieved, but the signifi-cance of this circumstance may be no deeper than that the best-developed analytical techniques are those relevant to Lagrangian models. Theo-retical physicists, that is to say, still lean heav-ily on their experience with Lagrangians when thinking about the S matrix.

Among the above examples of supplementary principles, baryon number conservation presents an especially severe challenge for the bootstrap hypothesis; to "derive" this principle an approach totally unrelated to Lagrangians seems required. Second-degree analyticity also deserves special mention. In the past the author's personally-favored version of bootstrap hypothesis has re-flected the Lagrangian influence by including second-degree analyticity as a distinct and sepa-rate constraint on the S matrix. My interest now in the simpler hypothesis stems from esthetics, coupled with the striking continued nonexistence of models having demonstrably acceptable space-time characteristics, with or without second-degree analyticity.

The hadron bootstrap hypothesis will be judged in large measure according to the success achieved in deriving "non-Lagrangian" supplementary principles from the requirement of S-matrix compatibility with the real world. The current rapid growth of analytical techniques relevant to Regge asymptotic behavior suggests that, among "non-Lagrangian" global principles, second-degree analyticity will be among the first to have its status clarified.

V. EXPERIMENTAL IMPLICATIONS; MODELS

Implied by the hadron bootstrap hypothesis is the theoretical possibility, not only of explaining global hadronic attributes such as baryon number, but of calculating in the $\alpha \to 0$ limit *all* hadronic masses[13] and reaction amplitudes without any input parameters. Implied at the same time, however, is that the properties of no selected particle or subset of particles are more amenable to calculation than those of any other. Since all hadrons are mutually interdependent in a bootstrap, an attempt to completely understand any individual strongly interacting particle requires an understanding of all. This "all or nothing" character of the hypothesis makes its experimental predictive content extraordinarily elusive. Given, that is, the richness of observed hadronic phenomena, it is manifestly beyond human capability ever to predict everything from nothing, even if everything flows uniquely from self-consistency. It is, nonetheless, a historical fact that important encouragement for the bootstrap idea has arisen from experimental observations of hadron properties. How is this paradoxical situation to be understood?

It is to be understood in terms of approximate and limited extrapolation schemes or "models," based on general S-matrix principles. Each scheme (model) accepts a certain increment of experimental information about the hadron S matrix and then attempts to predict as much as possible about "neighboring" hadron properties. There has by now been sufficient variety and success for such models as to make apparent the deep dynamical content of unitarity when combined with analyticity and Poincaré invariance. It has, in particular, been established that the predictive content associated with traditional equations of motion for specified degrees of freedom is at least matched by the content of general S-matrix principles, without any need to identify definite "degrees of freedom." The recognition that equations of motion are unnecessary for predicting hadron behavior has been a powerful spur to

the bootstrap idea.

S-matrix extrapolation schemes (models) are never sharply defined with respect to either input or output and are inevitably characterizable by the derogatory term "phenomenology," since they represent no more than an application of widely accepted general principles. The hadron bootstrap hypothesis, nevertheless, implies that the predictive power of these principles is limited only by human ingenuity and dedication. By working harder and (or) by exercising more powerful mathematical techniques, physicists are supposed, by the hypothesis, always to be able to reduce the experimental input and increase the predictive output of S-matrix extrapolation schemes. There is supposed to be no irreducible minimum ratio of input to output.

It may be noted that once a selection has been made of the experimental input and of the approximations to be tolerated in a particular S-matrix extrapolation model, the technique employed to accomplish the extrapolation may resort to the same type of equation used to evaluate "fundamenton" models. The term "fundamenton" is used here to characterize any arbitrarily assignable component in a theory, such as an elementary particle or a field in a Lagrangian. By definition a bootstrap theory contains no fundamentons, but in an approximate S-matrix extrapolation model the experimental input in effect plays the role of fundamenton.

The potential model of classical nuclear physics provides an excellent example. From the S-matrix standpoint, as shown by Charap and Fubini[14] following the more general work of Mandelstam,[15] the experimental input consists of certain conservation laws, such as baryon number and isotopic spin, plus the position and residues of pion and nucleon poles, together with the knowledge that these poles are relatively isolated from their neignbors. It is then possible to use general S-matrix principles to extrapolate from this input to predict (approximately) a wide variety of phenomena involving nucleons of low kinetic energy. The Charap-Fubini extrapolation technique employs a differential equation formally identical to a Schrödinger equation for a nucleon wave function under the influence of Yukawa potential, the strength and range of the latter being determined by a pion-pole position and residue. Nucleons and pions thus appear as fundamentons in this particular extrapolation model; they are accepted, that is to say, as arbitrary input.

Other S-matrix models assign a fundamenton role to other types of experimental input, and attempt to cover (approximately) other ranges of phenomena by extrapolation. Because the ranges of different models may partially overlap each other, the fundamenton of one model may be part of the predicted output from another. Considered collectively, therefore, the use of such models to investigate the bootstrap hypothesis is compatible with the possibility that *no* fundamentons are tolerable in a completely self-consistent hadron S matrix.

VI. CONCLUSION

If human limitation allows no hope for extrapolation from nothing to everything, can one even imagine what might constitute "verification" of the hadron bootstrap hypothesis? Certainly not in the sense of a fundamenton theory where all predictions flow from unambiguous arbitrary input. Nonetheless, increasingly remarkable theoretical correlations of experimental facts about hadrons may come to be accomplished purely through general properties of the analytic S matrix. If at the same time no example of a fundamenton-containing S matrix is constructed, and if observed global hadronic attributes such as baryon number conservation and Regge asymptotic behavior can be deduced from general principles, it may gradually become plausible that the only uniquely necessary input is the requirement of self-consistency.

---

*Research supported in part by the Air Force Office of Scientific Research under Contract No. AF 49 (638)-1545, Joseph Henry Laboratories, Princeton University, Princeton, N. J. 08540, and in part by the U. S. Atomic Energy Commission.

[1]D. Iagolnitzer and H. P. Stapp, Commun. Math. Phys. 14, 15 (1969).

[2]J. Coster and H. P. Stapp, J. Math. Phys. 11, 1441 (1970); 11, 2743 (1970).

[3]H. P. Stapp, Phys. Rev. D 3, 1303 (1971).

[4]We share the point of view of Stapp, expressed in Ref. 3, that the state vector concept is to be understood as a nonrelativistic approximation, not as the basic vehicle for expressing quantum superposition. We take the description of the quantum world to be realized through scattering amplitudes between asymptotic states.

[5]By "zero" we mean a mass so small that the associated Compton wavelength is much larger than all relevant macroscopic distances.

[6]For references, see H. P. Stapp, Nuovo Cimento (to be published).

[7]We ignore gravitation in this discussion, so little being known about the relation to the quantum world of this extraordinarily weak interaction.

[8]W. Heisenberg, Z. Physik 120, 513 (1943); 120, 673 (1943).

[9]The line of thought pursued here suggests that the conventional description of superposition through scattering amplitudes may not be "an absolute truth" but only an approximation, somehow related to the smallness of the fine-structure constant. In such an event one may feel less perplexed at the well-known philosophical absurdities that result from an attempt to apply quantum superposition to the real world. The point of view of this paper, already stated several times, is that, in a scientific framework, quantum and classical worlds should be regarded as separate.

[10]A partial bootstrap may be incapable of explaining why the electromagnetic coupling of hadrons should be related, as observed, to the conserved hadronic attributes called isospin and hypercharge. As discussed in Secs. IV and V, the hadron bootstrap hypothesis is supposed to explain all hadronic symmetries and conservation laws, but the relation of these attributes to electromagnetism may not be understandable until the origin of the latter has been comprehended.

[11]Without leptons, weak interactions seem unlikely to be represented within a scientific bootstrap. From the bootstrap viewpoint, a more than superficial theory of weak interactions thus promises to be as formidable a task as understanding the origin of electromagnetism.

[12]For a review of generally accepted S-matrix properties, see R. Eden, P. Landshoff, D. Olive, and J. Polkinghorne, *The Analytic S Matrix* (Cambridge Univ. Press, Cambridge, England, 1966); G. F. Chew, *The Analytic S Matrix* (Benjamin, New York, 1966).

[13]Only ratios of hadron masses are meaningful in the absence of contact between the S matrix and the real world. There is no mechanism, in other words, for setting the scale of momentum space.

[14]J. M. Charap and S. Fubini, Nuovo Cimento 14, 540 (1959); 15, 73 (1959).

[15]S. Mandelstam, Phys. Rev. 112, 1344 (1958).

MULTIPLE-PRODUCTION THEORY VIA TOLLER VARIABLES*

Naren F. Bali

Lawrence Radiation Laboratory, University of California, Berkeley, California

and

Geoffrey F. Chew

Lawrence Radiation Laboratory and Department of Physics, University of California, Berkeley, California

and

Alberto Pignotti†

Lawrence Radiation Laboratory, University of California, Berkeley, California

(Received 5 July 1967)

Toller's group-theoretical analysis of kinematics is exploited to define a complete set of variables, each of independent range, for particle production of arbitrary multiplicity. In terms of these variables, the generalized Regge-pole hypothesis leads to a simple, unambiguous, and experimentally accessible prediction for high-energy multiple-production cross section. A flat Pomeranchuk trajectory is shown to violate the Froissart bound.

A variety of multiperipheral models for inelastic reactions at high energy has been discussed in the literature,[1-7] but the implementing variables have been incomplete or imperfectly matched to the factorizability which characterizes such models. In this paper we exploit the work of Toller[8] to define a complete set of variables for particle production of arbitrary multiplicity, the range of each variable being independent of the others. The new variable set is natural for the implementation of any multiperipheral model, leading to a phase space that factors asymptotically in the same manner as does the amplitude. We apply our variables to the (unique) generalization of the Regge-pole hypothesis, achieving a simple, unambiguous, and experimentally accessible prediction for multiple-production cross sections at high energy which maintains the factorization property. One important aspect of the result is the exclusion of the possibility of a flat Pomeranchuk trajectory.

For the N-particle production reaction $a+b \rightarrow 1+2+\cdots+N$, we begin by selecting a particular ordering of final particles so as to define a set of $N-1$ momentum transfers Q_{mn} according to the diagram of Fig. 1. Each different ordering leads to a different set of variables; any of these sets is complete, the choice between them being a matter of convenience usu-

614

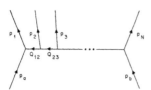

FIG. 1. Diagram defining the momentum transfers Q_{nm}.

ally resolved by appeal to the multiperipheral concept. That is to say, for describing a particular region of final-particle momenta, one generally chooses that set of variables for which all Q_{mn}^2 in this region are small, while all $s_{mn} = (p_m + p_n)^2$ are large.

The number of variables needed to describe an amplitude with a total of $N+2$ ingoing and outgoing particles is well known to be $3N-4$, once Lorentz invariance is included. We divide the total variable set into three categories, a set of $N-1$ t variables, a set of $N-1$ ξ variables, and a set of $N-2$ ω variables. This choice is motivated in detail by Bali, Chew, and Pignotti[9] on the basis of Toller's group-theoretical analysis.[8] The t variables are obvious: $t_{mn} = Q_{mn}^2$. Less obvious but still recognizable are the ξ_{mn}, which are also in one-to-one correspondence with the Q_{mn}, $i\xi_{mn}$ being the

analytic continuation of the angle in the rest system of Q_{mn} between the direction of $\vec{p}_m$ and that of $\vec{p}_n$. In the region of interest here the Q_{mn} are spacelike (the t_{mn} are negative), and Toller has shown that each ξ_{mn} is real, ranging from 0 to ∞ independently of the other variables.

The ω_n are the least familiar components of our variable set. The members of this subset are in one-to-one correspondence with the internal vertices of Fig. 1. To understand ω_n, go into the rest frame of p_n, where the spatial components of the two adjacent momentum transfers point in the same direction. Then consider the rigid rotation about this axis of all momenta standing on the left of the vertex n and the independent rigid rotation of all momenta standing on the right. The difference of these two-rotation angles is ω_n, which thus has a range 0 to 2π.

In Ref. 9 it is shown explicitly how to pass by a succession of Lorentz transformations from the variables t_{mn}, ξ_{mn}, ω_n to the ordinary momentum variables or to the channel invariants $s_{mn} = (p_m + p_n)^2$ and $s = (p_a + p_b)^2$. In the interest of brevity we confine ourselves here to the observation that s_{mn} is a linear function of $\cosh \xi_{mn}$, with coefficients that depend only on the t's adjacent to the m and n vertices:

$$s_{mn} = [\lambda_m^{1/2} \lambda_n^{1/2} / 2(-t_{mn})] \cosh \xi_{mn} + \text{function of } t\text{'s}, \tag{1}$$

with $\lambda_n = \lambda(m_n^2, t_{n-1,n}, t_{n,n+1})$ (for the end vertices, one of the t's should be replaced by m_a^2 or m_b^2), where

$$\lambda(t_i, t_j, t_k) = t_i^2 + t_j^2 + t_k^2 - 2t_i t_j - 2t_i t_k - 2t_j t_k. \tag{2}$$

Thus a large value of ξ_{mn}, with adjacent t's small, implies a large value of s. It turns out that s depends on all $3N-4$ variables, but when all the $\cosh \xi_{mn}$ are large,

$$s \sim \frac{\lambda_1^{1/2} \lambda_N^{1/2}}{2(-t_{12})^{1/2}(-t_{N-1,N})^{1/2}} \cosh \xi_{12} \cosh \xi_{23} \cdots \cosh \xi_{N-1,N} \prod_{i=2}^{N-1} (\cos \omega_i + \cos q_i), \tag{3}$$

where

$$\cosh q_n = \frac{m_n^2 - t_{n-1,n} - t_{n,n+1}}{2(-t_{n-1,n})^{1/2}(-t_{n,n+1})^{1/2}}. \tag{4}$$

The $3N-4$ dimensional phase space

$$d\Phi_N = d^4 p_1 \delta(p_1^2 - m_1^2) d^4 p_2 \delta(p_2^2 - m_2^2) \cdots d^4 p_N \delta(p_N^2 - m_N^2) \delta^4(p_1 + p_2 + \cdots + p_N - p_a - p_b),$$

VOLUME 19, NUMBER 10   PHYSICAL REVIEW LETTERS   4 SEPTEMBER 1967

in terms of the new variables becomes

$$d\Phi_N = \frac{2}{m_a^2 m_b^2\, 2^{3N}} \frac{\lambda_1^{1/2}\lambda_2^{1/2}\cdots\lambda_N^{1/2}}{(-t_{12})(-t_{23})\cdots(-t_{N-1,N})} dt_{12}dt_{23}\cdots dt_{N-1,N}$$

$$\times d\cosh\xi_{12}d\cosh\xi_{23}\cdots d\cosh\xi_{N-1,N}d\omega_2 d\omega_3\cdots d\omega_{N-1}d\psi\,\frac{\delta(\cosh\eta - p_a\cdot p_b/m_a m_b)}{\sinh\eta}, \qquad (5)$$

the angle ψ describing rigid rotations of the entire final set of N momenta about the common direction of p_a and p_b in a frame where these initial momenta are parallel. Evidently the spin-averaged matrix element will not depend on ψ, but it will in general depend on all the other variables appearing in formula (5). For a target at rest, $\cosh\eta$ is the energy of the incident particle in units of its own rest mass. In terms of s,

$$\cosh\eta = \frac{s - m_a^2 - m_b^2}{2m_a m_b} \sim \frac{s}{2m_a m_b}.$$

The single constraint interlocking our variables arises through the delta function in $\cosh\eta$. It follows, however, from formula (3) that when all the $\cosh\eta_{mn}$ are large,

$$\cosh\eta \sim (\text{factorizable function of } t_{mn} \text{ and } \omega_n)\cosh(\xi_{12} + \xi_{23} + \cdots \xi_{N-1,N}), \qquad (3')$$

so that for a fixed set of t's and ω's the constraint is only on the sum of the ξ's. It is typical of multiperipheral models that when η is large, most of the production occurs in regions where every ξ_{mn} is large. Thus the approximation (3') can be used to simplify the phase space:

$$d\Phi_N \approx \frac{\text{factorizable function of } t\text{'s and }\omega\text{'s}}{\sinh\eta}\delta(\xi_{12} + \xi_{23} + \cdots + \xi_{N-1,N} - \xi^{(+)})$$

$$\times dt_{12}\cdots dt_{N-1,N}d\xi_{12}d\xi_{23}\cdots d\xi_{N-1,N}d\omega_2 d\omega_3\cdots d\omega_{N-1}d\psi, \qquad (6)$$

where, from formula (3),

$$\cosh\xi^{(+)} \approx s\,\frac{2(-t_{12})^{1/2}(-t_{N-1,N})^{1/2}}{\lambda_1^{1/2}\lambda_N^{1/2}}\,2^{N-2}\prod_{i=2}^{N-1}(\cos\omega_i + \cos q_i)^{-1}. \qquad (7)$$

We are now in a position to write down a cross-section formula. Suppose, for example, that the Regge-pole hypothesis is adopted for the absolute square of the amplitude, summed over final spins and averaged over initial spins[3-7,9]:

$$\langle |A(t_{mn}, \xi_{mn}, \omega_n)|^2\rangle_{Av} \sim f_1(t_{12})f_2(t_{12}, \omega_2, t_{23})\cdots f_N(t_{N-1,N})(\cosh\xi_{12})^{2\alpha_{12}(t_{12})}(\cosh\xi_{23})^{2\alpha_{23}(t_{23})}\cdots. \qquad (8)$$

The internal "vertex functions" f_n describe the coupling of two Regge trajectories to a physical particle, while f_1 and f_N couple two physical particles to a single Regge trajectory. Taken together with formula (6) and the flux factor, and integrating over $d\psi$, the behavior (8) leads to

$$s^2 d\sigma_N \sim F_1(t_{12})F_N(t_{N-1,N})\prod_{i=2}^{N-2}F_i(t_{i-1,i},\omega_i,t_{i,i+1})\exp[2\alpha_{12}(t_{12})\xi_{12}]$$

$$\times \exp[2\alpha_{23}(t_{23})\xi_{23}]\cdots dt_{12}dt_{23}\cdots d\xi_{12}d\xi_{23}\cdots d\omega_2\cdots d\omega_{N-1}\delta[\xi_{12} + \xi_{23} + \cdots - \xi^{(+)}], \qquad (9)$$

a result containing a wealth of physically interesting predictions, especially if one exploits the correlations between different reactions flowing from the universality of the vertex functions F_n. (The function f_n differs from F_n only by a factor that depends in a known manner on $t_{n-1,n}$, $t_{n,n+1}$, and ω_n.)

VOLUME 19, NUMBER 10   PHYSICAL REVIEW LETTERS   4 SEPTEMBER 1967

We make no attempt here to exhaust the content of formula (9), but three of the most obvious features are:

(a) Consider a reaction in which all leading trajectories are the Pomeranchuk and suppose this trajectory to be perfectly flat (a fixed pole) at $\alpha = 1$. Then

$$s^2 d\sigma_N \sim \prod_{i=1}^{N} F_i e^{2\xi_i^{(+)}} dt_{12} dt_{23} \cdots d\xi_{12} d\xi_{23} \cdots d\omega_2 \cdots d\omega_{N-1} \delta[\xi_{12} + \xi_{23} \cdots - \xi^{(+)}],$$

or, integrating over the $d\xi$'s and remembering (7),

$$d\sigma_N \sim (\text{function of } t\text{'s and } \omega\text{'s})(\ln s)^{N-2} dt_{12} dt_{23} \cdots dt_{N-1,N} d\omega_2 \cdots d\omega_{N-1}. \qquad (10)$$

The limits on the t and ω intervals become independent of s for large s; so there is a conflict with the Froissart limit[10] for $N > 4$, showing that peak shrinkage, such as that associated with a moving pole, is essential to the consistency of the model.

(b) Assuming all poles to move, if formula (9) is integrated over the $d\xi$'s, we find that

$$s^2 d\sigma_N \sim \frac{\prod_i F_i}{2^{N-2}} \left\{ \frac{\exp[2\alpha_{12}\xi^{(+)}]}{(\alpha_{12} - \alpha_{23})(\alpha_{12} - \alpha_{34}) \cdots (\alpha_{12} - \alpha_{N-1,N})} \right.$$
$$\left. + \frac{\exp[2\alpha_{23}\xi^{(+)}]}{(\alpha_{23} - \alpha_{12})(\alpha_{23} - \alpha_{34}) \cdots} + \cdots + \frac{\exp[2\alpha_{N-1,N}\xi^{(+)}]}{(\alpha_{N-1,N} - \alpha_{12}) \cdots} \right\} dt_{12} dt_{23} \cdots d\omega_2 d\omega_3 \cdots . \qquad (11)$$

The energy dependence of this differential cross section is

$$\propto s^{2[\alpha_{max} - 1]},$$

where α_{max} is the highest trajectory in the chain. Such a dependence was conjectured by Zachariasen and Zweig.[7]

(c) For processes in which one trajectory in the chain lies well below the others, the differential final-particle spectrum will favor low subenergies for the particle pair corresponding to the low-lying trajectory. That is to say, even at a fixed incident energy it is possible to investigate the characteristic Regge structure by studying ratios of final subenergies. The logarithmic distribution in the ratio of two subenergies, keeping all other ratios fixed, is predicted by formula (9) to be a straight line whose slope is determined by the difference of the corresponding trajectory heights. In particular, for $N = 3$ after integrating over $d\omega_2$,

$$d\sigma_3 \sim (\text{function of } t\text{'s}) \, s^{\alpha_{12}(t_{12}) + \alpha_{23}(t_{23}) - 2} (s_{12}/s_{23})^{\alpha_{12}(t_{12}) - \alpha_{23}(t_{23})} dt_{12} dt_{23} d\ln(s_{12}/s_{23}). \qquad (13)$$

A concluding remark is that clusters of final particles with low total mass can replace any or all of the single outgoing particles in Fig. 1, so long as the experimenter sums over degrees of freedom within a given cluster—except for the cluster mass m_n. The general problem is analyzed in detail in Ref. 9. It turns out that all formulas given in this paper continue to hold, with single-particle masses replaced by cluster masses.

Questions involving total cross sections and the over-all multiplicity of production require integration over the dt's and $d\omega$'s as well as a summation over final-particle combinations. These matters will be considered in another paper.

*This work was supported in part under the auspices of the U. S. Atomic Energy Commission.

†On leave from Consejo Nacional de Investigaciones Científicas y Técnicas.

[1]D. Amati, S. Fubini, A. Stanghellini, and M. Tonin, Nuovo Cimento 22, 569 (1961).

[2]S. Frautschi, Nuovo Cimento 28, 409 (1963).

[3]K. A. Ter Martirosyan, Zh. Eksperim. i Teor. Fiz. 44, 341 (1963) [translation: Soviet Phys.−JETP 17, 233 (1963)] ; Nucl. Phys. 68, 591 (1965).

VOLUME 19, NUMBER 10   **PHYSICAL REVIEW LETTERS**   4 SEPTEMBER 1967

[4]T. W. B. Kibble, Phys. Rev. 131, 2282 (1963).
[5]M. S. K. Razmi, Nuovo Cimento 31, 615 (1964).
[6]Chan Hong Mo, K. Kajantie, and G. Ranft, CERN Report No. Th.719, 1966 (to be published).
[7]F. Zachariasen and G. Zweig, California Institute of Technology Reports Nos. 68-116 and 117, 1967 (unpublished).
[8]M. Toller, Nuovo Cimento 37, 631 (1965).
[9]N. F. Bali, G. F. Chew, and A. Pignotti, Phys. Rev. (to be published).
[10]M. Froissart, Phys. Rev. 123, 1053 (1961).

AN INTEGRAL EQUATION FOR SCATTERING AMPLITUDES*†

G. F. Chew‡

Department of Physics and Lawrence Radiation Laboratory, University of California, Berkeley, California

and

M. L. Goldberger§

Palmer Physical Laboratory, Princeton University, Princeton, New Jersey

and

F. E. Low‡

Department of Physics, Massachusetts Institute of Technology, Cambridge, Massachusetts

(Received 9 January 1969)

We outline a dynamical scheme for studying Regge behavior of scattering amplitudes. It is based on unitarity and the concept of short-range momentum-space correlations in high-energy processes.

We describe here a type of integral equation that generalizes the 1962 model of Fubini and collaborators (ABFST model)[1] to more realistic multiperipheral mechanisms such as the multi-Regge model.[2-4] The essential requirement is that interparticle correlations in the production amplitude should involve a finite number of "links" in the multiperipheral "chain". We restrict ourselves, that is to say, to a certain type of "short-range order" in momentum space, but the details of this order are not crucial.

Virtues of the equation are these: (a) It operates entirely inside the physical region, both the kernel and the inhomogeneous term being obtainable through analysis of actual reactions. (b) It is a dynamical equation in the same sense as the Schrödinger equation, rather than a consistency condition imposed by analyticity. In fact, no reference is made to analyticity. (c) Regge poles are generated, and the consistence of input and output poles provides a natural bootstrap mechanism. (d) The dynamical relation of Regge cuts to poles is illuminated. Being an expression of unitarity, the equation includes absorptive effects.

To illustrate the method, let us suppose that the most general reaction initiated by two spinless particles a and b is $a+b \rightarrow a+b+(n$ spinless particles of mass $\mu)$. If the initial and final momenta, as well as a set of $n+1$ momenta transfers, are denoted as in Fig. 1, a simple multiperipheral model for the production amplitude assigns to it the form

$$G_a(p_0,p_1)G(p_0,p_1,p_2)G(p_1,p_2,p_3)\cdots G_b(p_n,p_{n+1},p_{n+2}) + \text{exchange terms}. \tag{1}$$

In a multi-Regge model with a single type of input trajectory and without dependence on "vertex angles", for example, we would have

$$G(p_{i-1},p_i,p_{i+1}) = \beta(p_i^2,p_{i+1}^2)[(p_{i-1}-p_{i+1})^2]^{\alpha in(p_i^2)}, \tag{2}$$

with α_{in} the "input" trajectory and β the coupling constant for an internal vertex. The factor $G_a(p_0,p_1)$ would be the coupling constant for the end vertex connecting particle a to the chain, while G_b would have the form (2) but with the external coupling constant appropriate to particle b. More complicated models might include several input trajectories and input cuts, different kinds of produced particles,

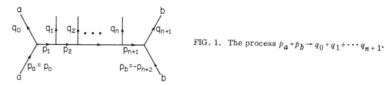

FIG. 1. The process $p_a + p_b \rightarrow q_0 + q_1 + \cdots q_{n+1}$.

VOLUME 22, NUMBER 5   PHYSICAL REVIEW LETTERS   3 FEBRUARY 1969

dependence on vertex angles, etc. The associated correlation functions G would be more complicated but each would depend on a finite number of "adjacent" variables in the multiperipheral sense.

Continuing with our simple example, let us denote the absorptive part of the elastic amplitude for $a+b \to a'+b'$ by

$$\sum_{n=0}^{\infty} A_n(p_a',p_a,p_b',p_b),$$

where

$$A_n = \int d^4 q_0 \cdots d^4 q_{n+1}\, \delta(p_a+p_b-\sum_{i=0}^{n+1}q_i)\delta^+(q_0^2-m_a^2)\delta^+(q_0^2-\mu^2)\cdots\delta^+(q_{n+1}^2-m_b^2)$$

$$\times G_a^*(p_0',p_1')G_a(p_0,p_1)G^*(p_0',p_1',p_2')G(p_0,p_1,p_2)\cdots G_b^*(p_n',p_{n+1}',p_{n+2}')$$

$$\times G_b(p_n,p_{n+1},p_{n+2}) + \text{exchange terms},\qquad(3)$$

with

$$p_1'=p_0'-q_0,\, p_2'=p_1'-q_1,\, \cdots,\, p_{n+2}'=p_{n+1}'-q_{n+1}.$$

Here δ^+ denotes the positive-energy part of the delta function. If the coupling constant $\beta(t_i,t_{i+1})$ is small except for small values of t_i and t_{i+1}, the exchange terms tend to be small. We shall here ignore them completely, although exchange can be partially included by an appropriate complication of the correlation function.

In the ABFST model $G(p_{i-1},p_i,p_{i+1})$ contains no correlation between p_{i-1} and p_{i+1} and in fact depends only on p_i^2, being the elementary propagator for the line labeled by p_i. A recursion relation between A_{n+1} and A_n is then immediate. To accommodate our more complicated correlation we must back up one rung along the chain and undo the integration over q_0 in Formula (3). We thus define a new function B_n from which A_n is obtained as follows:

$$A_n(p_a',p_a,p_b',p_b)=\int d^4 q_0\,\delta^+(q_0^2-m_a^2)G_a^*(p_a',p_1')G_a(p_a,p_1)B_n(p_a',p_a,q_0,p_b',p_b),\qquad(4)$$

where $p_1=p_a-q_0, p_1'=p_a'-q_0$. The quantity B_n satisfies a recursion relation of the ABFST type, namely

$$B_n(p_0',p_0,q_0;p_b',p_b)=\int d^4 q_1\,\delta^+(q_1^2-\mu^2)G^*(p_0',p_1',p_2')G(p_0,p_1,p_2)B_{n-1}(p_1',p_1,q_1;p_b',p_b)\qquad(5)$$

with $p_2=p_1-q_1, p_2'=p_1'-q_1$. Summing Eq. (5) from $n=1$ to ∞ and defining

$$B=\sum_{n=0}^{\infty}B_n,$$

we obtain the integral equation

$$B(p_0',p_0,q_0;p_b',p_b)=B_0(p_0',p_0,q_0;p_b',p_b)$$

$$+\int d^4 q_1\,\delta^+(q_1^2-\mu^2)G^*(p_0',p_1',p_2')G(p_0,p_1,p_2)B(p_1',p_1,q_1;p_b',p_b),\qquad(6)$$

where B_0 corresponds to the two-particle unitarity contribution:

$$B_0(p_0',p_0,q_0;p_b',p_b)=\delta^+((p_0+p_b-q_0)^2-m_b^2)G_b^*(p_0',p_0'-q_0,-p_b')G_b(p_0,p_0-q_0,-p_b).\qquad(7)$$

The complete absorptive part is obtained through the linear operation (4), i.e., integration over q_0, performed on B rather than on B_n.

A more economical notation is achieved through the change of variables, $p_0=P_0+\frac{1}{2}Q, p_0'=P_0-\frac{1}{2}Q, p_1$

209

$= P_1 + \frac{1}{2}Q, \, p_1{}' = P_1 - \frac{1}{2}Q$, etc., plus $p_b = P_b - \frac{1}{2}Q, \, p_b{}' = P_b + \frac{1}{2}Q$. The familiar invariant variables are then

$$s = s_0 = (p_0 + p_b)^2 = (p_0{}' + p_b{}')^2 = (P_0 + P_b)^2,$$

$$s_1 = (p_1 + p_b)^2 = (p_1{}' + p_b{}')^2 = (P_1 + P_b)^2, \text{ etc.},$$

$$t_0{}^{(\pm)} = (P_0 \pm Q/2)^2, \, t_1{}^{(\pm)} = (P_1 \pm Q/2)^2, \text{ etc.},$$

while the actual negative momentum transfer t is

$$t = (p_a - p_a{}')^2 = (p_b{}' - p_b)^2 = Q^2.$$

We may then rewrite our fundamental equation (6) as

$$B(P_0, P_1; P_b, Q) = B_0(P_0, P_1; P_b, Q) + \int d^4 P_2 \, \delta^+((P_1 - P_2)^2 - \mu^2) B(P_1, P_2; P_b, Q) H(P_0, P_1, P_2; Q), \qquad (8)$$

where

$$H(P_0, P_1, P_2; Q) = G^*(P_0 - \tfrac{1}{2}Q, P_1 - \tfrac{1}{2}Q, P_2 - \tfrac{1}{2}Q) G(P_0 + \tfrac{1}{2}Q, P_1 + \tfrac{1}{2}Q, P_2 + \tfrac{1}{2}Q). \qquad (9)$$

More complicated correlations can be accommodated by undoing two or more integrations in defining B. The kernel H would then depend on a correspondingly longer sequence of adjacent P's, but the form of the integral equation would be similar.

Although the driving term B_0 corresponds in the multi-Regge model to the well-known AFS cut in the angular momentum plane (the A_0 formed from B_0 is precisely this cut), the iteration of B_0 through the integral equation generates an infinite sequence of cuts that sums up to Regge poles. The underlying basis for these poles is the same as in the ABFST model: invariance of the kernel $H(P_0, P_1, P_2; Q)$ under the little group of fixed-Q simultaneous Lorentz transformations of P_0, P_1, and P_2. This invariant operation does not involve any transformation of P_b, and when s_0, s_1, and s_2 are asymptotic the invariance can be translated into the statement that the kernel depends only on the ratios s_0/s_1 and s_1/s_2, together with t, $t_0{}^{(\pm)}$, $t_1{}^{(\pm)}$, and $t_2{}^{(\pm)}$. Using the same reasoning as ABSFT, one may then conclude that if B is regarded as a function of the invariants s_0, s_1, $t_0{}^{(\pm)}$, and t, the solution of the homogeneous equation as $s_0, s_1 \to \infty$ is of the form

$$B(s_0, s_1, t_0{}^{(\pm)}, t_1{}^{(\pm)}; t) \sim s_1{}^{\alpha(t)} b(s_0/s_1, t_0{}^{(\pm)}, t_1{}^{(\pm)}, t_1{}^{(\pm)}; t), \qquad (10)$$

where $\alpha(t)$ is the largest eigenvalue of the homogeneous equation. Forming A by integration then leads to

$$A(s, t_0{}^{(\pm)}; t) \sim s^{\alpha(t)} a(t_0{}^{(\pm)}; t), \qquad (11)$$

the physical absorptive part to be obtained by setting $t_0{}^{(\pm)} = m_a{}^2$.

Provided the eigenvalue equation for α has a solution, there is thus a coherence leading to Regge poles in the infinite sum of cuts. The original AFS cut will not disappear but will be modified by the equation. Symbolically, if the equation is written as $B = B_0 + KB$, the solution $B = [1 - K]^{-1} B_0$ contains the cut both in B_0 and in K, so there is a damping. Questions of this type we defer to a subsequent paper.

In terms of invariants the asymptotic homogeneous equation becomes

$$B(s_0, s_1; t_0{}^{(\pm)}, t_1{}^{(\pm)}; t) = g_0{}^2 \int \frac{ds_2}{s_1} dt_2{}^{(+)} dt_2{}^{(-)} K\left(\frac{s_0}{s_1}, \frac{s_2}{s_1}, t_0{}^{(\pm)}, t_1{}^{(\pm)}, t_2{}^{(\pm)}; t\right)$$

$$\times B(s_1, s_2; t_1{}^{(\pm)}, t_2{}^{(\pm)}; t), \qquad (12)$$

where we have explicitly extracted a factor $g_0{}^2$ to characterize the strength of an internal vertex. The eigenvalue equation corresponding to (10) is

$$b\left(\frac{s_0}{s_1}; t_0{}^{(\pm)}, t_1{}^{(\pm)}; t\right) = g_0{}^2 \int_0^1 dx \, x^{\alpha(t)} \int dt_2{}^{(+)} dt_2{}^{(-)} K\left(\frac{s_0}{s_1}, x, t_0{}^{(\pm)}, t_1{}^{(\pm)}; t\right) b\left(\frac{1}{x}; t_1{}^{(\pm)}, t_2{}^{(\pm)}; t\right); \qquad (13)$$

the range of $t_2^{(\pm)}$ being restricted by

$$\Delta(t, t_2^{(+)}, t_2^{(-)}) = t^2 + t_2^{(+)2} + t_2^{(-)2} - 2[tt_2^{(+)} + tt_2^{(-)} + t_2^{(+)}t_2^{(-)}] < 0.$$

Now, for large s_0/s_1 and small $s_2/s_1 = x$ the multi-Regge model has the kernel behavior

$$K \sim \left(\frac{s_0}{s_1}\right)^{\alpha_{\text{in}}(t_1^{(+)}) + \alpha_{\text{in}}(t_1^{(-)})} \frac{F(t_1^{(\pm)}, t_2^{(\pm)}; t)}{[-\Delta(t, t_2^{(+)}, t_2^{(-)})]^{1/2}}. \tag{14}$$

In the weak-coupling limit, we may employ this form throughout (13) and deduce

$$b = \left(\frac{s_0}{s_1}\right)^{\alpha_{\text{in}}(t_1^{(+)}) + \alpha_{\text{in}}(t_1^{(-)})} \beta(t_1^{(\pm)}, t), \tag{15}$$

with

$$\beta(t_1^{(\pm)}, t) = \frac{g_0^2}{\alpha(t) - [\alpha_{\text{in}}(t_1^{(+)}) + \alpha_{\text{in}}(t_1^{(-)}) - 1]} \int dt_2^{(+)} dt_2^{(-)} \frac{F(t_1^{(\pm)}, t_2^{(\pm)}, t)\beta(t_2^{(\pm)}, t)}{[-\Delta(t, t_2^{(+)}, t_2^{(-)})]^{1/2}}. \tag{16}$$

Assuming F to be factorizable,

$$F(t_1^{(\pm)}, t_2^{(\pm)}, t) = F_1(t_1^{(\pm)}, t)F_2(t_2^{(\pm)}, t), \tag{17}$$

we find the eigenvalue equation

$$1 = g_0^2 \int \frac{dt^{(+)} dt^{(-)}}{[-\Delta]^{1/2}} \frac{F_2(t^{(\pm)}, t)F_1(t^{(\pm)}, t)}{\alpha(t) - [\alpha_{\text{in}}(t^{(+)}) + \alpha_{\text{in}}(t^{(-)}) - 1]}. \tag{18}$$

The corresponding Chew-Pignotti model[5] may be obtained immediately by assuming $\alpha_{\text{in}}(t)$ to be independent of t, i.e., $\alpha_{\text{in}}(t^\pm) \to \bar{\alpha}_{\text{in}}$. Then

$$\alpha = 2\bar{\alpha}_{\text{in}} - 1 + g^2(t), \tag{19}$$

where

$$g^2(t) = g_0^2 \int \frac{dt^{(+)} dt^{(-)}}{[-\Delta]^{1/2}} F_2(t^{(\pm)}, t)F_1(t^{(\pm)}, t). \tag{20}$$

We close with some comments on the possibility of a self-consistent weakly coupled Pomeranchuk trajectory. If we write

$$\alpha(t) = 1 - g_0^2 X(t), \tag{21}$$

and set $\alpha_{\text{in}}(t) = \alpha(t)$, then Eq. (18) becomes

$$1 = \int \frac{dt^{(+)} dt^{(-)}}{[-\Delta]^{1/2}} \frac{F_1(t^{(\pm)}, t)F_2(t^{(\pm)}, t)}{X(t^{(-)}) + X(t^{(-)}) - X(t)}, \tag{22}$$

independent of g_0^2. Assuming Eq. (22) to have a solution, it is tempting to infer a connection between the two properties $\alpha \approx 1$ and $\alpha' \approx 0$, both properties following from the smallness of g_0^2. Unfortunately, in a more realistic model, with additional input trajectories lying below the Pomeranchuk and strongly coupled, such a simple inference may not be possible. The lower lying input trajectories may play an important role in determining the properties of the output Pomeranchuk. The above example, nonetheless, illustrates the bootstrap potentialities of our equation.

*A preliminary version of this work was reported at the Vienna Conference on High Energy Physics by W. Frazer.

†Since this work was completed, we have received a preprint by I. G. Halliday and L. M. Saunders containing very similar ideas.

‡Work supported in part by the U. S. Atomic Energy Commission.

§Work supported in part by the U. S. Air Force Office of Scientific Research, under Contract No. AF49(638)-1545, and the U. S. Atomic Energy Commission.

VOLUME 22, NUMBER 5   PHYSICAL REVIEW LETTERS   3 FEBRUARY 1969

[1]L. Bertocchi, S. Fubini, and M. Tonin, Nuovo Cimento 25, 626 (1962); D. Amati, A. Stanghellini, and S. Fubini, Nuovo Cimento 26, 6 (1962).

[2]T. W. B. Kibble, Phys. Rev. 131, 2282 (1963).

[3]K. A. Ter Martirosyan, Zh. Eksperim. i Teor. Fiz. 44, 341 (1963) [translation: Soviet Phys.—JETP 17, 233 (1963)].

[4]F. Zachariasen and G. Zweig, Phys. Rev. 160, 1322, 1326 (1967). See also N. Bali, G. F. Chew, and A. Pignotti, Phys. Rev. Letters 19, 614 (1967).

[5]G. F. Chew and A. Pignotti, Phys. Rev. 176, 2112 (1968).

PHYSICAL REVIEW D   VOLUME 32, NUMBER 10   15 NOVEMBER 1985

Single-surface basis for topological particle theory

G. F. Chew

Lawrence Berkeley Laboratory and Department of Physics, University of California, Berkeley, California 94720

V. Poénaru

Département de Mathématiques, Université de Paris-Sud, 91405 Orsay Cedex, France
and Institut des Hautes Études Scientifiques, 35 route de Chartres, 91440 Bures sur Yvette, France
(Received 26 June 1985)

Orientations of the embellished two-dimensional bounded manifold that embeds the Feynman graph in topological particle theory (TPT) are shown capable of representing *all* discrete particle properties; there is no need for a second surface. By recognizing more fully than heretofore the patch structure of the surface, a representation is found for quark generation and lepton generation as well as for spin-chirality. "Color" is given a representation similar to that found earlier for isospin. Previously developed representations of baryon number, electric charge, and "quark" chirality remain essentially unchanged. The proposed modifications of topological particle theory leave hadron dynamics unaltered while facilitating TPT extension to electroweak interactions.

I. INTRODUCTION

In evolution since 1978 has been a continuing effort to associate all discrete particle properties with orientations of one- and two-dimensional Feynman-graph "embellishments," with the aim of removing arbitrariness from particle theory.[1-6] Rather than starting with a Lagrangian, a graphical S-matrix expansion is taken as fundamental in this approach—which we call "topological particle theory" (TPT). Convergence of the topological expansion to a unitary S matrix must eventually be verified, but a generalization of Feynman's rules for topological amplitudes satisfies unitarity order by order within the expansion (Ref. 7 and Appendix C).

How are expansion components ordered? TPT extends the Feynman graph from a one-dimensional entity to something two-dimensional by embedding the graph in a surface. The reason for so "thickening" is to endow the embellished graph with an unambiguous *complexity* or "entropy." Each embellished graph is characterized by a set of non-negative integers that describe its entropy (Appendix A); when graphs are combined through connected sums, entropy cannot decrease, so components of the topological expansion may be arranged in a sequence of increasing entropy. Entropy integers such as genus stem from the topology of the embedding two-dimensional manifold; numbers of embedding dimensions higher than 2 are unsuitable for entropy bookkeeping.

The hope for convergence of the topological expansion is based on the tendency, recognized[8,9] in the early 1970s and now familiar in so-called "$1/N$ expansions," for Feynman amplitudes to diminish in multiplicity with increasing entropy. For strong-interaction topologies the effective value of N has turned out in some cases to be as large as 2^{10}, corresponding to ten two-valued orientations of Feynman-graph embellishments[10] (Appendix C). This paper will describe the most economical and consistent oriented embellishments so far recognized.

Two different categories of consistency requirements may be identified.

(1) Although Feynman's original expansion did not contemplate graph *contraction*—whereby certain graphs are recognized as equivalent to other graphs (contraction is often characterized as "duality")—a single term of the topological expansion represents all embellished graphs that can be contracted to the same structure.[3] Contractions maintain orientations together with entropy indices and control dynamics at the level of zero entropy, where amplitudes are self-generating ("bootstrap," Ref. 11 and Appendix B). A consistent set of entropy indices and associated contraction rules place severe demands on those Feynman-graph embellishments that connect with the Poincaré group; Poincaré embellishments will remain almost unaltered in this paper, although we shall slightly extend those established earlier so as to describe spin as well as chirality. But "extra-Poincaré" embellishments are less constrained by consistency with contraction, and it is here that we shall be proposing modifications.

(2) A second category of constraints arises from the need to provide a basis for classical momentum measurement. Elementary hadrons are self-generated at zero entropy, but it is necessary, in order to achieve meaning for the "asymptotic" momentum measurements on which the S matrix is based, to postulate massless photons not generated by hadrons but coupled to a conserved electric charge that can be carried by hadrons. (Reference 13 has described the generation from TPT hadron dynamics of a nonhadronic weakly interacting scalar boson that shares some characteristics of photons and thereby gives promise of a future TPT connection between the fine-structure constant and the dimensionless hadronic coupling constant.) (Stapp[12] has shown how classical electromagnetism emerges from coherent soft-photon collections generated through Feynman rules.) Photons must be accommodated with the same two-dimensional oriented manifold that describes hadrons, and this requirement con-

strains embellishments related to electric charge. We shall, in this paper, tamper only slightly with earlier charge-related embellishments of the Feynman graph, which required elementary photons to be accompanied by seven other elementary massless "gauge" bosons[14] and endowed many terms of the topological expansion with an SU(2) isospin symmetry.[15]

Our tampering will relate chiefly to "quark color" and "quark generation." The oriented two-dimensional manifold has no room for SU(3) color gauge symmetry; TPT accommodates no vector gluons in the sense of QCD. [Dependence of TPT on zero-entropy contractions excludes massless gauge-vector strongly interacting analogues of photons—such as the gluons of QCD. Massless gauge-vector bosons require contraction-forbidding *nonzero entropy* in spin-chiral topological structure (Appendix A).] Nevertheless, as will be seen in Sec. III, "topological quarks" have a three-valued degree of freedom reminiscent of color in pre-QCD quark models.

Both quark color and quark generation have previously been associated in TPT with a second surface[3]—transverse to the Feynman graph—but this paper will find a natural place for these degrees of freedom within the surface that houses all other particle features. Certain lines and patches within this surface, that previously were unoriented, will be given orientations of the kind already associated with electric charge[4] and with chirality.[6] The location of these newly oriented lines and patches decouples their orientations from Poincaré transformations and allows their interpretation as "color" and generation.

Why was it felt for so many years (since 1978) that a second, transverse, surface was needed? Historically the TPT need for color was inferred from contraction consistency *before* introduction of the lines that are now proposed as carriers of color. (These lines were introduced in 1981, not to carry color but to control chirality,[6] in the sense reviewed in Sec. V A.) Consistent contraction rules required an ordering of line segments incident on cubic vertices along the "belt"—the boundary of the surface that embedded the Feynman graph. A natural way to achieve such ordering was to thicken the belt into a transverse two-dimensional manifold called the "quantum surface."[3] A triangulation of the quantum surface, dual to the belt, then allowed association of elementary hadrons with triangulated disks on the transverse surface, and orientations of hadron-disk perimeters became interpreted as quark generation. We have thus far found no inconsistency in such a quantum-surface representation of "quark" "color" and generation and, indeed, purely strong-interaction Feynman rules remain unchanged by the representation to be proposed in this paper.

Why then might elimination of the quantum surface constitute an advance for TPT? Our motivation relates to electroweak interactions of hadrons. For "gauge" bosons the quantum surface is irrelevant, a situation that extends to leptons and to a nonhadron family of "horizontal" (H) scalar bosons predicted by TPT (Ref. 16). As a consequence there is no visible source of "quark"-generation symmetry breaking so long as this degree of freedom resides on the quantum surface. Our proposal to locate quark generation on the same surface as leptons and H

bosons opens the door to quark-generation mixing and symmetry breaking.

The proposals of the present paper, furthermore, are economical. Beyond representing color and quark generation, the quantum surface has never proved to be very useful. All interesting action has occurred on the "classical" surface housing the Feynman graph. Entropy indices, for example, have derived exclusively from the classical surface. The threeness of "color," which heretofore has been associated with quantum-surface triangles,[3] equally well associates with the classical-surface feature that all nontrivial vertices along the belt are cubic. We shall recognize in Sec. III that representation of "color" gives a new meaning to quark triality: $3 = 1 + 2$, where 2 is the number of orientations of a "color" carrying line and 1 is the number of Feynman lines close to a belt cubic vertex.

Also worth anticipating in this introduction is that the representation to be proposed for quark generation is parallel to that for lepton generation while at the same time being distinct. There are four "quark" generations in one to one correspondence with four lepton generations and mechanisms for symmetry breaking are similar. Mixing lepton generations will nevertheless be impossible, while mixing quark generations appears as an option.

II. FEYNMAN AND FINKELSTEIN LINES; MOMENTUM AND ELECTRIC CHARGE

In TPT any open connected Feynman graph F with N external lines (carrying momentum in the usual sense) is embedded in a two-dimensional bounded, connected, oriented surface Σ with F ends located along the surface boundary $\partial\Sigma$ at separate points e_i, $i = 1, \ldots, N$. We shall denote an F line as f. The surface Σ has in previous literature been called the "classical surface." The (closed) one-dimensional $\partial\Sigma$ decomposes into disconnected "external" and "internal" parts. The external part we shall refer to as the "belt," following the terminology of Ref. 3. Internal parts of $\partial\Sigma$ touch no F ends but relate to internal f's in a way to be described in Sec. IV. The belt decomposes exhaustively into connected particle portions π_i, each π housing within its interior (not at a π end point), exactly one F end, i.e., an elementary particle i belongs to a π_i which includes e_i.

Forthcoming embellishments of Σ will decorate each π so as to uniquely to establish a particle type. This section will review the charge decoration of π's for elementary mesons and gauge bosons as an illustration of general TPT principles related to momentum and electric charge.

Figure 1 shows how a three-meson, single-vertex F is embedded in an oriented disk whose circular belt (here the entire $\partial\Sigma$) divides into 3 π's, each π containing one e. Notice that F embedded within an oriented Σ implies a cyclic ordering of f's incident on any F vertex.

When a graph F is built by a connected sum of two other graphs F' and F'', there is a corresponding connected sum of the embedding surfaces

$$\Sigma = \Sigma' \# \Sigma''$$

made by identifying and erasing certain segments of the π_i' and π_i'' (within $\partial\Sigma'$ and $\partial\Sigma''$) that belong to an inter-

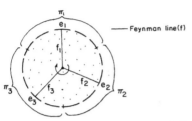

FIG. 1. Single-vertex, 3-meson Feynman graph embedded in a disk.

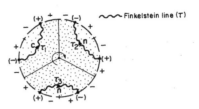

FIG. 3. Finkelstein-line embellishment of the Fig. 1 disk. (The shading will be explained by Fig. 15.)

mediate Feynman line f_i. The identification, called a "particle plug," is made so as to perpetuate the global orientation of Σ and also those local orientations, to be described in the course of this paper, that establish particle identity. Figure 2 shows how global Σ orientation is perpetuated when two cubic meson vertices are connected by a single intermediate meson $\pi_i' \leftrightarrow \pi_i''$. In this example the entire π_i' identifies with the entire π_i''. Generally we require some segment from each "half" of a π to be identified and erased in a particle plug, recognizing that e_i divides any π_i into two halves. (The possibility that certain segments of π_i *not* be identified and erased in a particle plug will be related in Sec. V to Lorentz transformations and the attendant nonconservation of particle spin and chirality.) The plugging rule is to *mismatch*, in identified segments of $\partial\Sigma'$ and $\partial\Sigma''$, the orientations *induced* by the respective Σ' and Σ'' orientations. Feynman end points e_i' and e_i'' are always identified and erased in a plug of particle i.

The global orientation of Σ implies alternating $+,-$ indices on successive belt segments, as we shall see immediately below. The plugging mismatch rule then translates into a rule that $+$ segments of the belt are to be identified with $-$ segments. The relation between $+$ and $-$ boundary segments is like that between particle and antiparticle or between an "ingoing" and "outgoing" particle. Both these notions in TPT connect with the global Σ orientation.

We have noted how Feynman's momentum-transporting graph F divides each π into two halves; further boundary segmentation results from oriented lines, beginning and ending on the belt and not crossing F or each other, that were proposed by Finkelstein and collaborators[4] in order to represent electric charge. Here we shall denote a Finkelstein line by the symbol τ, because these lines subsequently become recognized as the TPT repository of isospin.[15] Figure 3 adds τ's to Fig. 1 and

thereby carves the belt into a total of 12 segments. We have attached $(+,-)$ indices to these segments according to the rule that, following the sense of the global Σ orientation, any boundary segment immediately following a point of F contact has a $(-)$ index. This rule is general and we shall generally use the symbol β for any segment of $\partial\Sigma$. In Fig. 3 there are 6 β^-'s and 6 β^+'s.

The orientation of each τ allows algebraic characterization by a two-valued index defined by comparing τ orientation to $(+,-)$ labels induced on the τ ends by Σ orientation in conjunction with F. Following the sense of Σ orientation along the belt, the first τ end encountered after an e carries a $(-)$ label, with subsequent alternation of $(+,-)$ labels on τ ends along the belt. If the two ends of a τ locate on the *same* belt component (as in Fig. 3), they necessarily carry *opposite* $(+,-)$ labels, because of the rule that no τ crosses any other line within Σ. If the two ends of a τ locate on *different* belt components (as in Fig. 4), we *require* location such that the $(+,-)$ labels be opposite. A c index is attached to a τ whose orientation is directed from its $(-)$ end toward its $(+)$ end; otherwise we give the τ an index n. In particle plugs τ orientation is continuous, so the (c,n) index corresponds to two conserved quantum numbers.

The (c,n) index is transferrable to certain β's along the belt which touch τ ends. The (c,n) index transfers from a τ end to the contacting β whose $(+,-)$ label agrees with

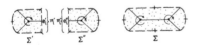

FIG. 2. Single-meson connected sum of two single-vertex disks.

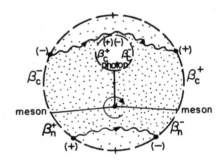

FIG. 4. Embellished Feynman graph for photon interaction with pair of charge mesons.

that of the τ end. Any $\beta_c^{\pm}$ can be said to carry ± 1 unit of electric charge while $\beta_n^{\pm}$ is electrically neutral; total electric charge is thus the number of β_c^{+} minus the number of β_c^{-}.

The reason a $\beta_c^{\pm}$ can be described as "charged" is that only such β's couple to photons; a photon π consists of β_c^{+} and β_c^{-}, as shown in Fig. 4 where a photon couples to a pair of charged mesons. Here the photon occupies an entire (closed) two-segment disconnected component of the belt while each meson occupies a four-segment (open) belt portion. Appendix A reviews the status of Fig. 3 within the topological expansion, and Ref. 17 deals with Fig. 4. Here we invoke these special topologies, without explanation, as simple examples of general principles.

Reference 15 explains that the orientation of each τ in Figs. 3 and 4 is independently reversible, so each meson and gauge boson carries an ordered pair of (c,n) labels—each such particle corresponding to the element of a 2×2 matrix in charge space. One may say more generally that isospin has been topologized; Ref. 15 shows SU(2) isospin symmetry to be a feature of any Feynman amplitudes whose topology permits reversibility of individual Finkelstein orientations. In TPT *all* internal symmetries and quantum numbers are analogously to be understood—as will emerge from what follows.

A topological quark or antiquark is usefully defined as a 2-β continuous belt interval divided at its midpoint by a τ and touched at exactly one end by F, as shown in Fig. 5. Each "quark" or "antiquark" carries a c or n label (i.e., has isospin $\pm\frac{1}{2}$). (The meaning of fractional electric charge for TPT "quarks" has been discussed most recently in Ref. 18.)

An elementary meson, such as appears in Figs. 3 and 4, belongs to a π which is a quark attached to an antiquark by an e. See Fig. 6. The belt of Fig. 3 is composed of three such meson π's while one of the two disconnected components of the belt in Fig. 4 consists of two meson π's.

We have associated the electric charge (isospin) of a quark with *one* of its two halves—the half that touches F. Such a segment of $\partial\Sigma$—touching an f at one end and a τ at the other—we shall designate as a "fermion unit," called ϕ_i, in anticipation of Sec. V which shows that any ϕ, isolated in Fig. 7, carries spin $\frac{1}{2}$ as well as isospin.

The gauge boson of Fig. 4 (discussed at length in Ref. 17) combines ϕ^{+} and ϕ^{-} units together with the intervening e; a quark (Fig. 5) combines a ϕ with one other β (to

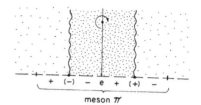

FIG. 6. Four-segment, $q\bar{q}$, belt portion corresponding to a meson.

be given a name in Sec. IV) but does *not* include an e. Because they do not carry momentum, quarks are not particles. Any particle has an e *inside* its π.

We remark that the τ bisecting a quark may be called a "quark line," although it does not carry momentum. Readers may have noted how in Fig. 3 τ's occupy the position of Harari-Rosner quark lines.[19]

Because of f and τ continuity, ϕ^{+}'s always plug into ϕ^{-}'s, and we may identify a conserved fermion number f which is the number of ϕ^{+}'s minus the number of ϕ^{-}'s. Fermion number is zero for mesons and gauge bosons but will not be zero for baryons and leptons. For all hadrons, the quark number is equal to the fermion number since fermion units in hadrons appear exclusively as "halves of quarks." We further remark in anticipation of Sec. IV that fermion (ϕ) units of $\partial\Sigma$ constitute the exclusive TPT repository of spin and chirality. All other boundary units carry zero spin and may be described as "bosonic."

This section has introduced no changes in previous TPT but merely given some new language to describe existing features. The new language is appropriate to changes that will follow. In anticipation of one such change we now independently orient each f as well as each τ. Such an orientation has previously been thought to be physically meaningless—merely telling the direction of flow of four-momentum p_{μ}^{i} attached to the line. Reversing the orientation of f_i can be compensated by changing p_{μ}^{i} to $-p_{\mu}^{i}$. Nevertheless, with a p_{μ}^{i} *and* an orientation attached to each f_i, continuity of Feynman orientation in particle plugs relates to momentum conservation in a sense similar to that by which continuous Finkelstein orientation relates to charge conservation. (An important difference is that no vertices occur along τ's.) Furthermore simultaneous reversal of *all* f orienta-

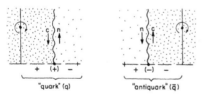

FIG. 5. Two-segment belt portions corresponding to "quark" and "antiquark."

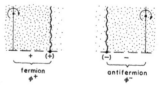

FIG. 7. Fermionic belt segments.

tions corresponds to TCP; one may now identify this symmetry as topological (Appendix D). Henceforth every f as well as every τ is to be understood as oriented.

III. JUNCTION LINES AND "COLOR" LINES; BARYON NUMBER AND LEPTON NUMBER

The surface Σ generally is "feathered"—locally a bounded smooth surface but with a finite number of *junction lines* J_l. Appendix C reviews the TPT consistency consideration that requires junction lines. A junction line can be either a (bounded) segment, with ends on the belt, or a circle within the Σ interior, and along such a line three pieces of smooth surface meet. When Σ is disconnected along its junction lines it becomes decomposed into a number of connected smooth "sheets" S_k; junction lines lie along the sheet boundaries ∂S_k which generally are built from portions of $\partial\Sigma$ plus junction lines. Each S_k has an orientation unambiguously correlated with that of other sheets by demanding a unique *induced* orientation for each junction line. Figure 8(a), which omits Finkelstein lines, shows the global orientation of a three-sheeted Σ with a single junction-line segment, embedding a baryon-antibaryon-meson single-vertex Feynman graph. Figure 8(a) can be misleading either by implying that Σ is embedded in a three-dimensional manifold, or that there is cyclic ordering of the three sheets meeting at a junction line. Figure 8(b) is more accurate—showing that J segments along the boundaries of each of three sheets are to be identified, but without any immediate cyclic ordering. (A cyclic ordering will be effectively provided by "color" lines described below.) Figure 8(b) also locates Finkelstein lines; in this example there are four (Harari-Rosner) "quark" lines together with one additional τ "close to" the junction line and touching the ends of the baryon and antibaryon f's. This extra Finkelstein line, an example of a type to be denoted τ_J, is motivated immediately below; Appendix B gives a further reason for this τ_J while explaining the different shading of patches in Fig. 8(b). Figure 8 illustrates the general rule that no lines (within the Σ interior) ever touch a junction line.

Those four oriented lines in Fig. 8(b) that each connect a boundary point to a *trivial* vertex inside Σ, are called "momentum-copy" lines f_c. Each junction-line end j is accompanied by the ends of exactly two f_c's and exactly one e, as shown in Fig. 9. A belt segment β separates j from the ends of each of these accompanying lines. Notice how, in the j neighborhood, distinction between f and f_c is provided by the τ_J—which contacts f.

Momentum-copy lines transport neither momentum nor electric charge but each f_c plays a role parallel to that of an associated f. Almost all statements about an f in Sec. II above apply also to an f_c. The exception is the rule that each particle π contains exactly one e. A π may contain 0, 1, or 2 j's and, correspondingly, 0, 2, or 4 f_c ends. In Fig. 8 the meson π contains no j's or f_c ends, while baryon and antibaryon portions of $\partial\Sigma$ each contain a single j and 2 f_c ends. The reason no π can contain more than 2 j's is that each j is separated by a single boundary segment from exactly 1 e. An e can be adjacent to at most 2 j's.

Let us pause here to expand on the useful notion implied by the foregoing that the belt is a closed *graph* with cubic and trivial vertices joined by β's. The cubic vertices coincide uniquely with j's. Any end of a τ, f, or f_c lying along the belt coincides with some trivial belt vertex. Additional trivial vertices occur at dividing points between adjacent hadron belt portions where a "quark" connects with a "mated antiquark." The belt graph corresponding to Fig. 8(b) is shown in Fig. 10, where line ends are indicated by line symbols. Here there are 2 cubic vertices and 19 trivial vertices.

Along any belt branch which lies along the boundary of a single sheet, $(+,-)\beta$ indices alternate according to the rule of Sec. II, but the 3 β's touching the same cubic vertex all carry the same $(+,-)$ label, which associates with the j located at that vertex. The induced orientation of a junction line points from a $j^{(-)}$ toward a $j^{(+)}$, as shown in Fig. 10 together with the induced belt orientation. The reader should remember that the junction line is *not* part of the belt but lies within the interior of Σ.

Section II introduced the notion of "quarks" and of fermion units of $\partial\Sigma$. Let us here add the notion of a *Y unit* (see Fig. 9)—which includes 3 β's joined by a j, together with end points of each β where either a τ or an f_c impinges. As in defining quark and fermion units, we do *not* include an e within a Y unit. The $\partial\Sigma$ of Figs. 8 and 10 may then be described as built from a Y^+ and Y^-, 4 q's, and 3 e's.

Figure 11 shows the $(3q, Y-)-\pi$ belonging to an elementary baryon. Here we do not show the (two-dimensional) neighboring interior of Σ, as we did in Fig. 6 for the meson β, but indicate the global orientation of Σ by the $(-)$ label on the j.

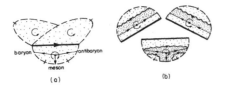

(a)   (b)

FIG. 8. (a) Single-vertex, meson-baryon-antibaryon Feynman graph embedded in a feathered surface. (b) Embellishment of the surface of (a) with Finkelstein and "color" lines.

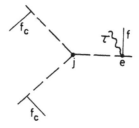

FIG. 9. Location of "color" line ends along the belt.

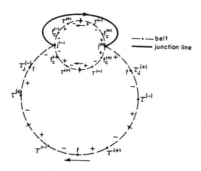

FIG. 10. The belt of graph for Fig. 8(b), with attached junction line.

As explained at the end of Sec. II, the f orientation in Fig. 11 could be assigned in either sense, but we require *as a new feature of TPT* (exemplified by Fig. 11) that the two f_c's arriving near the end of any junction line be *oppositely* oriented. A two-valued index generally attaches to each f_c in a sense similar to that of the (c,n) index on a τ; this index will be transferrable to certain "quarks" where it will be described as "quark color." Let us now make precise the definition of topological color.

Lines f_c are admitted only in continuously oriented sequences that end on the belt adjacent to f's. We shall call such a sequence of f_c's a "color line." The end of a color line inherits a $(+,-)$ label from the j which it accompanies and we require the two ends of the same color line always to carry opposite $(+,-)$ labels. An index No. 2 is attributed to a color line whose orientation is directed from $(-)$ toward $(+)$. Otherwise the color line carries index No. 3. Each j is accompanied by the ends of exactly one No. 2 and one No. 3 color line, as in Fig. 12.

If a quark touches F we say it carries "color" No. 1. If it touches an $f_{c,2(3)}$ we say the quark carries color No. 2

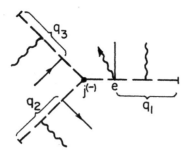

FIG. 11. Baryon belt portion (9 segments, comprising 3 "quarks" plus a Y^-).

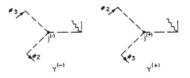

FIG. 12. Y units of the belt.

(No. 3). These "quark color" assignments are illustrated in Fig. 11. Every quark carries one of three colors.

The consistency of the foregoing requires that f_c orientation be continuous in particle plugs as well as at (trivial) vertices along color lines inside Σ. Conserved color quantum numbers are implied, but we shall see below that they are redundant with baryon and lepton number conservation. It is never possible independently to reverse f_c orientations, so there is no SU(2) [or SU(3)] color symmetry. However, simultaneous reversal of all f_c orientations, amounting to a global $2 \leftrightarrow 3$ color transformation, is a symmetry of strong-interaction topologies.

The Y plugging rule formulated in Ref. 5 is to be maintained. This rule recognizes a Y^+ or Y^- belt unit attached to each end of a junction line, as shown in Fig. 12, with the three legs of each Y being distinguishable. Leg No. 1 touches the coincident ends of two lines from the interior of Σ while opposite orientations of single incident lines distinguish legs No. 2 and No. 3. A plug identifies Y^+ with Y^- so as to achieve continuity of all four line orientations. At the end of leg No. 1, where two oriented lines impinge, the Y plug maintains orientations of tangent lines in the sense of Fig. 13.

Although never crossing f's or f_c's, τ's may thus have points of tangency. Reference 5 describes how $q\bar{q}$ plugs are made independently of Y plugs, not necessarily preserving quark color. Quark color switching is allowed.

Consistent contraction rules require distinguishability for all three legs of a Y. Heretofore TPT has employed the transverse quantum surface to define colors No. 2 and No. 3; no orientation has previously been given either to f's or f_c's. Adding these orientations and the Finkelstein-Feynman points of contact near junction lines are the only changes of Feynman-graph embellishment so far presented in this paper.

The number of Y^+ belt units minus the number of Y^- units is an absolutely conserved quantum number which has been called "boson number" and designated by the

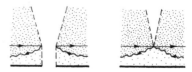

FIG. 13. Identification of Y legs in a connected sum.

symbol b (Ref. 20). Baryon number B is related to fermion number f and boson number b by the formula

$$B \equiv \tfrac{1}{2} f + \tfrac{1}{2} b . \tag{1}$$

Inspection of Fig. 11 shows a baryon to have $f = +3$ and $b = -1$, so $B = +1$, while according to Fig. 4, mesons and gauge bosons have $b = f = 0$ and hence $B = 0$. It is proposed in Ref. 17 that a lepton π is $Y^+ \phi^-$, Fig. 14, with $f = -1$ and $b = +1$, so leptons also have a zero baryon number. Figure 14 corresponds to the definition of lepton number

$$L \equiv \tfrac{1}{2} b + \tfrac{1}{2} f . \tag{2}$$

The π of Fig. 14 is seen to have $L = +1$, while all hadronic π's have $L = 0$.

Continuity of "color" lines implies conservation of "color" quantum numbers but each unit of boson number b is accompanied by exactly one unit of each "color." Therefore conservation of

$$b = L - B \tag{3}$$

is redundant with "color" conservation. (Reference 15 points out that b is proportional to hypercharge—the difference between electric charge and the 3-component of isospin—so topological "color" relates to hypercharge.) This feature of "color" has remained unchanged from the quantum-surface representation, as have Feynman rules for "quark-color" switching. In fact all purely strong-interaction Feynman rules[7] are unaffected by the TPT changes proposed in this paper.

We close this section with three remarks.

(1) As with f_c's, TPT freezes the orientation of any τ_J—a τ with at least one end adjacent to a j. Freezing in the c direction of hadronic τ_J orientations has been long established.[3,4] Reference 17 will explain why nonhadronic τ_J's have orientation frozen in the n sense.

(2) "Color" lines, although similar in some respects to τ's, always have vertices *inside* Σ; an f_c never has *both* ends on the belt, although it may have both ends inside Σ. τ's never terminate inside Σ although they may have points of tangency with other lines.

(3) The continuous sequence of f_c's building a "color" line is in 1-to-1 correspondence with some collection of f's. (In strong-interaction topologies, this collection forms a continuous sequence.) Even though momentum is

not transported by a "color" line, a momentum p_μ^l attaches to each momentum-copy line. We shall characterize the union of F with all momentum-copy lines and their linking vertices as the extended momentum graph M (with lines designated μ).

IV. ORIENTED SPINOR AND SCALAR PATCHES

The collection of Feynman lines, "color" lines, and isospin (Finkelstein) lines divide the surface Σ into patches with may be differentiated as either "spinor" or "scalar." What is the topological meaning of such adjectives, which usually associate with behavior of indices in Lorentz transformations? We have seen in Secs. II and III how each color or isospin line inherits a "sense"—expressed through $(+, -)$ labels on its ends—from the orientation of Σ in conjunction with the Feynman graph. Facing along any τ or f_c in a direction from $(-)$ to $(+)$ there is a patch of Σ immediately to the "left" and another immediately to the "right," where left and right are defined by the orientation of Σ. Rather than using these adjectives we choose to employ the terms spinor and scalar, as shown by Fig. 15. Our choice of terminology here stems from the consequence of our rule for $(+, -)$ indices [e.g., Figs. 3, 4, and 8(b)] that any patch touching a Feynman line is spinor according to Fig. 15. We shall find in Sec. V A that patches adjacent to F generate indices which transform as spinors under Lorentz transformation. Patches which are scalar according to Fig. 15 never touch F and do not generate indices involved in Lorentz transformations. Scalar patch-generated indices are discussed in Secs. V C and V D. It is a general TPT principle that the Feynman graph always locates inside a spinor region of Σ. Figure 8(b) (also Figs. 19 and 23) illustrate this principle.

The boundary segments designated by ϕ in Sec. II (Fig. 7) are seen to locate exclusively along the boundaries of spinor patches. We shall find all other β's to locate exclusively along the boundaries of scalar patches. A characterization of ϕ as "fermionic" and all other β's as "bosonic" is appropriate when one recalls the S-matrix connection between spin and statistics established by Stapp.[21] As used in this paper the two adjectives spinor and fermion are to be understood as synonymous—both implying spin-$\tfrac{1}{2}$ and Fermi statistics. The two adjectives scalar and boson here both imply spin-0 and Bose statistics.

A long-standing TPT postulate has associated chirality with a ϕ label induced by an orientation of the adjacent spinor patch. Section V A will extend the chirality label on ϕ's to a double label which includes spin, after the present section has given general rules for which Σ

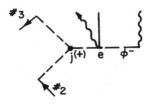

FIG. 14. Lepton Y^+, ϕ^- belt portion.

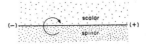

FIG. 15. Separation of scalar from spinor patch by a Finkelstein or "color" line.

patches are to be oriented and how. The notion of ϕ as a fermionic "boundary unit" will be accompanied by two bosonic boundary units, either a single β (Sec. V B) or a Y-shaped β triplet (Sec. V C). We shall associate quark and lepton generation, respectively, with patch-induced labels on these two types of bosonic boundary unit.

Any patch whose boundary includes a β we shall designate as σ_i with boundary $\partial\sigma_i$. We orient each σ_i and, independently, each disconnected component $(\partial\sigma_i)_d$ of $\partial\sigma_i$, with these orientations to propagate coherently to the new patch formed when a $\beta^+\beta^-$ pair is identified and erased in a particle plug. That is, if the symbol σ_i is understood to mean an oriented patch together with its oriented boundary, such a plug amounts to a connected sum

$$\sigma = \sigma' \# \sigma''.$$

Figure 16 shows the connected sum of a patch σ' that has a single boundary component and a patch σ'' that has two boundary components. The indicated orientations are local; the global Σ orientation must simultaneously be perpetuated, as in Fig. 2 (and Fig. 20, below). In this example σ' carries two (local) orientations while σ'' and σ both have three orientations. All patches within purely strong-interaction topologies are disks like σ', with single boundary components and thus two orientations, but Ref. 17 will show purely electroweak topologies to require cylindrical patches like σ'', with two disconnected boundary components and thus three orientations associated with the patch.

Particle quantum numbers will stem from σ-induced labels on $\partial\Sigma$ "units;" each unit belongs to exactly one boundary component of some σ. The orientation of its σ will endow any $\partial\Sigma$ unit with one two-valued label and the orientation of its boundary component will supply a second.

All 6 patches of Fig. 3 (3 spinor and 3 scalar) and all 5 patches of Fig. 4 (3 spinor and 2 scalar) are disks, each with a pair of $\partial\Sigma$ units among boundary segments; each of these σ's is "doubly oriented." Figure 8(b) at first sight might appear to contain 12 patches, but one of these (spinor, in the center, bounded entirely by 1 τ and 2 μ's) has no β along its boundary. Such a patch is not a σ and will not be oriented; such a patch will be characterized as "white." (We shall describe elsewhere how white spinor patches become σ's when gauge bosons are inserted therein.)

The junction line in Fig. 8(b) further influences the patch structure so as to reduce the number of σ's, a related effect is the occurrence, at each of the J ends, of a Y-shaped boundary unit. The requirement that no line inside Σ ever touch a J has always in TPT been accompanied by a rule that local areas of Σ connected by a J, as well as the boundaries of these areas, are coherently oriented, in the sense of Fig. 17. We maintain these rules and correspondingly speak of a (single) σ_J as any collection of patches (on separate Σ sheets) that touch the same J. The boundary of any σ_J contains, on a single component, "mated" Y^+ and Y^- $\partial\Sigma$ units for each bounded J "within" the σ_J. These two units—Y^+ and Y^-—inherit common labels from the σ_J. The two notions of σ_J and mated Y boundary units may be captured by thinking of a junction line as lying inside σ_J even though it places its mated Y^+Y^- pair on a single boundary component. All σ_J's are scalar. Examining Fig. 8(b) with the foregoing in mind reveals a total of 9 doubly oriented σ's, 4 spinor and 5 scalar, one of the latter being a σ_J with a single boundary component.

At this stage we need to describe the "internal" part of $\partial\Sigma$, which is entirely fermionic, consisting of $\phi^+\phi^-$ circular components belonging to oppositely oriented spinor $\partial\sigma$ components. An internal component of $\partial\Sigma$ contains no e's (no π's) but is tangent to a μ and also to a τ as shown in Fig. 18; a "gauge hole" here has appeared within Σ. Reference 17 will explain this term. A strong-interaction example of a gauge hole is given in Fig. 19—where 4 external mesons and 1 intermediate meson line appear in a 2-vertex embellished Feynman graph. All 9 patches here are σ's (5 spinor and 4 scalar) and we have designated by σ_1 and σ_2 the two spinor patches that touch the internal component of $\partial\Sigma$. These two patches are independently oriented but their boundaries are not. The ϕ^+ and ϕ^- bounding a gauge hole in Σ always belong to oppositely oriented $(\partial\sigma)_d$'s; otherwise the hole would be contracted to zero—with two patches becoming a single patch, as in the upper part of Fig. 19.

The patch structure of Fig. 18 arises when the meson plug of Fig. 2 involves a quark-antiquark joining where ϕ^+ and ϕ^- belong to oppositely oriented patch boundaries (one clockwise and one anticlockwise) and so may not be identified. See Fig. 20. As reviewed in Sec. V A, the possibility of such mismatches for the fermionic halves of quarks is required by unitarity. In electroweak topologies, as explained by Ref. 17, gauge holes relate to conserved currents (to the maintenance of a zero photon mass).

In Ref. 6 gauge holes were collapsed to "chiral-switch lines" within Σ. (Chiral-switch lines were called "topological gluons" in Ref. 6.) For describing chirality in strong interactions, switch lines and gauge holes are equivalent, but to represent electroweak interactions gauge holes are more convenient.

It has been remarked above that all strong-interaction σ's are doubly oriented; the orientations of σ and $\partial\sigma$ here

FIG. 16. Connected sum of two patches. Orientation of patch-boundary components is indicated by arrows along β's (segments of $\partial\Sigma$—which lack intrinsic orientation).

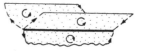

FIG. 17. A junction-line patch σ_J.

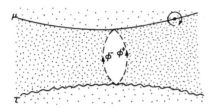

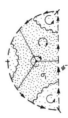 

FIG. 18. Gauge hole.

FIG. 20. Two single-vertex, 3-meson embellished Feynman graphs whose connected sum generates the gauge hole of Fig. 19.

behave in a parallel fashion under connected sums. Either can be represented by an orientation *induced* in a $\partial\Sigma$ unit. In Ref. 6 only one of the two orientations was recognized and was physically identified for spinor patches with chirality. The chosen orientation was that of a patch but could just as well have been the patch boundary. No attempt was made in Ref. 6 to topologize spin.

Labels attached to a $\partial\Sigma$ unit depend on *comparing* its σ and $(\partial\sigma)_d$ orientations to orientations already introduced in Sec. II. We compare σ orientation to global Σ orientation: if these two orientations agree (disagree), we attach a U (D) label to the $\partial\Sigma$ unit. Global Σ orientation further induces in any $\partial\Sigma$ unit an orientation to which a $(\partial\sigma)_d$ orientation can be compared; if these orientations agree (disagree) we attach an O (P) label to the $\partial\Sigma$ unit. Every boundary unit thereby acquires both a (U,D) and an (O,P) label.

V. INTERPRETATION OF LABELS ON BOUNDARY UNITS

A. Spin and chirality

Section II defined a "fermionic" β to be a segment of $\partial\Sigma$ lying along the boundary of a spinor patch; such a β we have called a ϕ. The present section explains the alge-

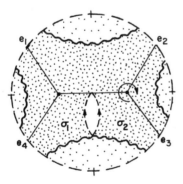

FIG. 19. Gauge hole inside an embellished two-vertex, four-meson Feynman graph.

braic sense in which a ϕ is a fermion unit.

The (U,D) and (O,P) labels on a belt ϕ we propose to interpret, respectively, as spin and chirality. On any ϕ these labels correspond to a Dirac 4-spinor index. (U,D) associates with spin $(+\frac{1}{2}, -\frac{1}{2})$ carried by a belt ϕ in the sense of a Stapp M-function index,[21] while O [P] corresponds in the Weyl basis for Dirac 4-spinors to the "upper" ["lower"] projection $(1+\gamma_5)/2$ [$(1-\gamma_5)/2$]. Under Lorentz transformations the 4-valued $(U,D)\times(O,P)$ index on any ϕ^- transforms like ψ_α while that on any ϕ^+ transforms like $\bar{\psi}_\alpha \equiv (\psi^\dagger\gamma_0)_\alpha$.

In hadronic belt portions, ϕ units appear exclusively as "halves" of "quarks." We shall explain in the following paragraphs how the topological proposals of this paper lead to the quark-line Feynman rules of Ref. 7 and give them a detailed representation. As a preliminary we note that a "quark" invariably appears "mated" with an "antiquark" in a continuous 4-β interval along $\partial\Sigma$ that begins and terminates at the M ends touching the mated pair of ϕ's, but the ϕ^+ and ϕ^- may or may not belong to the same σ. In Figs. 3 and 8(b) both members of every mated (ϕ^+, ϕ^-) pair belong to the same component of a $\partial\sigma$, while in Fig. 19 one of the (ϕ^+, ϕ^-) mated pairs along the belt is divided between two oppositely oriented $\partial\sigma$'s by a gauge hole in Σ.

If (the belt) ϕ^+ and ϕ^- of a quark line belong to the same disk σ, then both carry the *same* (O,P) and (U,D) labels; hence the 4×4 Dirac matrix connecting indices on ϕ^+ and ϕ^- is diagonal. Symmetry under reversal of σ and $\partial\sigma$ orientations then implies a *unit Dirac matrix* in the corresponding Feynman rule. This unit matrix, as discovered independently by Mandelstam[22] and by Stapp,[23] allows topological contraction ("duality") in the presence of quark spin and chirality; the unit Dirac matrix is central to the algebra of hadronic supersymmetry (Appendix B).

On the other hand, when the ϕ^+ and ϕ^- at the ends of a quark line belong to two *different* disks with oppositely oriented boundaries as in Fig. 19, their (U,D) labels are independent while (O,P) labels *disagree*. The Feynman rule of Ref. 7, in conformity with Lorentz invariance, uniquely represents the connection between ϕ^+ and ϕ^- indices for such a topology by the Dirac matrix (Weyl basis)

2692   G. F. CHEW AND V. POÉNARU   32

$$\not{p}/m_0 = \begin{bmatrix} 0 & \sigma \cdot p/m_0 \\ \tilde{\sigma} \cdot p/m_0 & 0 \end{bmatrix}, \tag{4}$$

where p is the momentum carried by the intermediate Feynman (or momentum-copy) line tangent to the gauge hole separating the two patches (see Fig. 19). The parameter m_0 is the elementary-hadron mass. Reference 17 will generalize such association of a gauge hole with a 4-vector, associating the pair of spin indices on the (nonbelt) ϕ^+, ϕ^- bounding the gauge hole with the four values of a Lorentz tensor index.

The full residue of any elementary hadron pole thus gets a familiar Feynman factor $1 + \not{p}/m_0$ for each quark line that accompanies an intermediate momentum line. Quark lines thereby satisfy the Dirac equation on shell $[p^2 = m_0^2$, chirality here being suppressed because $\frac{1}{2}(1 + \not{p}/m_0)$ is a projection operator], but within the Feynman factor we recognize the unit matrix as representing single disk-patch quark propagation, while $\not{p}/m_0$ represents two-patch propagation. Unless chirality reverses, spin cannot change. If chirality reverses, spin may or may not change. The foregoing unsymmetrical posture of spin and chirality, implicit in the Dirac "switch" matrix (4), parallels the inequivalent (two- and one-dimensional) manifolds underlying these particle attributes.

Historically, before any topology for spin was introduced, it was understood that Lorentz invariance and unitarity require the switch matrix (4). Once Feynman rules are given (including phase factors for closed loops[23,24]) strong-interaction theory is complete, so readers may wonder why it has been thought necessary to topologize spin. There are two reasons. (1) The theory must be extended *beyond* strong interactions to situations where the Feynman rules remain to be discovered. We anticipate the sole topological repository of *any* particle's spin and chirality to be its ϕ content. (2) Topologizing spin as well as chirality may be a step toward eventually topologizing the full Poincaré group. A continuing defect of TPT is that, although momentum flows through the Feynman graph, there so far has appeared no way to give topological meaning to *quantity of* momentum. Presence of spinor patches on *both* sides of each Feynman line and association of momentum factors with gauge holes invite developments beyond those of this paper.

Reference 17 will discuss a general requirement of Lorentz invariance (going beyond strong interactions) that the boundary of any patch, whether connected or not, either includes no ϕ's or exactly one $\phi^+ \phi^-$ pair carrying a single (U, D) label. Furthermore any $\partial \sigma$ that includes a ϕ must *not* include β's other than ϕ's. We have here another statement that each σ must be purely fermionic or purely bosonic.

B. Quark generation

A "quark" as defined by Fig. 5 is a 2-β connected belt interval, a (4-spinor, 2-isospinor) ϕ plus a companion β lying along the boundary of a scalar patch and never touching the Feynman graph. Such a belt segment appears only within "quarks" and will be designated δ. A δ carries no

(c, n) index [no isospin, because its $(+, -)$ label disagrees with that of the contacting τ] but has a 4-valued $(U, D) \times (O, P)$ label. We shall designate the label on a δ by an index $G = 1, \ldots, 4$ according to Table I.

Disconnection of δ's from F we interpret as disconnecting G from Poincaré transformations. In other words we interpret G when attached to a δ as an internal quark degree of freedom, and we call it "topological quark generation." The δ boundary unit is bosonic.

The door is open to mixing of topological quark generations when hadrons interact with nonhadrons. In all examples presented in this paper, each δ^- is mated with a δ^+ along the boundary of a disk σ whose orientation is reversible and whose boundary orientation is reversible. The indices on δ^+ and δ^- are connected in Feynman rules for such topologies by a 4×4 unit matrix; there is no breaking of G symmetry and no G mixing. But if the two members of a mated $\delta^+ \delta^-$ pair belong to *different* σ's, there will be nondiagonal 4×4 matrices acting on the quark-generation space and "physical" quark generations will be those (generally inequivalent) linear superpositions of the four topological quark generations that diagonalize these matrices. G mixing for quarks is probably inconsistent for pure strong interactions and for gauge-boson coupling to hadrons, as will be explained elsewhere, but is plausible in hadron coupling to the H neutral scalar bosons predicted by TPT (Ref. 25). There is furthermore expectation of coupling between topological quark generation and the lepton generation to be described in Sec. V C.

Notice how the isospin-carrying Finkelstein line in Fig. 5 divides any quark into a fermion half—carrying a 4-valued Dirac index—and a boson half—carrying a 4-valued generation index. Because the fermion half touches M, the quark Dirac index interacts with momentum and is affected by Lorentz transformations. A lepton (Fig. 14) also divides into a fermion half and a boson half, although the boson half here is a Y (not a δ), and Sec. V C will reveal four lepton generations corresponding to $G = 1, \ldots, 4$. Leptons and quarks in TPT bear certain similarities but the parallelism does not go as deep as in standard theory. Because the two halves of a TPT quark are not separated by an e, a TPT quark does not carry momentum, and because lepton generation associates with a junction line, there will be four separately conserved lepton numbers. G mixing for quarks is not accompanied by G mixing for leptons.

The earlier topological representation of quark generation, based on the transverse surface Σ_Q, also corresponded to a 4-valued index because two independent orientations were responsible.[3] Since the transverse surface attached only to hadrons and never played any role in elec-

TABLE I. Topological generation index.

G		
1	O	D
2	P	D
3	P	U
4	O	U

troweak interactions, however, no Σ_Q motivation ever was found for quark-generation switching.

C. Lepton generation

Finally we turn attention to the Y-shaped boundary unit (Fig. 12) that touches the end of a junction line J. We have already noted that each mated Y^+Y^- pair carries a single 4-valued $(O,P)\times(U,D)$ label. We may designate this label by the symbol G already defined by Table I; when attached to a Y as to a δ, G cannot interact with momentum or undergo Lorentz transformations and therefore represents an internal particle degree of freedom. One reason a Y, even though contacting M, is bosonic (scalar) rather than fermionic is because G is continuously preserved along any junction line; fermionic attributes—spin and chirality—must have capacity to switch.

Defining a quantum number b_G as the number of Y_G^+ minus the number of Y_G^-, each of 4 b_G's is exactly conserved. Our previously defined boson number b is equal to $\sum_{G=1}^{4} b_G$. According to Fig. 14 each lepton contains one Y^+, so each lepton carries $+1$ unit of a b_G. To connect b_G with lepton number for generation G, it is necessary to recall the "freezing" of all labels on hadronic Y's. From the beginning of TPT no degrees of freedom have attached to hadronic Y's (Ref. 3), and we do not propose now to tamper with this principle, which will be reviewed below. We extend the convention of Ref. 3 that hadronic Y's carry a frozen O label by requiring them also to carry a frozen U label. In other words, according to Table I, hadronic Y's always carry $G=4$. In tabulating elementary-hadron quantum numbers one may ignore the single-valued G index carried by hadronic Y's while recognizing each quark as carrying a 4-valued G index. At the same time Ref. 17 will stress that lepton coupling to hadrons through junction lines breaks lepton-G symmetry because this coupling is possible only for $G=4$.

Even though leptons couple to hadrons through G, there remains for each G an absolutely conserved lepton number:

$$L_G \equiv b_G + B\delta_{G4} . \tag{5}$$

Remembering that $B=0$ for all elementary particle except baryons, and that for hadrons $b_G = -B\delta_{G4}$, we see that $L_G = b_G$ for leptons while $L_G = 0$ for all hadrons. The previously defined lepton number $L = b + B$ [Eq. (3)] is equal to $\sum_{G=1}^{4} L_G$.

Notice that the label G corresponds to physical lepton generation but not to physical quark generation. Physical quark generations are superpositions of different G's, and there is no absolutely conserved "quark" number that carries a G label.

The conserved quantum number L_G is nonzero not only for leptons but for H bosons,[25] whose closed Y^+Y^- belt portion is shown in Fig. 21. Each of these two (nonmated) Y's carries a separate G label. If Y^+ carries G' and Y^- carries G'', then the H has $L_{G'}=+1$ and $L_{G''}=-1$.

For reasons defined in Ref. 17, the Finkelstein orientation of any nonhadronic Y is frozen in the n sense—nonhadronic Y's being electrically neutral. Both hadronic and nonhadronic

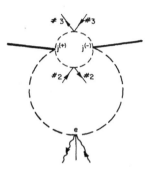

FIG. 21. Y^+Y^- belt portion of an H (horizontal) boson. The $j^{(+)}$ and $j^{(-)}$ are *not* mated by a junction line.

Y's touch the extended Feynman graph M at the ends of all three Y "legs," with the end which touches F also contacting a τ_J. The difference between hadronic and nonhadronic Y lies partly in the opposite τ_J orientation and partly in the freezing of hadronic (O,P) and (U,D) labels. The latter difference relates to the contractability of zero-entropy hadronic topologies; intermediate elementary hadrons may be indistinguishable from "bound states" of several elementary hadrons while collections of nonhadrons are never equivalent to a single nonhadron. (A nonhadron always is accompanied by nonzero entropy.) Zero-entropy contractions begin by collapsing mated "quark-antiquark" $\partial\Sigma$ intervals together with the attached σ's. Rules for such collapse depend entirely on "quark" orientations; hadronic Y's are invisible in strong interactions because Y^+Y^- mated pairs are not adjacent to each other along $\partial\Sigma$. Boundary contact between different hadrons occurs exclusively through "quark" content (see Figs. 3 and 8).

No principle has here been proposed forbidding a scalar-patch boundary from including both a Y and a δ. Such inclusion could couple lepton and quark generations and appears natural when H bosons interact with hadrons; Ref. 17 will provide examples.

Although we now forbid patches of mixed bosonic-fermionic character, such as were employed by Ref. 18 in discussing how parity symmetry can arise from junction lines, a source of parity asymmetry from junction lines has survived through our tying the meaning of (O,P) indices to orientations of $\partial\sigma$'s that may be built partially from τ_J's. Freezing of τ_J orientation not only breaks isospin symmetry but now has the potential to break parity symmetry.

VI. SUMMARY

This paper has simplified topological particle theory (TPT) by restricting Feynman-graph embellishments to a single two-dimensional surface. Strong-interaction Feynman rules have remained unaffected by this change, while quark and lepton generations have been placed on a basis that, although common, forbids lepton-generation mixing at the same time as allowing generation mixing for "topological quarks." The four values taken by a generation index have been connected to the four values of a Dirac spinor index, while preserving the scalar (internal) character of the generation index.

Topological representation of isospin has remained unchanged while that of "color" has become similar to that of isospin. [There continues, however, to be no SU(3) "color" symmetry.] The form of TPT proposed here recognizes each elementary particle as "built" from a momentum-carrying Feynman-line end surrounded by three types of "unit"—a fermionic unit ϕ and two bosonic units, δ and Y. Each unit carries a $(+,-)$ index, which alternates within a particle. Gauge bosons are ϕ^+, ϕ^-, leptons (antileptons) are Y^+, ϕ^- (ϕ^+, Y^-), and H bosons are Y^+, Y^-. (The comma here locates the Feynman-line end.) Hadrons are built from Y units plus "quarks" q, a q $(\bar{q})$ being $\phi^+\delta^-$ $(\delta^+\phi^-)$. Mesons are $q, \bar{q}$, baryons (antibaryons) are q, Y^-qq $(Y^+\bar{q}\,\bar{q},\bar{q})$, and "hexons" are $qqY^-, Y^+\bar{q}\,\bar{q}$ (Ref. 10). A quark, like a lepton, now has a fermion half and a boson half, with generation located in the latter, but a quark continues, as in previous versions of TPT, *not* to carry momentum.

ϕ units carry, in addition to a Dirac 4-spinor index, a (c,n) isospin index. Y units carry a frozen c *or* n index (this freezing being a source of symmetry breaking) as well as a 4-valued, strictly conserved, G index. δ units carry only a G index whose conservation may be violated in hadron-nonhadron interactions.

ACKNOWLEDGMENTS

The proposals of this paper have been influenced by discussions with F. Capra, B. Dougherty, D. Issler, J. Finkelstein, and H. P. Stapp. The precise topology of spin, in particular, is the result of a suggestion from Stapp. We are indebted to D. Issler for a careful reading of the manuscript. This research was supported in part by the Director, Office of Energy Research, Office of High Energy and Nuclear Physics, Division of High Energy Physics of the U.S. Department of Energy under Contract No. DE-AC03-76SF00098.

APPENDIX A: ENTROPY INDICES

For completeness here and in the following appendices we reproduce certain essential TPT ingredients from Ref. 3, with minor adjustments required by the present paper.

An S-matrix connected part M is a superposition of amplitudes M_γ

$$M = \sum_\gamma M_\gamma , \tag{A1}$$

each M_γ belonging to an embellished Feynman graph γ. Each γ is characterized by a set $g_i(\gamma)$ of non-negative integers, called "entropy indices," which allow terms within the topological expansion (A1) to be ordered according to increasing topological complexity. (The first-appearing γ's not only have the smallest number of Feynman vertices but the lowest entropy.) Multivertex γ's are connected sums of single-vertex γ's

$$\gamma = \gamma' \# \gamma'' \# \cdots . \tag{A2}$$

In any connected sum an entropy index $g_i(\gamma)$ cannot decrease and usually increases. Entropy indices exhibit either "strong entropy"

$$g_i(\gamma) \geq g_i(\gamma') + g_i(\gamma'') + \cdots \tag{A3}$$

or, at least, "weak entropy"

$$g_i(\gamma) \geq \max[g_i(\gamma'), g_i(\gamma''), \ldots] . \tag{A4}$$

Four entropy indices g_i suffice for strong interactions. Two of these, g_1 and g_2, which we describe as "entropy of the first kind," were identified already in 1973 by Veneziano[9] and relate to th(F)—the infinitesimal thickening of the Feynman graph. The genus of th(F) is g_1 while g_2 also depends on genus but further involves the number b of disconnected components of ∂th(F) that contain ends e of F:

$$g_2 \equiv g_1 + b - 1 . \tag{A5}$$

The index g_1 is "strong" in the sense of (A3) while g_2 is "weak" in the sense of (A4).

An embellished graph γ with $g_1 = g_2 = 0$ is described as "planar" (graphs with $g_1 = 0$, $g_2 = 1$ are called "cylindrical"). Strong-interaction dynamics is dominated by planar γ because of high quark multiplicity. Each quark has a multiplicity 2^5 (2 spins, 2 chiralities, 2 isospins, and 4 generations), and only for planar γ can each closed Feynman loop be accompanied by two closed quark loops. The multiplicity of embellished strong-interaction Feynman graphs generally decreases rapidly with increasing g_2.

The remaining entropy indices, g_3 and g_4, we describe as "entropy of the second kind." For strong interactions the index g_3 is the number of gauge holes within Σ, while g_4 is the number of independent Möbius bands within Σ—closed paths which cross junction lines. Both g_3 and g_4 are "strong" and both relate to quark-line complexity. The index g_3 counts the number of "chiral switches" along quark lines while g_4 counts "color switches." Reference 10 has emphasized that indefinitely large values of g_3 and g_4 make important contributions to strong-interaction dynamics; graph multiplicity does not necessarily decrease with increasing g_3 and g_4. (According to Ref. 10, entropy of the second kind is responsible for dynamical development of a GeV strong-interaction scale starting from a zero-entropy TeV scale.)

Zero-entropy γ's—with $g_1 = g_2 = g_3 = g_4 = 0$—exhibit a 2^{10} (bosonic)$\times 2^5$ (fermionic) supersymmetry[26,27] which gradually becomes broken with increasing entropy. Elementary hadrons constitute a zero-entropy supermultiplet (Appendix B). Any embellished graph that includes nonhadrons (external or internal) necessarily has nonzero entropy.

Reference 17 discusses further entropy indices that become significant for nonhadrons. An example is g_5, defined as the number of nonhadronic Feynman loops.

APPENDIX B: ZERO ENTROPY

Contraction rules, unchanged from Ref. 3, imply that any multivertex zero-entropy γ is equivalent to a single vertex. Also unchanged is the rule that any zero-entropy

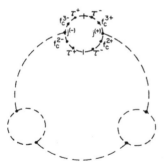

FIG. 22. A 3-beaded zero-entropy belt graph. Only those trivial vertices on one bead are shown.

FIG. 24. Shorthand quark-line diagram corresponding to Fig. 23.

belt is a "necklace of beads," a 3-beaded example is given in Fig. 22. Each bead has the same structure—containing no e's but along each branch, that connects the bead's mated j^+j^- pair, there is a mated pair of τ ends and a mated pair of f_e ends. A separate 3-patch (2 scalar, 1 spinor) sheet of Σ, without any portion of F, belongs to each branch of a zero-entropy bead, as shown by the single-bead zero-entropy example of Fig. 8(b); the "color" lines on the two sheets belonging to the same bead are oppositely oriented.

A typical Feynman-graph carrying zero-entropy sheet, S_F, corresponding to the three junction lines implied by Fig. 22, is the disk shown in Fig. 23, where spinor and scalar patches are distinguished by different shading. The shading in Fig. 8(b) has the same significance. There are no color lines on a zero-entropy S_F. It is immediately verifiable that all four entropy indices are zero for the example of Figs. 22 and 23.

The combined content of Figs. 22 and 23 is economically expressible by the shorthand quark-line diagram of Fig. 24, where each "colored" quark line carries a 2^5-valued index. (The color labels in Fig. 24 could be omitted.) Each quark line in Fig. 24 associates with a τ (not a τ_J) that

connects q and $\bar{q}$ intervals of $\partial\Sigma$. The arrows on a quark line correspond to the $(+,-)$ indices on the ends of a τ (pointing from $-$ toward $+$). The Feynman graph could be omitted from the quark-line diagram without loss of information, but we include F in Fig. 24 to remind readers that TPT quarks do not carry momentum.

Contraction rules imply a single mass m_0 for all elementary hadrons. (According to Ref. 10, m_0 lies in the TeV range.) Figure 25 gives the shorthand representation for the four categories of elementary hadron. Readers should compare Fig. 25 to Figs. 24 and 23 and note the meaning of "diquark" implied by these figures. The diquark notion is central to TPT hadron dynamics.[10] All strong-interaction embellished Feynman graphs are built by connected sum of zero-entropy γ's. Strong-interaction Feynman rules are given by Ref. 7 in terms of zero-entropy vertex functions.

APPENDIX C: EMBELLISHED LANDAU GRAPHS

An embellished Landau graph $L_l^{\gamma',\gamma''}, \ldots$ is a connected sum of embellished Feynman graphs $\gamma',\gamma'', \ldots$ not accompanied by erasure of identified $\partial\Sigma$ segments. Instead the identified segments remain within $L_l^{\gamma',\gamma''}, \ldots$ as discontinuity lines. (Discontinuity lines are not patch boundaries.) Figure 26 shows $L_l^{\gamma',\gamma''}$ for the connected sum of Fig. 20. This example, because of the gauge hole, has nonzero entropy, so the corresponding γ (Fig. 19) is not contractible, while any zero-entropy connected sum (of zero-entropy γ's) is contractible to a single-vertex γ. Nevertheless, the amplitude M_γ belonging to any γ, regardless of entropy content, is an analytic function of momentum with singularities which associate with that set of $L_l^{\gamma',\gamma''}, \ldots$ belonging to connected sums which by

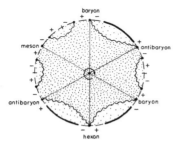

FIG. 23. A zero-entropy sheet embedding a 6-hadron Feynman graph. The belt is that of Fig. 22.

FIG. 25. Shorthand representation of elementary hadrons.

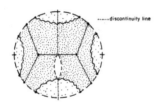

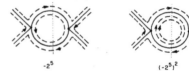

FIG. 27. Shorthand examples of quark and diquark closed loops.

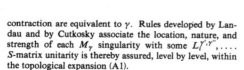

FIG. 26. Embellished Landau graph generated by the connected sum of Fig. 20.

contraction are equivalent to γ. Rules developed by Landau and by Cutkosky associate the location, nature, and strength of each M_γ singularity with some $L_{\bar{r}}^{\gamma',\gamma''}, \ldots .$ S-matrix unitarity is thereby assured, level by level, within the topological expansion (A1).

Landau-Cutkosky rules provide the dynamical equations of zero entropy. The equations are equivalent to planar discontinuity formulas for spinless elementary particles, but for each closed loop of a zero-entropy $L_{\bar{r}}^{\gamma',\gamma''}, \ldots$ there is a multiplicity factor N_0, where

$$N_0 = 2^{10} - 2^5 . \tag{C1}$$

Because of Fermi statistics[24] each zero-entropy quark loop brings a factor -2^5 while each diquark loop brings a factor $(-2^5)^2$. Figure 27 gives a sample shorthand representation of zero-entropy Landau graphs with quark and diquark loops.

Formula (C1) reveals the need for junction lines in TPT. Without junction lines (and attendant "diquarks") N_0 would be negative and a consistent S matrix probably could not be achieved, because strong-interaction vertex functions would not be Hermitian analytic.[21] Hermitian analyticity $M(p) = M^\dagger(p^*)$ is a property of the connected parts of an analytic unitary S matrix.

APPENDIX D: P, C, T

The effect of time reversal, charge conjugation, and parity inversion on a topological amplitude M_γ corresponds to certain orientation reversals. Let R denote the reversal of *all* orientations, both global and local. Under R every boundary unit undergoes $+ \leftrightarrow -$ and in $\leftrightarrow$ out interchange, but all labels, $(c,n), (O,P), (U,D), (1,2,3)$ remain unaltered. If R_p denotes reversal of all Feynman-line orientations, R_p^σ reversal of all spinor-patch orientations and $R_p^{\partial\sigma}$ orientation reversal of all spinor-patch boundaries (the subscript p is a reminder that these orientations relate to the Poincaré group), then

$$P = R_p^{\partial\sigma} , \tag{D1}$$

$$C = R R_p R_p^\sigma , \tag{D2}$$

$$T = R R_p^\sigma R_p^{\partial\sigma} . \tag{D3}$$

It follows that

$$CP = R R_p R_p^\sigma R_p^{\partial\sigma} \tag{D4}$$

corresponds to a reversal of all "non-Poincaré" orientations. It also follows that

$$PCT = R_p . \tag{D5}$$

[1]G. F. Chew, J. Finkelstein, J. P. Sursock, and G. Weissmann, Nucl. Phys. **B136**, 493 (1978).

[2]G. F. Chew, Nucl. Phys. **B151**, 237 (1979).

[3]G. F. Chew and V. Poénaru, Z. Phys. C **11**, 59 (1981).

[4]G. F. Chew, J. Finkelstein, R. M. McMurray, Jr., and V. Poénaru, Phys. Rev. D **24**, 2287 (1981).

[5]J. Finkelstein, Z. Phys. C **13**, 157 (1982).

[6]G. F. Chew and J. Finkelstein, Z. Phys. C **13**, 161 (1982).

[7]G. F. Chew and M. Levinson, Z. Phys. C **20**, 19 (1983).

[8]G. 't Hooft, Nucl. Phys. **B72**, 461 (1974).

[9]G. Veneziano, Nucl. Phys. **B74**, 365 (1974); Phys. Lett. **52B**, 220 (1974).

[10]G. F. Chew, D. Issler, B. Nicolescu, and V. Poénaru, in *Proceedings of the XIXth Rencontre de Moriond, la Plagne,*

France, 1984, edited by J. Tran Thanh Van (Editions Frontières, Gif-sur-Yvette, 1984), p. 143.

[11]R. Espinosa, Nuovo Cimento **88A**, 185 (1985).

[12]H. P. Stapp, Phys. Rev. D **28**, 1386 (1983).

[13]G. F. Chew, Phys. Rev. D **27**, 976 (1983).

[14]G. F. Chew and J. Finkelstein, Phys. Rev. Lett. **50**, 795 (1983).

[15]G. F. Chew and J. Finkelstein, Phys. Rev. D **28**, 407 (1983).

[16]G. F. Chew, J. Finkelstein, B. Nicolescu, and V. Poénaru, Z. Phys. C **14**, 289 (1982).

[17]D. Issler (in preparation).

[18]G. F. Chew and V. Poénaru, Phys. Rev. D **30**, 1579 (1984).

[19]H. Harari, Phys. Rev. Lett. **22**, 562 (1969); J. Rosner, *ibid.* **22**, 689 (1969).

[20]G. F. Chew, J. Finkelstein, and D. Issler, Lawrence Berkeley

Laboratory Report No. 17189, 1983 (unpublished).

[21]H. P. Stapp, Phys. Rev. **125**, 2139 (1962).

[22]S. Mandelstam, Phys. Rev. D **1**, 1745 (1970).

[23]H. P. Stapp, Phys. Rev. D **27**, 2445 (1983); **27**, 2478 (1983).

[24]C. E. Jones and P. Finkler, Phys. Rev. D **31**, 1393 (1985); **31**, 1404 (1985).

[25]G. F. Chew, J. Finkelstein, B. Nicolescu, and V. Poénaru, Z. Phys. C **14**, 289 (1982).

[26]P. Gauron, B. Nicolescu, and S. Ouvry, Phys. Rev. D **24**, 2501 (1981).

[27]G. F. Chew, Phys. Rev. Lett. **47**, 764 (1981).

September 23, 2012

Unification of Gravity with Electromagnetism

Geoffrey F. Chew

Theoretical Physics Group
Physics Division
Lawrence Berkeley National Laboratory
Berkeley, California 94720, U.S.A.

Principles of Maxwell, Lorentz, Milne, Dirac and Feynman are combined to unify gravity with electromagnetism. Special-relativistic settled reality (*SR*) evolves, as universe age increases, via cosmological Feynman paths. Although *SR* is 'classical', its *evolution* is 'quantum mechanical'. A *unitary* Hilbert-space Lorentz-group representation--a *lightlike*-fiber-bundle—allows definition of *divergenceless* Lorentz-tensor self-adjoint *retarded-potential* operators. Feynman-path action (real but *not SR*) is invariant under a 7-parameter group that augments 6-parameter SL(2,c) by a 1-parameter compact (Kaluza-Klein) group generated by discrete electric charge. *Sublightlike* charged-particulate matter (a component of 'objective reality') reflects 'zitterbewegung'—*fluctuation* of a lightlike-velocity direction specified by 2 angles of a 3-sphere fiber. The third fiber angle is Dirac conjugate to a discrete-spectrum electric-charge operator, whose commutation with *all* potential operators renders *discrete* the spatially-localized temporally-stable *charged-particle* component of *SR*.

1

Introduction

Physics at molecular or smaller scales has relied on the special-relativistic 'particle' concept, but the *absence* of *finite*-dimensional *unitary* Lorentz-group representations precludes a Dirac-theoretic [1] Fock space that houses finite-spin 'Lorentz particles'. [The S matrix is *not* a 'Dirac theory'.] The present paper employs Gelfand-Naimark (*GN*) infinite-dimensional *Hilbert-space* unitary representation of the Lorentz-group [2] to define *classical retarded* gravitational-electromagnetic fields. A special-relativistic settled reality (*SR*) reconciles Dirac principles with 'detectable particles'. Particle properties such as electron mass are determined by an invariant Feynman-path action [3] that combines gravity with electromagnetism.

Deferring path action to another paper, we here *locally* represent (classical) gravity and electromagnetism by *expectations* of self-adjoint *retarded* Lorentz-*tensor* potential operators that 'Dirac-extend' the classical Lienard-Wiechert 4-*vector* retarded electromagnetic potential. Maxwell's equations are satisfied by *SR*.

Dalembertians of potential-operator expectations are current densities of conserved electric charge, energy and momentum. A 'detectable particle' is a discretely-charged *temporally-stable* 'concentration' in classical Lorentz-tensor densities. Our theory admits no *Dirac* meaning for 'particle wave function'. Although the Lemaître-Hubble redshift displayed by our proposal [4] might seem incompatible with special relativity, *all* our reality is Lorentz-group based.

The Gelfand-Naimark Hilbert space, that unitarily represents the 6-parameter semisimple 'Lorentz' Lie group SL(2,c), comprises normed twice-differentiable functions of location in a 6-dimensional super manifold—a fiber bundle of compact, SU(2), 3-sphere *fiber* over an invariantly metricized (noncompact) 3-hyperboloid *base space*. Astonishingly, this Hilbert space *also* represents a 12-parameter Lie group isomorphic to transformations of a unimodular 2×2 complex matrix through right *or* left multiplication by other such matrices.

The 6 generators of SL(2,c)$_R$ commute with the 6 generators of SL(2,c)$_L$. Referring to the product group as 12L, we here show how electric-charge *discreteness*, accompanying action symmetry under a 7-parameter 12L *subgroup*, underpins the experimental-physics (special-relativistic) meaning of 'individual particle'. Our ('cosmologically-relativistic') unified quantum theory of gravity and electromagnetism defines 'particle' to be a discretely-charged temporally-stable component of 'settled' (classical) reality.

Within each of 6 12L commuting-generator 'matched' pairs, one pair member generates displacements along a (noncompact) real line while the other generates displacements around a (compact) circle. Eigenvalues of the former are continuous while, of the latter, discrete. The action that determines universe Feynman-path (quantum) dynamics is invariant under both SL(2,c)$_R$ *and* a 1-parameter compact *left* subgroup of 12L. The *septet* of symmetry generators— *GN*-represented by 7 self-adjoint (Dirac) operators—are 'universe constants of motion'.

2

Central to this paper is a pair of *left* generators we call '*quc* energy' (continuous) and '*quc* electric charge' (discrete). The abbreviation '*quc*' stands not for reality but for a lightlike 'quantum-universe constituent' that, in collaboration with other *quc*s, Dirac-defines reality. An individual *quc*, despite lacking the temporal stability essential to objective reality, carries conserved momentum, angular momentum and electric charge in a Dirac wave-function sense. 'Dirac particle reality' is achieved through *expectations* of self-adjoint operators acting on a finite (although currently-huge) tensor product of *GN* single-*quc* Hilbert spaces.

Any *quc* momentum component—a right Lorentz-group generator in a *single-quc* Hilbert space-- commutes with the corresponding (same direction) component of the *quc*'s angular momentum. Two *different*-direction *quc* momentum components do *not* commute with each other. Although a corresponding property of *angular*-momentum operators is familiar in previous quantum theory, *quc*-momentum eigenvalues are continuous. *Quc*-angular-momentum eigenvalues (unsurprisingly) are discrete. Momentum operators generate displacements in the (non-compact) 3-dimensional metricized *hyperbolic quc*-fiber-bundle base space.

For pedagogical reasons we avoid the term 'Lorentz boost' that associates to a change of velocity. The velocity of a *quc* is unchangeably lightlike (always *c* in magnitude). On the other hand, *quc* momentum, despite being a 'boost generator', is a component of a 3-vector set of *quc spatial*-displacement generators within a generator 6-vector. Our *classical SR* meaning for *particle* momentum is more familiar; status as 3-vector component of a *positive* 4-vector there prevails.

A *fixed*-direction component of *quc* momentum generates an *infinitesimal* displacement in the *curved* base space. But, because of the curvature, a *non*-infinitesimal displacement in fixed direction does *not* follow a geodesic. In contrast, a lightlike displacement generated by the *quc*'s *energy* operator follows a geodesic that parallels *quc* velocity. *All* displacements of *GN* fiber-bundle *quc* location are at fixed 'age' within a 'big-bang' spacetime—the interior of a forward lightcone where redshift-specifying age is Minkowski distance from lightcone vertex. [4]

Particle momentum accompanies (settled) positive-energy objective-reality. A particle's temporal stability requires its energy to be positive. Probabilistic (Copenhagen) interpretation of the S matrix reflects 'hidden'--inaccessible to 'observation'--*negative*-energy reality. [4] *Another* paper, dealing with Feynman-path action, will define a classical *quc-path* momentum that, although *not* a *positive*-energy particle momentum, *is* a component of a right-Lorentz 4-vector.

Quc-path momentum *differs* from the 6-vector-component self-adjoint operator we call '*quc* momentum', while *both* these concepts differ from our reality's 4-vector '*particle* momentum'--conserved by vanishing divergence of gravitational potential rather than by action invariance. *Quc-path* 4-vector momentum (in contrast) is conserved by the *definition* of 'cosmological Feynman path' (*CFP*). Neither in ray nor in path is '*quc*' to be understood as 'particle'. ('Hidden reality', manifesting *negative*-energy *quc*s, is addressed in Reference (4).)

The universe spacetime representing the group 12L is a forward-lightcone interior. 'Age' enjoys a purely-classical 'arrowed' and scale-setting (redshift-specifying) status. Neither a group Casimir nor other nontrivial self-adjoint operator, age is invariant under the *full 12L* group. *Age,* 'arrowed' with unambiguous origin, plays a role similar in many ways to that of the *single* (classical-Newtonian, Euclidean-group-invariant) time of nonrelativistic (Heisenberg-Schrödinger-Dirac) quantum theory, despite the arbitrariness of that (non-arrowed) time's origin.

Symmetry

The group *12L* is isomorphic to transformations of a unimodular 2×2 complex matrix through left *or* right multiplication by some (other) such matrix. Any left *12L* element commutes with any right element (6-parameter left and right semisimple subgroups). *12L* invariance of a Haar measure accords this group a *unitary GN* Hilbert-space representation. [4]

Although all *12L generators* are represented by self-adjoint (Dirac) operators, not all *12L* elements correspond to symmetries. Path action is invariant under a 7-parameter symmetry subgroup, whose generator septet comprises the universe's 'constants of motion'. A left-Lorentz symmetry generator-- electric charge--joins the six right-Lorentz generators of momentum and angular momentum (a Lorentz 6-vector). Zero total-universe angular momentum corresponds to Mach's principle.

GN unitary 12L Hilbert-space representation is by functions of location in a 6-dimensional fiber-bundle at *fixed* age. Each bundle attaches a 3-sphere fiber to every location in a *12L*-invariantly metricized 3-dimensional hyperbolic base space that (independently of fiber) provides 4-vector representation of left *and* right Lorentz groups. Location in base space is '*quc* spatial location'. Two fiber angles specify '*quc*-velocity direction'; the third angle is Dirac (Kaluza-Klein) conjugate to *quc* electric charge. (No *quc* attributes, either in Hilbert space or in Feynman path, correspond to *SR*.)

Definition of 'Reality' by *Expectation*

Exceptional ages, spaced by a Planck-scale age interval δ, each carry a universe ray—a vector in a *quc* quasi-Fock space. Meaning for 'settled reality', within (*quc*-less although *quc-CFP-traversed*) slices of continuous 'classical spacetime' (where age is *not* exceptional), is through *classical* retarded right-Lorentz-tensor fields. These fields are prescribed by *expectations*, with respect to that *earlier* quasi-Fock-space ray whose age is closest to the field-point age, of self-adjoint multi-*quc* right-tensor retarded-field operators. The finite number of *qucs* doubles with each successive step of exceptional universe age. The minimum number-- that at universe beginning--was 1. Prominent among the classical fields prescribing settled local reality is a divergenceless second-rank symmetric (right) Lorentz-tensor-- *conserved energy-momentum current density*. This tensor field is the Dalembertian of our retarded gravitational potential, divided by Newton's gravitational constant.

We caution readers against attribution of 'ordinary-language' significance to the Hilbert-space term 'expectation'-- a real number prescribed by pairing some ray with a self-adjoint operator. *Absent* from our meaning for 'expectation' is any probabilistic or consciousness aspect. There is no 'Schrödinger cat'—only a 'cat'.

4

Electromagnetic 'expectation reality' has two distinct components, matching the unambiguous decomposition of the Lienard-Wiechert retarded vector potential into a zero-Dalembertian 'radiation' component proportional to electric-charge 'accelerations' and a 'Coulombic' component proportional to charge 'velocities'. (The Dalembertian of the latter is the conserved electric-charge current density.) 'Photon reality' (as recognized by experimental physicists) resides *partly* in the former vector-potential component *and* partly in the energy-momentum tensor.

Unitary Hilbert-Space 12*L* Representation

We now introduce the *GN* Hilbert-space *unitary* Lorentz-group representation. A later section will define the self-adjoint operators whose expectations prescribe reality. We invoke Pauli's 2×2 matrices, a tool familiar in particle theory although unexploited in Reference (2). A lightlike *single-quc* rigged Hilbert space provides 'regular' *12L* representation. Single-*quc* Hilbert-space vectors are twice-differentiable normed complex functions of the coordinates of a 6-dimensional ('super') manifold. Three (noncompact) dimensions spatially locate a *quc*, two (compact) specify its velocity direction and one (compact) underpins its charge. The latter dimension *also* distinguishes lightlike 'bosonic' *qucs* from 'fermionic'. (Analytic S-matrix CPT symmetry entangles 'internal' particle quantum numbers with the *complex* Lorentz group which preserves energy-momentum complex 4-vector inner products.)

Gelfand and Naimark defined a Hilbert space of functions of a unimodular 2×2 complex matrix *a* through *three* complex variables s, y, z (six real variables) according to the following product of three unimodular 2×2 matrices, each of which coordinates the manifold of a 2-parameter *abelian* 12 *L* subgroup:

$$a(s,\ y,\ z) = exp(-\sigma_3 s)\times\ exp\ (\sigma_+ y)\times\ exp(\sigma_- z). \tag{1}$$

In Formula (1) the symbols $\sigma_1, \sigma_2, \sigma_3$ denote the standard (handed) set of Pauli hermitian traceless self-inverse 2×2 matrices (determinant −1), with $\sigma_\pm \equiv \frac{1}{2}(\sigma_1 \pm i\sigma_2)$. The matrix σ_3 is real diagonal while the real hermitian-conjugate matrix pair, σ_+ and σ_-, each has a unit off-diagonal element.

Three two-parameter subgroups are represented, respectively, by '*s*', '*y*' and '*z*' submanifolds of the 6-manifold *a*. The *z* manifold represents a 2-parameter 'velocity-direction' subgroup of *right* SL(2,c) while the *s* manifold represents the 2-parameter *left* subgroup of diagonal matrices that is central to this paper. The 2-dimensional *y* manifold, although associating neither to a right nor to a left subgroup, enjoys *geometrical* significance: A *directed* geodesic of Milne's hyperbolic 3-space [5] is specified by the pair y, z of complex variables.

The complex coordinate *z* specifies geodesic *direction* while *y* spatially locates the geodesic in a 2-dimensional surface *transverse* to this direction. Finally, a point *along* the z, y geodesic is *longitudinally* coordinated by *Re s*. Noteworthy is *absence* of any *geometrical* role for *Im s*; this absence is crucial to what follows.

Alternative to the unimodular-matrix factorization (1) is the factorization $a(s, y, z) = u(a)\times h(a_5)$, where $u(a)$ is *unitary* unimodular while $h(a_5)$ is *positive-hermitian* unimodular with $a_5 \equiv exp(-\sigma_3 Re\ s)\times exp\ (\sigma_+ y)\times exp(\sigma_- z)$. The matrix functions $u(a)$ and $h(a_5)$ we do not display here but are straightforwardly computable. The unitary $u(a)$ maps *a* onto a compact 3-dimensional fiber space (a 3-sphere) that covers a noncompact *12L*-invariantly-metricized 3-

5

dimensional base space onto which a_5 is mapped by $h(a_5)$. Base space will below alternatively be coordinated by a positive 4-vector of fixed invariant magnitude.

We employ Dirac's shorthand [1] of denoting, by a *single* symbol, *both* a (real classical) *quc coordinate and* a self adjoint *quc operator* whose spectrum comprises the possible values of this coordinate. An example is the symbol $Re\ s^\sigma$--linearly related to what might either be called 'the local time' of $Quc\ \sigma$ or its 'longitudinal coordinate'. The symbol E^σ will denote Quc-σ energy in (below-defined) 'σ local frame'. The 'canonically-conjugate' (when appropriately normalized) Quc-σ time and energy operators, $Re\ s^\sigma$ and E^σ, are below seen not to commute.

The 6-dimensional Haar measure,

$$da^\sigma = ds^\sigma\ dy^\sigma\ dz^\sigma, \tag{2}$$

is invariant under $a^\sigma \rightarrow a^{\sigma\Gamma} \equiv a^\sigma \Gamma^{-1}$, with Γ a 2×2 unimodular matrix representing a *right* SL(2,c) transformation of the coordinate a^σ. The measure (2) is *also* invariant under analogous left transformation. The 'volume-element' symbol $d\xi$ in (2), with ξ complex, means $d\ Re\ \xi \times d\ Im\ \xi$.

Any Hilbert-space *individual-quc* vector is a complex differentiable function $\psi(a^\sigma)$ with invariant (finite) norm,

$$\int da^\sigma |\ \psi(a^\sigma)|^2\ . \tag{3}$$

The integration in (3) spans the full y^σ and z^σ complex planes and the full $Re\ s^\sigma$ line but only a 2π interval of $Im\ s^\sigma$. Expectations of self-adjoint operators such as the Quc-σ discrete electric-charge,

$$Q^\sigma \equiv i\hbar^{1/2}g\ \partial/\partial Im\ s^\sigma, \tag{4}$$

with g an elsewhere-addressed universal dimensionless constant, are specified by the norm (3).

The Quc-σ continuous-spectrum energy-operator companion to (4) is

$$E^\sigma_\tau \equiv i(\hbar/2\tau)\ \partial/\partial Re\ s^\sigma, \tag{5}$$

the symbol τ here standing for some *discrete* and *exceptional* value of universe age that labels a *quc* Hilbert space. (Our velocity unit is such that $c = 1$.) The six generators of (right) SL(2,c) are *also* self-adjoint linear homogeneous superpositions of partial first derivatives (in the s,y,z basis)—derivative superpositions with coefficients dependent on a^σ, but that *all* commute with (4) and (5) while *each* commutes with exactly *one* other member of the right-generator sextet.

Two positive-trace self-adjoint hermitian 2×2 unimodular-matrix functions, bilinear in a^σ, $a^{\sigma\dagger}$ and linear in τ that, together, are *equivalent* to the coordinate quintet a_5^σ--$Re\ s^\sigma$, y^σ, z^σ--plus the age τ are

$$x^\sigma(a_5^\sigma, \tau) \equiv \tau\ a^{\sigma\dagger}a^\sigma, \tag{6}$$

$$v^\sigma(a_5^\sigma) \equiv a^{\sigma\dagger}(\sigma_0 - \sigma_3)\ a^\sigma. \tag{7}$$

The hermitian matrix $x^\sigma(a_5^\sigma, \tau)$ is a positive-timelike 4-vector with 'Lorentz magnitude' $x^\sigma \cdot x^\sigma = \tau^2$, while the hermitian matrix $v^\sigma(a_5^\sigma)$ is a positive-lightlike 4-vector ($v^\sigma \cdot v^\sigma = 0$) such that

$x^\sigma \cdot v^\sigma = \tau$. The dimensionful 4-vector x^σ locates *Quc* σ in spacetime (with respect to lightcone vertex) while dimensionless v^σ specifies this *quc*'s lightlike velocity 4-vector--whose timelike component equals 1 in the *Quc-σ local frame* where $x^\sigma = (\tau,0,0,0)$.

Our definition of 'reality' supposes present-universe age τ to be huge compared to an ('inflation'-interpretable) 'starting-age' τ_0 that in turn was huge compared to the Planck-scale exceptional-age spacing δ. The universe starting age τ_0 establishes the 'macro' scale of laboratory physics ($\sim$ 1 km)--in effect determining Avogadro's number. Physics (as pursued by humanity) recognizes a unique right-Lorentz local frame--the frame in which cosmic background radiation is isotropic, reflecting initial-universe spherical symmetry.

Failure of the commutator,

$$[E^\sigma(\tau), v^\sigma (a_5^\sigma)] = i\hbar/\tau \; v^\sigma (a_5^\sigma),$$ (8)

to vanish at finite τ reflects redshift.[4]

One might suppose '*quc* lightlikeness' to preclude massive particles, but we shall define reality via *expectations* of self-adjoint lightlike gravitational and electromagnetic potential *operators* that are functions of the (right) 4-vector *Quc-σ* operators $x^\sigma(a_5^\sigma, \tau)$, $v^\sigma (a_5^\sigma)$ and the invariant operators $E^\sigma(\tau)$ and Q^σ. *Fluctuation* of *quc* velocity-direction, at *fixed* momentum, helicity, energy and electric charge, is expected in the universe wave function. Conserved ('classical') current densities of energy-momentum and electric-charge are then generally *not* 'lightlike'. Although Dirac's lightlike electron-velocity operator, which led Schrödinger to coin the term, 'zitterbewegung', differs importantly from the *quc*-velocity operator (7), Schrödinger's language is appropriate for describing 'cosmological origin of rest mass'.

Classical Retarded Settled Reality

We now blend the *classical* Lienard-Wiechert (*LW*) *retarded-field* notion with *fixed-age* $\tau = N\delta$ Hilbert-space self-adjoint *quc* operators. The reader is warned against confusing the self-adjoint field operators defined here with the *quantum radiation fields* employed by the Standard Model-- the latter operators *not* being Dirac-associable to *SR*.

Let the positive-timelike 4-vector symbol x, with $(N\delta)^2 < x \cdot x < [(N+1)\delta]^2$, denote a spacetime location *between* the ages $N\delta$ and $(N+1)\delta$. We shall define a continuous set of x-labeled single-*quc* self-adjoint operators on the N (exceptional-age) Hilbert space housed at Age $N\delta$. Summing over *all qucs*, settled reality is then prescribed by interpreting a corresponding continuum of Ray N expectations as the divergenceless (retarded, classical) *LW* electromagnetic vector potential. First derivatives of this vector potential yield Maxwell's electric and magnetic fields; the potential's Dalembertian is the conserved electric-charge current density. Maxwell's (classical) equations, for electric and magnetic fields in terms of current density, apply throughout the interior of the spacetime slice.

In what follows the single superscript σ is to be understood as identifying *Quc* σ at age $N\delta$. For the retarded electromagnetic vector-potential operator $A^\sigma_\mu (x)$, associated to *Quc* σ and the spacetime-location x (the 'field-point'), we postulate

$$A^\sigma_\mu (x) \equiv \theta_{ret}(x, a_5^\sigma) \; Q^\sigma v^\sigma_\mu / v^\sigma \cdot (x - x^\sigma),$$ (9)

7

the retardation step function $\Theta_{ret}(x, a_5{}^{\sigma})$--defined two paragraphs below--*not* depending on *Im* s^{σ}. All operators in (9) commute. The (right) 4-vector operators x^{σ} and v^{σ} have been defined, respectively, by Formulas (6) and (7). Because the *quc*-velocity 4-vector v^{σ} is lightlike, the Lorentz-divergence of $A^{\sigma}{}_{\mu}(x)$ vanishes—a consideration that will render electric-charge conservation an aspect of 'classical reality'.

     The x dependence of the potential is seen to reside in the step function $\Theta_{ret}(x, a_5{}^{\sigma})$ and in the invariant *LW*-denominator operator, $v^{\sigma} \bullet (x - x^{\sigma})$. Because $v^{\sigma} \bullet v^{\sigma} = 0$, this denominator has the same value at *all quc* spacetime locations (not only that of age $\mathcal{N}\delta$) along the lightlike trajectory with *Quc-σ* velocity that passes through x^{σ}. If the $a_5{}^{\sigma}$ trajectory *intersects* the x backward lightcone, *classical LW* language refers to that intersection's location as *the* spacetime location of the 'retarded source' for the electromagnetic potential $A^{\sigma}{}_{\mu}(x)$. Age $\mathcal{N}\delta$ *quc*s whose *spatial* locations are *far* from that of the slice-interior point x, thereby admit *LW* description as being located in the 'distant past' of x.

     The symbol $\Theta_{ret}(x, a_5{}^{\sigma})$ in (9) denotes a function equal to 1 *iff* the $a_5{}^{\sigma}$ trajectory (passing with velocity v^{σ} through the *Quc-σ* spacetime location x^{σ}) intersects the x *backward* lightcone. Otherwise $\Theta_{ret}(x, a_5{}^{\sigma})$ vanishes. (*Any* lightlike trajectory not located *on* the x lightcone intersects the x forward-backward lightcone exactly *once*.) Summed over all sources, the Ray-$\mathcal{N}$ expectation of (9) prescribes the classical electromagnetic vector potential $A^{\mathcal{N}}{}_{\mu}(x)$ within the Ray-$\mathcal{N}$ immediate future.

     Although the symbol $A_{\mu}(x)$, *without* superscript $\mathcal{N}$, may be employed to designate the retarded vector potential *almost everywhere* in spacetime, exclusion must be remembered of the exceptional ray ages where $\tau = \mathcal{N}\delta$. Classical reality is not defined *on* the hyperboloids that house rays. Second-order differential equations connecting (classical) potentials to current densities, that are meaningful *inside* any 'spacetime slice', only *approximately* extrapolate these fields from one slice to the next. Universe reality evolution is *quantum-mechanically* determined by cosmological Feynman paths (*CFP*s) --via action-specified *phases* of *complex* unimodular numbers. [*CFP* prescription is *not* addressed here.]

     We now attend to a settled *gravitational* reality founded on the self-adjoint *quc* energy operator (5) rather than the *quc* electric-charge operator (4). The corresponding pair of commuting left-Lorentz-group generators, invariant under the right Lorentz group and representable by individual *quc*s, are Dirac conjugate to real and imaginary parts of the complex *quc* coordinate s^{σ}.

     Paralleling the electromagnetic divergenceless vector potential $A_{\mu}(x)$ is a gravitational divergenceless symmetric-tensor potential $\Phi_{\mu\nu}(x)$. When divided by G, the Dalembertian of $\Phi_{\mu\nu}(x)$ prescribes (without Heisenberg uncertainty) the 4-vector current density of conserved energy-momentum. We anticipate qualitative difference between electromagnetic and gravitational objective reality, symptomized by positivity of *Maxwell-field* energy density. We expect to categorize *all* electromagnetic reality as 'objective' whereas gravitational reality comprises not only objective but also 'hidden' negative-energy components that have necessitated a probabilistic interpretation for Copenhagen quantum theory.

     The (Newton-*LW*-Dirac) gravitational-potential operator

$$\Phi^{\sigma}_{\mu\nu}(x) = G\,[E^{\sigma}V^{\sigma}_{\mu\nu}(x) + V^{\sigma}_{\mu\nu}(x)E^{\sigma}], \tag{10}$$

where the divergence-less symmetric-tensor retarded-field self-adjoint operator $V^{\sigma}_{\mu\nu}(x)$, defined by

$$V^{\sigma}_{\mu\nu}(x) \equiv \Theta_{ret}(x, a_5{}^{\sigma})\, v^{\sigma}_{\mu}\, v^{\sigma}_{\nu} / v^{\sigma} \bullet (x - x^{\sigma}), \tag{11}$$

right transforms as a symmetric second-rank Lorentz tensor of zero invariant trace. Paralleling the electromagnetic vector potential, the tensor classical gravitational potential $\Phi^{\mathcal{N}}{}_{\mu\nu}(x)$ is the Ray $\mathcal{N}$ expectation of (10), summed over all *quc*s (whose *total* number is finite). The Dalembertian of $\Phi^{\mathcal{N}}{}_{\mu\nu}(x)$, when divided by G, is the energy-momentum tensor --presumed to manifest both massless and massive particles (e.g., photons, electrons, atoms) *and* electromagnetic-field energy-momentum, all positive-energy components of objective reality. 'Hidden reality', associated to negative energy but also manifested by $\Phi_{\mu\nu}(x)$, is elsewhere addressed. [4]

Conclusion

We have unified gravity and electromagnetism through a succession of *quc* quasi-Fock-space rays that unitarily represent a 12-parameter 'right-left doubling' of SL(2,c). Self-adjoint *quc* energy (source of gravity) and *quc* electric charge (source of electromagnetism) are, respectively, Dirac conjugate to real and imaginary parts of a complex Gelfand-Naimark coordinate for a 6-dimensional (super) fiber-bundle manifold. Sustained has been 'Dirac reality'-- via self-adjoint Hilbert-space operators--as well as principles of Maxwell and Feynman.

A 7-parameter symmetry group (with a 6-parameter Lorentz subgroup) is generated by 7 conserved self-adjoint operators— momentum, angular momentum and electric charge —whose expectations separately aggregate to zero for the universe as a whole. Elsewhere described is a *single-quc* spherically-symmetric *initial* condition that specifies 'zero total-universe energy'. Classical 'settled' reality resides in electromagnetic and gravitational fields within invariant spacetime slices of Planck-scale age width. Each universe ray is separated from its successor by such a slice.

Any ray (after the first at Age τ_0) is determined from its predecessor by the actions of cosmological Feynman paths that traverse the intervening slice. Gravitational and electromagnetic line-integral path action, proportional to the potentials here defined, is specified in a separate paper. Although no spacetime-slice-inhabiting particle (an *SR* component for some age interval where $\tau \gg \tau_0$) lives forever, a *quc* never dies. The number of *qucs* at age $N\delta$ -- equalling 2 raised to the power, N - N_0, where $N_0 = \tau_0/\delta$, increases monotonically with universe age.

Acknowledgements

Intense discussions over two decades with Henry Stapp are reflected in this paper. Help has also been received from Jerry Finkelstein, David Finkelstein, Dave Jackson, Don Lichtenberg, Stanley Mandelstam, Ivan Muzinich, Ralph Pred, Ramamurti Shankar, Eyvind Wichmann and Bruno Zumino.

References

1. P. A. M. Dirac, *Quantum Mechanics*, 3rd ed., Oxford University Press, New York (1947).
2. M. Naimark, *Linear Representations of the Lorentz Group*, MacMillan, New York (1964).
3. R. P. Feynman, *Rev. Mod. Phys.* **20**, 267, (1948).
4. G. F. Chew, *arXiv* 1107.0492 (2011).
5. E. A. Milne, *Relativity, Gravitation and World Structure*, Clarendon Press, Oxford (1935).

August 23, 2013

Extended-Lorentz Quantum-Cosmology Symmetry Group

U(1) × SD(2,c)L × SL(2,c)R

Geoffrey F. Chew

Theoretical Physics Group
Physics Division
Lawrence Berkeley National Laboratory
Berkeley, California 94720, U.S.A.

Summary

Unitarily representable by transformations of Milne quantum-universe (MQU) Hilbert-space vectors is a 9-parameter 'extended-Lorentz' Lie group whose algebra comprises 9 conserved MQU-*constituent* ('*quc*') attributes: electric charge, energy, spirality, 3-vector momentum and 3-vector angular momentum. Commutation with the full symmetry algebra by the 3-element Lorentz-*extending* sub-algebra identifies any *quc* by its (*permanent*) trio of charge, spirality and energy integers.

Milne's redshift-specifying 'universe age' is invariant under the MQU symmetry group. Also invariant is the (elsewhere specified) universe hamiltonian--a self-adjoint age-dependent Hilbert-space operator (*not* a symmetry-algebra member) that generates universe evolution with increasing age through a 'Schrödinger' (first-order) differential equation.

Remark: Ontological (*not* mathematical) language recognizes the 'objective reality' of a 'particle'—e.g., a photon. An MQU particle is an approximately-stable 'relationship' between *different quc*s—a 'marriage' whose persistence allows 'recognition' by *intermediate*-scale *quc* aggregates endowed with 'consciousness'. 'Intermediate scale' means *particle-scale-huge* while *Hubble-scale-tiny*.

Introduction

Dirac quantum theory--based on self-adjoint Hilbert-space operators--has for physics been impeded by the absence of unitary *finite*-dimensional Lorentz-group representations. The Gelfand-Naimark (GN) unitary Hilbert-space *infinite-dimensional* representation, [1] although *unusable* by a physics founded on positive-energy objective reality, applies *cosmologically* to a Milne (redshifting, 'big-bang') [2] *quantum universe* (MQU} where *constituent* energies may be positive *or* negative and where *total* energy vanishes. [3]

Milne's cosmological application of the Lorentz group--to a universe spacetime whose hyperbolic 3-space is *invariantly* metricized, is *profoundly* different from the physics application of the Lorentz group--to a spacetime where 3-space is Euclidean with *non-invariant* metric. Widely-misunderstood Milne cosmology associates 'redshift' so directly to universe age as to render redshift 'trivial'.

Four-dimensional *Milne spacetime* occupies the *interior* of a *forward* lightcone, where the positive (arrowed) 'age', τ, of any location is its 'Minkowski distance' from the lightcone vertex. MQU is governed by the symmetries of a here-specified 9-parameter Lie group that we designate as $U(1) \times SD(2,c)_L \times SL(2,c)_R$—a product of *four* individually-semisimple subgroups, three of which, each 1-parameter, commute with the entire group. Universe age (*not* a Hilbert-space operator) and universe hamiltonian [3] (a self-adjoint operator, although not a member of the group algebra) are both invariant under the 9-parameter MQU symmetry group.

A 2×2 unimodular-matrix meaning for the subscripts L and R in the foregoing notation will here be explained, as well as use of the symbol D in a notational location familiarly occupied by symbols such as U or L. Each of the nine members of the MQU algebra is (Stone) representable by a self-adjoint Hilbert-space operator. The (conserved) algebra corresponds to electric charge, 'spirality', energy, and a 6-vector (nonabelian) combination of momentum and angular momentum.

Particle physics is a local positive-energy restricted-scale approximation to MQU that ignores (redshifting) universe expansion. Not yet established but plausible is an *approximate* relation between charge, spirality and baryon number that accompanies approximate CP and CPT particle-physics symmetries. Reference (3) addresses both the meaning of spirality and the physics 'spatial-flattening' of MQU symmetry to the 10-parameter group that Wigner associated to the name of Poincaré.

The present paper and Reference (3) both employ the (pronounceable) acronym '*quc*' to denote an 'MQU constituent'. Each *quc* separately represents the symmetry group. The total number of *quc*s, although huge, is finite and constant--age independent because the hamiltonian lacks *quc* annihilation or creation operators. Each of the nine group generators represents a 'nameable' conserved *quc* attribute. The integer-valued spirality attribute distinguishes odd-integer 'fermionic' *quc*s from even-integer 'bosonic'.

Objective reality—*positive*-energy temporally-stable spatially-localized 'measurement-accessible' current density--involves at least *two qucs*. A single *quc* cannot represent 'matter'—the definition of which requires some stable self-sustaining *relationship* between *different qucs*.

Fiber-Bundle Factorization of a *Quc*'s 6-Dimensional Unimodular 2×2 Matrix Coordinate

The 6-dimensional complex-unimodular 2×2 matrix coordinate of a *quc* (uncovered by GN in a purely-mathematical non-cosmological context [1]) admits factorization into a *product* of individually-unimodular 3-dimensional *unitary* and *positive-hermitian* 2×2 coordinate matrices. The unitary factor represents the unmetricized ('directional') fiber of a bundle whose metricized (geometric) base space is represented by the positive-hermitian factor. Either order of the two factors is possible with the *same* unitary factor, but when the hermitian factor stands on the right it is *different* from, although unitarily equivalent to, the hermitian factor when standing on the left.

Three 'Euler' angles,

$$0 \leq \varphi' < 4\pi, \, 0 \leq \vartheta < \pi, \, 0 \leq \varphi < 2\pi, \tag{1}$$

specify, collectively, the unitary 2×2 matrix,

$$\boldsymbol{u} = exp \, (i\boldsymbol{\sigma}_3\varphi'/2) \, exp \, (i\boldsymbol{\sigma}_1\vartheta/2) \, exp \, (i\boldsymbol{\sigma}_3\varphi/2). \tag{2}$$

The matrix (2) is isomorphic to to a unit-radius 3-sphere.

In Formula (2) the symbols $\boldsymbol{\sigma}_1$ and $\boldsymbol{\sigma}_3$ denote (Pauli) hermitian traceless self-inverse *real* 2×2 matrices, each with determinant -1, $\boldsymbol{\sigma}_3$ being diagonal and $\boldsymbol{\sigma}_1$ off-diagonal. (*Any* boldface symbol in the present paper denotes a 2×2 matrix.) If the full 6-dimensional unimodular *quc*-coordinate matrix is denoted by the symbol $\boldsymbol{a}$, then

$$\boldsymbol{a} = \boldsymbol{u} \, \boldsymbol{h}_R = \boldsymbol{h}_L\boldsymbol{u}, \text{ with } \boldsymbol{h}_L = \boldsymbol{u} \, \boldsymbol{h}_R \, \boldsymbol{u}^{-1}. \tag{3}$$

A positive-trace hermitian 2×2 matrix may represent either a right or left Lorentz positive 4-vector. A right 4-vector transforms under the group SL(2,c)$_R$ by multiplication from the right by the unimodular matrix representing this group *and* multiplication from the left by the hermitian conjugate of this same matrix. A left 4-vector transforms under the group SL(2,c)$_L$ by multiplication from the left by the unimodular matrix representing the latter and multiplication from the right by this matrix's hermitian conjugate.

Hyperbolic Base-Space Metric

Milne's Lorentz-invariantly-metricized hyperbolic 3-space is isomorphic to the 'base space' of a (classical) fiber bundle that requires reference neither to Hilbert space nor to '*quc*'. Reference (3), on the other hand, represents the invariant MQU hamiltonian's *quc* kinetic-energy as a 'hyperbolic laplacian'—a self-adjoint Dirac operator acting on complex normed functions of *quc* base-space location.

The MQU kinetic-energy operator, a GN-uncovered positive function of extended-Lorentz-group Casimirs, [1] maintains such fiber-ignoring base-space meaning when fiber space has dimensionality 2 rather than 3. Reduction of fiber dimensionality will below be related to the meaning of 'spirality'. The discrete meaning of both electric charge and spirality involves Hilbert space.

The combination of *quc* Hilbert space with *quc* classical fiber-bundle we call '*quc* fiber package'. This paper will identify a package with 2-dimensional fiber. The present section, however, addresses two alternative coordinations of a 6-dimensional 'classical' bundle where the fiber occupies a 3-sphere. Here, despite our use of Pauli matrices and complex numbers, we are ignoring Hilbert space.

Unitary equivalence of right and left hermitian base-space coordinate is conveniently representable in Pauli-matrix notation through 3-vector inner products. The imaginary Pauli matrix,

$$\sigma_2 = -i\sigma_3\sigma_1, \tag{4}$$

also hermitian (σ_3 and σ_1 anticommute), self-inverse, traceless and with determinant -1, combines with σ_1 and σ_3 to define a 'handed matrix 3-vector'. Defining a unit-magnitude 'direction 3-vector' *n* by the ordered set of components

$$n_1 = sin\Theta\ cos\phi,\ n_2 = sin\Theta\ sin\phi,\ n_3 = cos\Theta\ , \tag{5}$$

with

$$0 \le \Theta < \pi,\ 0 \le \phi < 2\pi, \tag{6}$$

the symbol $\sigma \cdot n$ represents the 3-vector inner product

$$\sigma \cdot n \equiv n_1\sigma_1 + n_2\sigma_2 + n_3\sigma_3, \qquad (n \cdot n = 1). \tag{7}$$

The positive-hermitian (base-space) factor of the 6-dimensional *quc*-coordinate unimodular matrix, when standing to the right (left) of the unitary (fiber) factor *u,* will be denoted by the symbol h_R (h_L). *Any* positive-hermitian unimodular 2×2 matrix may be written in the form $exp(-\tfrac{1}{2}\beta\ \sigma \cdot n)$, with $\beta \ge 0$.

We shall denote the left (right) fiber-bundle base-space (positive-hermitian) factor by the symbol: $exp(-\frac{1}{2}\beta\, \boldsymbol{\sigma}\cdot n_{L\,(R)})$. It follows from (3) that $\boldsymbol{\sigma}\cdot n_L = \boldsymbol{u}\, \boldsymbol{\sigma}\cdot n_R\, \boldsymbol{u}^{-1}$—the left and right directional (unit) 3-vectors being related by a rotation that leaves β unchanged. The symbols $\beta_{R(L)}$ will sometimes be employed to denote the 3-vectors $\beta n_{R(L)}$.

Right-left invariance of (non-negative) β reflects this symbol's interpretability as 'shortest distance' between two different *quc* locations in the 3-dimensional hyperbolic base space, *one* of the two locations being at this space's *origin*.

The hyperbolic 3-dimensional fiber-bundle base space, occupied by a (fixed) finite although huge set of *qucs* at some fixed value of age, is invariantly metricized by

$$(d\beta)^2 + sinh^2\beta\, (dn_L \cdot dn_L = dn_R \cdot dn_R), \tag{8}$$

with

$$dn_{R(L)} \cdot dn_{R(L)} = (d\Theta_{R(L)})^2 + sin^2\Theta_{R(L)}\, (d\phi_{R(L)})^2 \,. \tag{9}$$

This paper will later denote simply by n a unit 3-vector equal to n_R, but *any* coordination of 3-dimensional base-space maintains Milne's (redshift-stipulating) hyperbolic (*not* elliptic) curvature.

Any 'Lorentz boost', either from right or left, when applied to *all quc* coordinates, merely shifts base-space origin to a new location, without altering the relationship between different *qucs* (such as the spatial distance between members of a pair). More generally, *any* element of $SL(2,c)_L \times SL(2,c)_R$, applied to *all quc* coordinates of MQU, leaves the universe unchanged.

Although the present paper considers only right 4-vectors, it attends to a 2-parameter abelian left-acting group SD $(2,c)_L$, a subgroup of $SL(2,c)_L$ that transforms a *quc*'s coordinate matrix $\boldsymbol{a}$ by multiplication from the left by a *diagonal* 2×2 complex unimodular matrix. The reader may understand the meaning of the symbol D either as 'diagonal matrix' or as 'displacement of a complex number'. (In the following section the displaced complex *quc* coordinate will be denoted by the symbol s.)

The nonabelian 6-parameter group $SL(2,c)_R$ transforms the coordinate $\boldsymbol{a}$ through *right* multiplication by a unimodular complex 2×2 matrix. The groups SD $(2,c)_L$ and $SL(2,c)_R$ commute with each other. The factorization $\boldsymbol{a} = \boldsymbol{u}\, h_R$ of the 6-dimensional *quc* coordinate matches our later Hilbert-space representation of a 9-parameter group that is isomorphic to $U(1) \times SD(2,c)_L \times SL(2,c)_R$, rather than to this group's L↔R equivalent. (In their representation of $SL(2,c)$, GN [1] made the arbitrary choice between L and R that we employ here. The reader, if a physicist, is warned that in representing SU(2) Wigner made the *opposite* choice.)

Alternative Factorizations of the *Quc*-Coordinate-Matrix; Spirality

An alternative to Formula (3) is a factorization of the (6-dimensional) unimodular 2×2 complex *quc*-coordinate matrix *a* into a product of *three* unimodular 2×2 matrices, each coordinating the manifold of a 2-parameter abelian SL(2,c) subgroup (acting from *either* right or left):

$$a(s, y, z) = exp(-\sigma_3 s) \times exp(\sigma_+ y) \times exp(\sigma_- z),\qquad(10)$$

where the real-matrix pair $\sigma_\pm$ is defined as $\frac{1}{2}(\sigma_1 \pm i\sigma_2)$ and each of the symbols *s, y, z* denotes a complex variable.

The leftmost factor in (10) is a *diagonal* 2×2 matrix which may itself be written as the product, $exp(-i\sigma_3 Im\, s) \times exp(-\sigma_3 Re\, s)$, of a commuting pair of unitary and positive-hermitian unimodular diagonal 2×2 matrices. It is useful to define a 5-parameter *quc*-coordinate matrix

$$b \equiv exp(i\sigma_3 Im\, s) \times a$$

,

$$= exp(\sigma_3 Re\, s) \times exp(\sigma_+ y) \times exp(\sigma_- z),\qquad(11)$$

that depends on *Re s, y, z* but *not* on *Im s*.

One dimension of the 3-sphere unitary factor in (3) thereby becomes recognized as enjoying status *distinct* from that of a remaining dimension *pair*. [The 3-sphere of Formula (1) is the product of an 'ordinary' 2-sphere and a 4π-circumference circle.] The distinguished fiber coordinate, *Im s*, is Dirac conjugate to the self-adjoint operator representing the symmetry-group generator we call 'spirality'. [3] The (ordinary) 2-sphere 'remainder' of *quc*-fiber 3-space we call '*quc* velocity-direction space'.

Positive Right 4-Vectors that Specify *Quc* Location in Base and Velocity-Direction Spaces

Two right 4-vectors, one positive-timelike and one positive lightlike, are equivalent to a quintet of real *quc* coordinates (the *y, z* complex-coordinate pair plus *Re s*) that specify the 2×2 unimodular matrix *b*. In the corresponding '*quc* package' neither a *quc*'s location in 3-dimensional metricized package base space nor its location in a 2-dimensional *quc* velocity-direction space depends on *Im s*. Locations in base-space and velocity-direction space are collectively equivalent to *b*.

Formulas (3) and (11) together expose the *positive-hermitian* unimodular 2×2 matrix,

$$B \equiv b^\dagger b\qquad(12)$$

$$= e^{-\beta\, \sigma\cdot n},\qquad(13)$$

as a dimensionless positive-timelike right 4-vector. The dimensionful positive factor τ (age) then allows the symbol $x \equiv \tau B$ to denote a 4-vector which locates a *quc* within Milne spacetime—the interior of a forward lightcone—by prescribing the *quc*'s displacement from the lightcone vertex. In a more familiar notation the 4 components of x are $\tau \cosh \beta$, $\tau n \sinh \beta$.

Complementing dimensionless B, which coordinates a fiber-bundle's metricized base space, is a second dimensionless positive right 4-vector—this one lightlike--to be denoted by the symbol v and coordinating a 2-dimensional unmetricized fiber. The 4-vector pair B, v is equivalent to b—an equivalence related below to invariant 4-vector inner products. (The inner product of any two *positive* 4-vectors is non-negative.)

The invariant inner product of two right 4-vectors will be denoted by the symbol $\bullet$. The inner product of *any* two right 4-vectors may be shown equal to the inner product of the unitarily-equivalent *left* 4-vector pair, so *either* product is invariant under $SL(2,c)_L \times SL(2,c)_R$ and, thereby, under the extended Lorentz group.

The *quc velocity-direction* right 4-vector is defined to be the dimensionless zero-determinant positive-hermitian matrix

$$v \equiv b^{\dagger}(\sigma_0 - \sigma_3)b, \qquad (14)$$

where the symbol σ_0 denotes the *unit* 2×2 matrix. Equivalence of the 5-dimensional coordinate matrix b to the positive 4-vector *pair* B, v accompanies the trio of inner products, $B \bullet v = 1$, $B \bullet B = 1$ and $v \bullet v = 0$, deducible by right-transforming to a special frame where $b = \sigma_0$. Explicit evaluation of Formula (14) via (10) and (11) reveals v *independence* of y—the 4-vector v being determined *entirely* by z and $\mathrm{Re}\, s$.

Fiber-Package Unitary Representation of Extended-Lorentz Group

An MQU ray is a sum of ('tensor') products, each with a below-discussed age-independent number of factors, of single-*quc* Hilbert-space vectors that each unitarily represents the 9-parameter extended-Lorentz group. A *regular*-basis electric-charge-Q Hilbert-space single-*quc* vector for Age τ is a complex differentiable function $\psi_Q^{\tau}(a)$ with invariant (finite) norm,

$$\int da\,|\psi_Q^{\tau}(a)|^2 . \qquad (15)$$

The invariant 6-dimensional volume element (Haar measure) da we now express through the trio (10) of complex-variable coordinates equivalent to a.

In the interest of notational simplicity we shall henceforth omit the age superscript τ. (Already omitted is a label to distinguish the *quc* in question from others with the same electric charge.) Further to be ignored except in *Eq.* (21) is the integer-Q *quc*-charge subscript; U(1) transformation (Kaluza-Klein) will be seen merely to shift wave-function *phase* by an increment proportional to Q.

The 6-dimensional Haar measure,

$$da = ds\, dy\, dz, \qquad (16)$$

is invariant under $a \to a^r \equiv a\Gamma^{-1}$, with Γ a 2×2 unimodular matrix representing a *right* Lorentz transformation of the coordinate a. The measure (16) is also invariant under analogous left transformation. The 'volume-element' symbol $d\xi$ in (16), with ξ complex, means $d\, Re\, \xi \times d\, Im\, \xi$.

The Hilbert-vector norm-defining integration (15) is, *wrt Im s*, over any continuous 2π interval of *Im s*. Below we shrink the Hilbert space so that *Re s* and *Im s* enjoy similar status in vector-norm definition. The norm (16) is then not invariant under the *full left* nonabelian group of Lorentz transformations but only under the 2-parameter abelian diagonal-matrix (D) left subgroup. Invariance under $SL(2,c)_R$ is unaffected.

A symmetry transformation specified by the 2×2 complex unimodular *right*-acting matrix Γ is *unitarily* Hilbert-space represented by

$$\Psi(a) \to \Psi(a\Gamma^{-1}).\tag{17}$$

Calculation shows $a\Gamma^{-1}$ to be equivalent to

$$z^r = (\Gamma_{22}z - \Gamma_{21})\,/(\Gamma_{11}-\Gamma_{12}z),\tag{18}$$

$$y^r = (\Gamma_{11}-\Gamma_{12}z)[(\Gamma_{11}-\Gamma_{12}z)y - \Gamma_{12}],\tag{19}$$

$$s^r = s + \ln(\Gamma_{11}-\Gamma_{12}z).\tag{20}$$

We now make explicit the single-*quc* Hilbert-space representation of $U(1)\times SD(2,c)_L\times SL(2.c)_R$, an element of which is specified by a U(1)-representing angle ω, $0 \le \omega < 2\pi$, an $SD(2,c)_L$-representing complex displacement Δ and an $SL(2.c)_R$-representing 2×2 complex unimodular matrix Γ. Under an ω, Δ, Γ-specified extended Lorentz-group element, a regular-basis single-*quc* Hilbert-space vector transforms to

$$\Psi_Q^{\omega,\Delta,\Gamma}(s, y, z) = e^{iQ\omega}\Psi_Q(s^r+\Delta, y^r, z^r).\tag{21}$$

Under the 9-parameter group the 2-dimensional volume element ds within the Haar measure (16) is invariant, as also is the 4-dimensional volume element $dy\, dz$. Such Haar-measure factorizability dovetails with the Formula (11) factorization of *quc* coordinate space.

Periodicity in Quc Hilbert-Vector Dependence on *Re s*

Displacement in the coordinate *Re s*, at fixed *Im s, y, z and τ*, displaces what we loosely call the 'time' of an individual *quc* at fixed values of 'everything else'. *Quc* energy, as a self-adjoint Hilbert-space operator representing a well-defined member of the *left* symmetry-subgroup algebra, is canonically-conjugate in Dirac sense to what we choose (with a dimensionality-endowing factor of τ) to call the 'time' of this *quc*. In the QMU hamiltonian [3] each *quc*'s gravitational potential energy is proportional to the *quc*'s energy—paralleling proportionality of *quc* electromagnetic potential energy to *quc* electric charge.

Although lightlikeness of the *quc* velocity-direction 4-vector **v** allows confusion (especially with *c* = 1) between 'temporal' and 'spatial' *quc* displacement, the group algebra unambiguously distinguishes right-invariant *quc* energy from a *quc*-momentum (3-vector) component of a right 6-vector—a component that generates (fixed-age) infinitesimal *quc* displacement in some direction through curved metricized base-space.

Because the infinitesimal-displacement direction must be specified in some *fixed* right-Lorentz frame, whereas a geodesic follows a curved path, the Reference (3) invariant self-adjoint *quc* hyperbolic laplacian—a positive Casimir function of 'geodesic-following' *second* derivatives to which a *quc*'s *kinetic* energy is proportional [1] —is *not* (as in Schrödinger's flat-space Hamiltonian) proportional to the inner product with itself of a *quc*'s 3-vector momentum.

Already noted has been the explicit confirmation by Formula (20) that (fixed-τ,*y*,*z*) displacements in *s* are right invariant (both real and imaginary parts). They further are invariant under the 3-parameter symmetry subgroup (with energy, spirality and electric charge as generators) that defines *quc* type, despite failure to be invariant under the full left-Lorentz subgroup. A Hilbert-space *shrinkage* specifying ray *periodicity* in dependence on *Re s* (periodicity for the hand of a '*quc* timepiece') maintains *quc* capacity to represent D-left-extended right-Lorentz transformations.

We therefore diminish each *quc*'s Hilbert space by the periodicity constraint,

$$\Psi(s, y, z) = \Psi(s + 2\pi, y, z), \tag{22}$$

with a matching redefinition of Hilbert-vector norm as integration in (15) over any (single) 2π interval of both *Im s and Re s*. The constraint (22) specifies *integer* eigenvalues for the self-adjoint operator *M* that is Dirac-conjugate to *Re s*.

Each of the three members of the extension subalgebra that complements the 6-member SL(2,c)$_R$ subalgebra is then Hilbert-space represented by a self-adjoint operator with (integer-specified) *discrete* spectra. The *quc* energy integer *M* is joined by the charge integer *Q* and the spirality integer *N*. (The spacing between neighboring *quc* energies is $\hbar/2\tau$—minuscule in the present universe on *any* of the scales of particle physics.)

Conclusion

Any *quc* is distinguished from all others by its conserved-integer trio, *Q, N, M*—the first integer, *Q*, specifying *quc* electric charge, the second, *N*, specifying *quc* spirality (distinguishing fermionic *qucs* from bosonic) while the third, *M*, specifies *quc* energy (in units of $\hbar/2\tau$). The finite total number of different *qucs*—the population of the set of allowed integer trios—is addressed in Reference (3). Because, within this set, any positive integer is accompanied by the corresponding negative integer, the total value *vanishes* of universe electric charge, spirality and energy.

Also vanishing is total-universe angular momentum—QMU's version of Mach's principle. Because *quc* momentum is continuous, Hilbert-space meaning for vanishing total-universe-momentum is more subtle; the meaning is addressed in an appendix to Ref. (3).

A 'type-basis' Hilbert-space vector $\Phi_{Q,N,M}(y, z)$, for the *quc* identified by the integer trio *Q, N, M*, is a differentiable normed complex function of two complex variables. In terms of the complex displacement (linear in Δ although *z*-dependent via Γ) that is specified by the formula

$$\delta\,(\Delta,\,\Gamma,\,z\,)\equiv\Delta+\ln\,(\Gamma_{11}-\Gamma_{12}z),\tag{23}$$

any *quc*-type-basis Hilbert-space vector represents the group element specified by ω, Δ, Γ through the transformation

$$\Phi_{Q,N,M}\,(y,\,z)\rightarrow e^{i[Q\omega\,+N\mathrm{lm}\,\delta\,+\,M\,\mathrm{Re}\,\delta\,]}\,\Phi_{Q,N,M}\,(y^{\Gamma},\,z^{\Gamma}).\tag{24}$$

The role of the *quc* coordinate *Re s*, absent from (24), deserves attention in our conclusion. This coordinate is resurrectable by defining the Hilbert vector

$$\Psi_{Q,N,M}\,(\boldsymbol{b})\equiv e^{-iM\,\mathrm{Re}\,s}\,\Phi_{Q,N,M}\,(y,\,z),\tag{25}$$

that recalls equivalence of the matrix coordinate $\boldsymbol{b}$ to the quintet *Re s, y, z* which coordinates the fiber bundle of 3-dimensional metricized hyperbolic base space and 2-dimensional velocity-direction fiber.

Established has been (classical) $\boldsymbol{b}$ equivalence to the right positive 4-vector pair $\boldsymbol{B}$ (base-space location) and $\boldsymbol{v}$ (velocity direction). Although both these 4-vectors depend on *Re s* and *z*, because $\boldsymbol{v}$ *fails* to depend on *y*, the latter may described *classically* as a '2-dimensional base-space *quc*-location coordinate'. Contrastingly, because *Re s* and *z* collaborate in 'fiber-package' roles that involve Hilbert space, this coordinate trio fails to admit a classical name.

Ontologically-justifiable names for 8 of 9 members of the extended-Lorentz algebra are unproblematic. The name 'spirality', assigned here to one algebra member, may or may not survive the test of usefulness.

Acknowledgements

Decades of discussions with Henry Stapp and Jerry Finkelstein have led to the quantum cosmology here proposed. Also contributing have been remarks by Korkut Bardakci, David Finkelstein, Eyvind Wichmann, Bruno Zumino, and Nicolai Reshetikhin.

References

1. M. Naimark, *Linear Representations of the Lorentz Group*. MacMillan, New York (1964).
2. E. A. Milne, *Relativity, Gravitation and World Structure*, Clarendon Press, Oxford (1935).
3. G. F. Chew, arXiv 1107.0492 (2011); *Schrödinger Equation for Milne Quantum Universe*, to be published (2013).